Periodic Table of the Elements

Light Metals · **Heavy Metals** · **Nonmetals** · **Inert Gases**

Period

Key / Legend:
- Atomic Number → 29
- Element Symbol → Cu
- Environmentally Important Element (●)
- Element Name → Copper

KEY: Solid □ · Liquid · Gas (dotted)

1 H Hydrogen																	2 He Helium
3 ● Li Lithium	4 Be Beryllium											5 B Boron	6 C Carbon	7 N Nitrogen	8 O Oxygen	9 F Fluorine	10 Ne Neon
11 Na Sodium	12 Mg Magnesium											13 ● Al Aluminum	14 ● Si Silicon	15 P Phosphorus	16 S Sulfur	17 ● Cl Chlorine	18 Ar Argon
19 K Potassium	20 Ca Calcium	21 Sc Scandium	22 Ti Titanium	23 ● V Vanadium	24 ● Cr Chromium	25 ● Mn Manganese	26 ● Fe Iron	27 ● Co Cobalt	28 ● Ni Nickel	29 ● Cu Copper	30 ● Zn Zinc	31 Ga Gallium	32 ● Ge Germanium	33 ● As Arsenic	34 ● Se Selenium	35 Br Bromine	36 Kr Krypton
37 Rb Rubidium	38 Sr Strontium	39 Y Yttrium	40 Zr Zirconium	41 Nb Niobium	42 ● Mo Molybdenum	43 Tc Technetium	44 Ru Ruthenium	45 Rh Rhodium	46 Pd Palladium	47 Ag Silver	48 ● Cd Cadmium	49 In Indium	50 ● Sn Tin	51 Sb Antimony	52 Te Tellurium	53 ● I Iodine	54 Xe Xenon
55 Cs Cesium	56 Ba Barium	57 La Lanthanum	72 Hf Hafnium	73 Ta Tantalum	74 W Wolfram	75 Re Rhenium	76 Os Osmium	77 Ir Iridium	78 Pt Platinum	79 Au Gold	80 ● Hg Mercury	81 Tl Thallium	82 ● Pb Lead	83 Bi Bismuth	84 ● Po Polonium	85 At Astatine	86 ● Rn Radon
87 Fr Francium	88 ● Ra Radium	89 Ac Actinium															

Lanthanide Series

58 Ce Cerium	59 Pr Praseodymium	60 Nd Neodymium	61 Pm Promethium	62 Sm Samarium	63 Eu Europium	64 Gd Gadolinium	65 Tb Terbium	66 Dy Dysprosium	67 Ho Holmium	68 Er Erbium	69 Tm Thulium	70 Yb Ytterbium	71 Lu Luteium

Actinide Series

90 ● Th Thorium	91 Pa Protactinium	92 ● U Uranium	93 Np Neptunium	94 Pu Plutonium	95 Am Americium	96 Cm Curium	97 Bk Berkelium	98 Cf Californium	99 Es Einsteinium	100 Fm Fermium	101 Md Mendelevium	102 No Nobelium	103 Lr Lawrencium

Geology and Hazardous Waste Management

Geology and Hazardous Waste Management

Syed E. Hasan

University of Missouri–Kansas City

Prentice Hall
Upper Saddle River, New Jersey

Dedicated to my wife, Farrukh—who first suggested the idea of writing this book—and to my children who lived without their Dad during the summer of 1994.

Library of Congress Catalog-in-Publication Data
Hasan, Syed E., 1939-
    Geology and hazardous waste management/Syed E. Hasan
      p.   cm.
    Includes bibliographical references and index.
    ISBN 0-02-351682 -8 (alk. paper)
    1. Hazardous wastes—Management.   2. Waste disposal in the ground—environmental
    aspects.   3. Environmental geotechnology.   I. Title.
    TD1064.H37   1995
    628.4'2—dc20
                                                    95–41347
                                                    CIP

Acquisitions Editor: Robert A. McConnin
Total Concept Manager: Kimberly P. Karpovich
Special Projects Manager: Barbara A. Murray
Assistant Vice President of Production and Manufacturing: David W. Riccardi
Manufacturing Buyer: Benjamin D. Smith
Cover Designer: Bruce Kenselaar
Text Designer: Custom Editorial Productions
Illustrations: Custom Editorial Productions

Cover photo: Over 7,000 drums containing hazardous wastes were found abandoned in South Carolina. Illustration from *Hazardous Waste Disposal Slide Set,* 1983 © James L. Ruhle & Associates, Fullerton, CA; reproduced with permission.

©1996 by Prentice-Hall, Inc.
Simon & Schuster Company/A Division of Viacom Company
Upper Saddle River, New Jersey 07458

Printed in the United States of America

10 9 8 7 6 5 4 3 2 1

ISBN: 0-02-351682-8

Prentice-Hall International (UK) Limited, *London*
Prentice-Hall of Australia Pty. Limited, *Sydney*
Prentice-Hall of Canada, Inc., *Toronto*
Prentice-Hall Hispanoamericana, S. A., *Mexico*
Prentice-Hall of India Private Limited, *New Delhi*
Prentice-Hall of Japan, Inc., *Tokyo*
Simon & Schuster Asia Pte. Ltd., *Singapore*
Editora Prentice-Hall do Brasil, Ltda., *Rio de Janerio*

Foreword

The impact of release of toxic chemicals, uncontrolled dumping of hazardous wastes, and lack of a regulatory framework have hit us hard, and we are paying a heavy price for our errors and omissions of the past. We can no longer afford to mistreat the Earth and its environment. We have to treat it with respect, use its resources diligently to sustain our lifestyle, and not repeat the mistakes of the past.

Environmental awareness is the key to all the efforts that are underway to protect the environment. An essential component of the comprehensive environmental efforts is environmental education. We have seen the introduction of environmental topics in the K–12 curricula. Beyond the school level, we are seeing more and more institutions of higher education offering courses in environmental disciplines. We at the University of Missouri-Kansas City feel proud to have envisioned this need nearly two decades ago and to have had our master's degree program in urban environmental geology operational since the fall of 1979. Even today, we hold the distinction of being the *only* university in North America that offers a master's level program in urban environmental geology. On a national level, though, full-fledged degree programs in fields like environmental geology are woefully lacking. At the bachelor's degree level we find only eight institutions that offer a major in environmental geology.

Educators are cognizant of the need to produce well-qualified people to manage the enormous environmental restoration jobs that await us in the twenty-first century. We see a growing number of course offerings in environmental sciences at American universities. Many geosciences and engineering departments have introduced courses in environmental geology, waste management, pollution control, and related areas. This trend will continue and we are likely to see a large increase in the number of environmental programs in the coming years.

Geology and Hazardous Waste Management is a well written and very timely book. The author has skillfully integrated geologic principles with the engineering, scientific, and regulatory aspects of hazardous waste management. The book will be extremely useful for students in environmental sciences, particularly those interested in learning about the complex nature of hazardous waste management. It will also serve as an important resource material to professionals in the burgeoning hazardous waste management industry and to the regulatory personnel.

James R. Durig
Dean, College of Arts & Sciences
University of Missouri–Kansas City

Preface

Management of hazardous waste became a major concern during the 1980s; it has assumed critical dimensions during the 1990s and will continue to demand our attention in the future. In the United States there are more than 1000 hazardous waste sites that have been placed on a priority list for cleanup by the U.S. Environmental Protection Agency, because of their imminent threat to human health and the environment. Added to these are nearly 15,000 contaminated sites at 1579 military installations scattered across the country. The estimated price tag for the latter alone is $30 billion, but could be as high as $200 billion. The cleanup itself will take decades. All this points to the fact that a sizable part of the future workforce will be engaged in the hazardous waste management industry. In order to ensure the availability of trained professionals to carry out this enormous task, some higher-education institutions have already introduced courses and degree programs in the hazardous waste management field; many others are in the process of doing so. It is anticipated that the number of colleges and universities offering programs in hazardous waste management and related areas will greatly increase in the next few years.

This book has been written to serve the educational need of students who will ultimately be trained to join the workforce in environmental fields in the coming years. Its purpose is to provide the reader with a thorough understanding of the principles of management of hazardous waste and to emphasize the critical relationship between geology and hazardous waste management. The comprehensive coverage of subject matter in the book will provide the knowledge students will need in order to confidently enter the fast-growing environmental job market.

The primary users of the book will be the seniors and beginning graduate students. The book is intended to be used as a text for courses in hazardous waste management offered in the geology, environmental sciences, and geotechnical and geological engineering degree programs, and will also serve as a good geological reference source for students in these programs.

In recent years, many geology and earth sciences departments have either introduced environmental geology and/or environmental science degree options or are actively considering doing so. For such programs, this book will be an excellent source for courses in waste and hazardous waste management.

No prior knowledge of geology is required, but a general familiarity with basic scientific disciplines is expected. However, to facilitate complete understanding, discussions of fundamental geologic concepts are included to provide the necessary background to individuals who have not had any coursework in the earth sciences. At the same time, discussions are carefully developed to avoid any repetition for those who have taken courses in the earth sciences.

Environmental firms and regulatory agencies frequently send their employees for short courses and workshops on various aspects of hazardous waste management. These courses, although providing a good overview of hazardous waste classification, treatment, and disposal technologies, generally do not emphasize the geologic and geotechnical aspects that are so critical in designing and executing cost-effective and successful remediation programs. The book strives to fill this gap and to provide a full understanding of the subject. The book will also serve as a useful reference for scientists, engineers, legal, and regulatory professionals working in the hazardous waste management industry.

I have used my training and experience in geology and geotechnical engineering to present a balanced approach to the subject of hazardous waste management. Some of the topics covered include site selection, regulatory framework, hydrogeological principles, groundwater contamination, and hazardous waste treatment and disposal methods. Appendices provide information on sources of geologic and related information and contact information for federal and state environmental agencies in the United States. Lists of environmental journals and magazines and of colleges and universities offering courses in waste and/or hazardous waste management will be found useful by both students and professionals alike. The list of acronyms in Appendix A will be especially useful for the reader as it deciphers the abbreviations and acronyms that have become an integral part of the rapidly evolving literature on hazardous waste management. Works used in the text are listed under *References Cited* at the end of each chapter. An additional supplemental reading list is provided for those who wish to pursue a particular topic in greater detail.

Discussions of regulatory aspects of hazardous waste management have been included at many places in the book. These discussions have been presented in a scientific and technical context, and not in a legal context; the discussions should not, therefore, be used for legal purposes.

I hope this book will meet the expectations of the reader. As the author, I greatly value the comments and opinions of the readers. No matter whether you are a student or an instructor, I would like to know how this book has served your needs. Please share your experience and any ideas you may have to improve the book, and send these to me either by mail (Department of Geosciences, University of Missouri, Kansas City, MO 64110-2499), by telephone (816) 235-2976, fax (816) 235-5535, or e-mail: shasan@cctr.umkc.edu.

ACKNOWLEDGMENTS

I owe special thanks to my long-time friends Richard L. Moberly, Associate and Senior Project Geologist, Woodward-Clyde Consultants; John E. Moylan, retired Chief of the Geotechnical Branch, U.S. Army Corps of Engineers, Kansas City District, and presently Senior Consultant, Woodward-Clyde, Inc., and David H. Stous, Senior Hydrogeologist, Burns & McDonnell Waste Consultants, Inc., for their valuable comments and suggestions in the early stages of development of the scope and contents of the book. Many other persons extended their willing cooperation that made this work possible. I cannot individually acknowledge their help, but I take this opportunity to thank them all.

To the Association of Engineering Geologists, I owe a measure of gratitude for publishing my request for case histories in its *AEG News*. Todd Beatty, Jacobs

Engineering Group, Inc., Santa Barbara, CA; Sandra Ihm, Warzyn, Inc., Addison, IL; Richard Rudy, Ecology & Environment, Tallahassee, FL; and Bill Shefchik and David Stous, Burns & McDonnell Waste Consultants, Inc., Overland Park, KS, deserve special thanks for providing materials for case histories used in the book.

I wish to thank Jeanne Brown for her assistance in word processing. To Erma Popek I am indebted for proofreading the manuscript.

The present-day information explosion makes obtaining the needed information from the burgeoning electronic databases a formidable task. An expert in library and information systems is a great asset to a textbook writer. The expertise and willing cooperation of professional librarians made my job of accessing the needed resources much easier and free of frustrations. I would like to take this opportunity to offer my special thanks to Janice Hood, Library Technician, Labet-Anderson, Inc., EPA Region VII, Kansas City, KS, for her assistance and patience during this project.

Students are the greatest asset that a professor has. Their perspectives on the subject have helped me in many ways, all of which enhanced the quality of the book. To all my past and current students, I say a hearty thank you.

Professional colleagues who went through the painstaking job of reviewing the manuscript have made many valuable suggestions and comments. I wish to thank James C. Lu, Amwest Environmental Group, University of Southern California; Gregory D. Boardman, Virginia Polytechnic Institute and State University; Charles W. Fetter, Jr., University of Wisconsin–Oshkosh; Eugene A. Glysson, University of Michigan; Allen H. Hatheway, University of Missouri–Rolla; Wendell H. Hovey, South Dakota School of Mines & Technology; Jeffrey L. Howard, Wayne State University; Chukwu Onu, Southern University and Agricultural & Mechanical College; and J. Jeffrey Price, Duke University.

Finally, my special thanks to Robert McConnin, Executive Editor, Prentice-Hall, Inc., for his continued support throughout the 18 months it took to complete the project. His expert advice and valuable suggestions, combined with his understanding of an author's difficulties and problems, made this arduous task more bearable and rewarding.

Syed E. Hasan

Contents

6 PHILOSOPHY AND APPROACHES TO HAZARDOUS WASTE MANAGEMENT   67

7 PHYSICAL, GEOTECHNICAL, AND GEOCHEMICAL PROPERTIES OF EARTH MATERIALS   78

8 HYDROGEOLOGY   115

9 CONTAMINANT TRANSPORT IN THE SUBSURFACE   137

10   CHEMICAL ANALYSIS AND QUALITY CONTROL   178

11   HAZARDOUS WASTE SITE SELECTION AND ASSESSMENT   183

12   PERSONNEL PROTECTION AND SAFETY AT HAZARDOUS WASTE SITES   215

13   TREATMENT TECHNOLOGIES FOR HAZARDOUS WASTES   243

14 METHODS OF HAZARDOUS WASTE DISPOSAL   264

APPENDICES

Introduction

HISTORICAL PERSPECTIVES

The threat to human health and life associated with naturally occurring toxins in plants, animals, and minerals has been known for a long time. In the past this knowledge was not only gainfully used—to prevent people from using such poisonous materials, effectively safeguarding their lives—it was also employed to kill people. A large body of literature on toxicology discusses the use of these toxic (poisonous) substances, including details of the antidotes used to counter the effects of specific poisons. Use of such toxic substances during the Roman period had become so widespread that by the fourth century B.C. even the general public found it distressing (Levey, 1966). Yet use of natural toxins hardly impacted the environment. With the invention of a large number of synthetic chemicals, however, the situation began to change in the late nineteenth and early twentieth century.

During the 1930s a large number of organic compounds were synthesized for use as pesticides and herbicides, resulting in the release of large quantities of harmful chemicals into the environment. In addition, post-World War II industrial and technological developments resulted in the establishment of many industries that generated wastes containing toxic chemicals. These wastes were conveniently disposed of anywhere and everywhere without any concern for their potential to adversely impact the environment and to afflict human beings with cancer and other chronic diseases. For nearly 40 years such careless disposal of hazardous wastes went on unabated—until 1976, when the U.S. Congress passed the Resource Conservation and Recovery Act (RCRA) to control the use and disposal of hazardous materials. This act was aimed at regulating hazardous waste throughout its entire life cycle, i.e., from the time of its generation, through transportation, storage, and treatment, to ultimate disposal. This approach

1

came to be known as the *cradle-to-grave* concept in the hazardous waste management industry.

Recent Events

There is frequent media coverage of the discovery of hazardous waste sites across the country. In fact, during the 1980s, the existence of a greater number of abandoned hazardous waste sites came to light than ever before; we may see the same trend during the 1990s. Why is it so? The answer lies in what may be termed the *latency effect*. Latency, in this context, relates to the unnoticed time gap that occurs between the time a hazardous waste is released into the environment and the time when the adverse effects associated with the release begin to manifest themselves. It is not uncommon for two to three decades to elapse before the extent of groundwater and soil pollution becomes noticeable, or before a person's health may be adversely affected. This latency might have led some to believe that disposing of hazardous materials in the environment does not pose problems. Of course, we know how fallacious this notion is; it is an illusion that we cannot live with any more. Regrettably it took the unfortunate events at Love Canal, Valley of the Drums, and Times Beach for us to realize the full magnitude and severity of the problems. On a positive note, it can be said that most nations of the world are now aware of the problems associated with careless disposal of hazardous waste and some are trying to implement measures that would ensure safe and environmentally acceptable management of hazardous wastes. This is the only option we have that will ensure survival of planet Earth and also guarantee a healthy and livable environment for the generations that will follow us.

GEOLOGY, THE SCIENCE OF EARTH

Earth is a unique planet. It is the only planet where conditions are just right to sustain various forms of life, including our own, that we know of. Keller (1992), in his popular book *Environmental Geology*, states: "Earth is our only suitable habitat." For us, Earth is our home. Despite advances made in space exploration and the possibility of "colonizing" other planets in the solar system, one can state with reasonable degree of certainty that it will not be possible in the near future. Even if we were successful in creating a giant dome on, let's say, the Moon, the limited space in the artificially created environment will not hold many people. Earth will still be our primary habitat.

There are many parallels between the habitat of human beings—the Earth—and an individual's home. In order to ensure that all systems function efficiently and that the house is safe and comfortable to live in, the owner of a home tries to get familiar with its various components, its foundation, and the heating, cooling, electrical, and plumbing systems. This knowledge helps in the detection of any problem at an early stage to avoid costly repairs or replacement. The same argument applies to the humans and the Earth. We only have to change our perspective. Instead of thinking of ourselves as a resident of a house on a piece of land, and being concerned only with the environment in our immediate vicinity, we have to broaden our view and look at the Earth from a much larger perspective and think globally.

As inhabitants of the Earth, it behooves each of us to know our "home" as best as we can. We need to be familiar with the processes that operate on and inside the Earth, to learn about its various systems and components, and to understand its origin and dynamics. The branch of science that deals with the study of earth is geology—literally the science of the Earth (*geo* = the Earth and *logia* = science or study). No other branch of human knowledge addresses questions relating to the origin and history of the Earth; the nature and composition of its materials; its internal make up; and the processes that operate unceasingly to bring about both external and internal changes. With this emphasis on geology, one may wonder: Why has it been that very few ever care to study geology, to learn about their home? The answer: For a long time geology was viewed largely in the context of the search for and exploitation of mineral resources, including coal and oil. Geologists were traditionally looked upon as scientists who spend their lives in the wilderness, searching for minerals and oil; nomad scientists who kept moving from place to place in mountains and valleys, at locations devoid of signs of the presence of humans. This stereotyping might have had some justification when geology was viewed as a science far removed from our daily lives. However, for the past 50 years and more, geologists have been active in many areas of human endeavors that affect our daily lives. It has now been recognized that geology plays a vital role in many of our activities. As early as 1973, Turner and Coffman, in their publication *Geology for Planning: A Review of Environmental Geology*, wrote: ". . . the geologist is an important member of the planning team. His (or her) role parallels that of the civil engineer, although the degree of involvement varies in several cases. The geologist also complements the roles of architect and landscape architect. The geologist has a definite and important role to play in planning activities" (Turner and Coffman, 1973, p. 25). Nowadays, it is very common to find geologists serving as staff members in organizations like land use planning commissions, waste management firms, city building code departments, hazard evaluation and emergency response teams, and in pollution assessment and abatement activities. Geologists also serve as managers and marketing personnel in fields related to energy and mineral resources. Recently geologists have joined with medical professionals, especially epidemiologists, to assess the possible link between the occurrence of certain chronic diseases and the geologic environment. With the liability associated with real estate transactions, geologists are being actively sought to carry out environmental audits of real estate. Many geologists are also involved in exploration, development, and management of groundwater resources.

HAZARDOUS WASTE

Simply speaking, hazardous waste is a waste material that has the potential to harm life forms and the environment. A *hazardous material* is not a *hazardous waste* until it is no longer useful, or has been abandoned or discarded. A toxic chemical (benzene, for example) is not a hazardous waste until it becomes part of a waste stream from which it cannot be separated for reuse. As long as the chemical is properly labeled and stored in the laboratory or factory, it is a *hazardous substance* but not a *hazardous waste*. Some waste material in itself may or may not be hazardous, but if it has the potential to become hazardous after interaction with the environment it is considered a hazardous waste. Hazardous wastes typically comprise organic and inorganic elements and compounds, many of which are known to be toxic to life forms—including humans. Regulatory agencies such as the U. S. Environmental Protection Agency (EPA) have developed an elaborate definition of hazardous waste, to comply with the laws. Specific tests have been developed to evaluate hazardous waste characteristics. Many substances have been included in a special list because of their known toxic or hazardous nature. Details of the EPA's definition of hazardous waste and its classification are discussed in Chapter 3.

Relevance of Geology to Waste Management

Geology in general, and environmental geology in particular, is of critical importance in management of conventional and hazardous wastes. Indeed, the specialty of environmental geology evolved in response to the environmental problems that began to emerge during the 1960s. Environmental geology was born in the early 1960s; and in the late 1960s and early 1970s the importance of this specialty became apparent as problems of contamination of our water systems began to surface. Very soon we started to look for ways to solve these problems. At about the same time came the recognition that for a land use plan to be sound and realistic, geologic information is vital and must be included in its formulation. These developments set the stage for recognition of environmental geology as a specialty in itself. The first textbook on this subject, *Environmental Geology*, written by Peter Flawn, was published in 1969. For quite some time it was debated whether the scope of environmental geology fell within the domain of a well-established branch of geology—engineering geology—or whether it should be treated as a new specialty. By the mid-1970s most geologists had agreed that there are various aspects of our physical environment—such as waste disposal, water and air pollution, study and prediction of hazardous natural events (such as earthquakes, flooding, volcanic activity), and geologic aspects of environmental health—that are outside the field of traditional engineering geology, and should be included within the domain of environmental geology.

Environmental geology made great strides in the 1970s and 1980s. Considering the state of our environment, the massive problems of cleanup of contaminated water supply sources, and waste management problems, it can be safely stated that environmental geology is here to stay. More and more trained environmental geologists will be needed to handle the astronomical job of cleaning up contaminated surface and groundwater systems and abandoned hazardous waste sites. According to the EPA, as of March 1992 there were about 30,000 sites nationwide containing hazardous wastes that pose threat to human life and the environment and need to be cleaned up. The EPA maintains a list, the National Priority List (NPL), that in October, 1992 included 1230 sites awaiting immediate cleanup actions (RCRA/Superfund Hotline, 1993). After preliminary investigations of a hazardous waste site indicate that it poses serious threat to human health and the environment, the site is assigned a priority ranking for cleanup and is added to the list. In a recent updating of the CERCLIS Data Base listing 37,851 sites, the EPA has classified 23,695 sites as not requiring further action. This leaves 14,156 hazardous waste sites that require remedial action (USEPA, 1995). Of these, 1,234 sites are on the NPL (Federal Register, 1995). The NPL does not include sites that are at establishments owned and operated by the

U.S. Department of Defense (DOD). A recent estimate (Environmental Protection, 1993) puts the number of DOD sites awaiting cleanup actions at 7300, and the cost to do the job at $25 billion. For 1993, DOD requested a budget of $5.3 billion to clean up its hazardous waste sites, which represents a 503 percent increase over its 1992 budget.

At *every* stage of hazardous waste management, geologic information provides the essential input that helps conceive, develop, test, and execute various plans in the complex, expensive, and time-demanding process of hazardous waste management. The relevance and significance of geology to hazardous waste management can be illustrated by the following example: A newspaper report draws the attention of the public to "... a strange smell coming from a certain area where a pile of dirt has been lying for some time. The liquid that comes out of the dump pile is characterized by orange and gray coloration. ...." The first, and perhaps the right, guess is that the site contains hazardous materials. In order to evaluate the nature, extent, and severity of the problem, a team of investigators begins study of the dump. In all cases, one of the members of the investigative team is the geologist. The leading role is played by him or her because it is the geologist who is trained to evaluate the piece of land, its topographic position, nature and characteristics of the earth materials occurring at the site, the groundwater, and the engineering geologic features. The geologist is also the expert to design a program of subsurface exploration and sampling that is dictated by the geology of the area. He or she analyzes the myriad of geologic data and synthesizes these to develop a workable plan to solve problems arising out of the presence of the hazardous waste at the location. A geologist's role does not end at this stage; indeed, geologic input is also very critical during the remediation stage.

PRESENT AND FUTURE TRENDS

Toxic effects of some commonly used synthetic chemicals have been known to scientists for a long time. It was not until the publication of the classic book *Silent Spring* by Rachel Carson (1962), however, that a broader awareness of the threat of man-made chemicals began to dawn upon the average person. The book triggered an ecological awakening that became the forerunner of many environmental movements that have occurred since the publication of the book. In more ways than one, Rachel Carson deserves the credit for exposing the hidden environmental threats of chemicals that had so innocently become part of our lives and were being used indiscriminately as pesticides, herbicides, and in other applications.

While the 1960s could be characterized as the period of awareness, the decade of the 1970s may be described as the years when the effects of careless use and disposal of chemicals began to manifest in many ways: The Cuyahoga River, in Ohio, catching fire (June 22, 1969) as a result of excessive contamination from flammable petroleum products that were dumped into it, and the Love Canal (1977) tragedy resulting in evacuation of its residents are just two of the many infamous examples. This period can also be characterized as the one when the United States began to take serious measures to prevent environmental degradation, and committed itself to preserving and maintaining quality of the environment. Several important pieces of legislation were enacted during the

1970s (see Chapter 4), and controls began to be imposed on producers, users, transporters, and disposers of hazardous materials.

During the 1980s all the industrialized and most of the developing nations became aware of the need to implement measures to control further degradation of the environment. The majority of the developed nations enacted legislation that led to the formulation of organizations for overseeing the environment. Laws to protect the environment were passed in ever-increasing numbers. During the late 1980s, however, a new realization dawned upon us: Careless use of chemicals is not only impacting the environment at places where the chemicals are used, or in the immediate vicinity, but on a much broader scale—across the globe. Depletion of the ozone layer and the greenhouse effects are two of the several examples of such global problems.

Also, during the 1980s we realized that there is not much time available to correct our past mistakes, and that if we do not take concrete steps to reverse the damage to our environment the problems will become irreversible. In 1989, *Time* instead of putting the usual Man or Woman of the Year on its first issue of the year, put Endangered Earth as Planet of the Year on the cover. The bulk of the issue was devoted to the Earth's environment. Lester Brown of the Worldwatch Institute, commenting on the severity of the problem and the limited time available, stated: "We do not have generations, we only have years, in which to turn things around" (*Time*, January 2, 1989, p. 30).

The 1990s, the last decade of the twentieth century, may be characterized as the period when human beings became well aware of the global significance of environmental degradation—ozone depletion, greenhouse effects, aquifer contamination, and air, land, and marine pollution. Environmental concern has now shifted from a local or regional scale to global scale. It was this concern that led to the convening of the United Nations Conference of Environment and Development (UNCED) in June 1992 at Rio de Janeiro, where an unprecedented 120 nations participated in a nonpolitical meeting. For the first time we witnessed concern for the environment becoming a part of the political process in both the developed and underdeveloped countries of the world.

Based on the environmental events of the past 35 years and the present trend, it can be stated that our survival is intimately tied to the quality of our environment. If we are not prepared to learn from past mistakes and are not willing to change our attitude of limiting environmental thinking to our immediate surroundings or country, and do not expand our views toward preservation of the global environment, the next century may initiate the demise of the most recent occupant of the Earth—*Homo Sapiens*. And despite the fact that humans were the last ones to arrive on the scene, they may also have the shortest tenure on the Earth and, more importantly, depart from the scene because of their own actions—an unparalleled instance in the 4.6-billion-year history of the Earth. On the other hand, a concerted effort by all nations of the world may turn the tide and allow us to remain on the planet for a far longer time.

INTERDISCIPLINARY NATURE OF THE SUBJECT

Our physical environment is complex and represents a close interaction between the biosphere, lithosphere, hydrosphere, atmosphere, and homosphere (the term includes human activities and their impact on the environment). In today's world, with its vast storehouse of knowledge, and in an age of specialization, it is clearly beyond the scope of one individual to acquire a thorough understanding of *all* aspects of the environment. One may become, for example, a zoologist with expertise in fresh water aquatic life forms, a hydrogeologist knowing everything about the occurrence and movement of groundwater, or an organic chemist having full knowledge of all chlorinated hydrocarbons, but these experts may not be able to assess the toxic effects of chemicals or how they are transported through the medium of soil or air, what kind of risk they pose to the environment, how such risk can be assessed, and how people can be saved from these risks. These aspects fall within the specialties of climatology, soil science, risk assessment, and toxicology. The point is that the solution of environmental problems is not within the domain of a single discipline: It is multidisciplinary and requires collaboration among experts drawn from various specialties.

The interdisciplinary nature of hazardous waste management can be illustrated by an example. The media carry a report revealing the presence of hazardous waste at a site without any information on its source, volume, or composition. Not knowing what lies in the dump, one has to proceed with extreme caution. This immediately calls for a personnel health and safety plan that is formulated by an industrial hygienist. Before the first person is even out to the site, a variety of background information needs to be gathered, such as past land use, climatic conditions, types of earth materials, ground and surface water regimes, native flora and fauna, and the like. While data-gathering may be done by a trained information specialist, the analysis and interpretation of the data require expertise of the earth scientist, climatologist, hydrologist, hydrogeologist, botanist, zoologist, and others. Upon review and analyses of the multitude of data, a site investigation plan is developed, with input from geologists, geotechnical and environmental engineers, surveyors, chemists, industrial hygienist, and other experts. After the field studies have been completed and samples taken, analytical chemists and toxicologists get involved in the assessment process. Depending upon the concentration of various chemicals and their toxic effects, the risks to human life and the environment are evaluated by experts in risk assessment. Their findings are reviewed by lawmakers, administrators, and public policy officials, who set laws and standards regarding exposure to the risk. At the same time as the situation is being evaluated, measures to clean up the contaminated site have to be formulated. Here again, the expertise of geologists, chemists, environmental engineers, managers, human resource experts, accountants (cleanup projects usually involve millions of dollars), and lawyers is needed.

A related aspect of hazardous waste site cleanup is finding the party or parties responsible for the environmental damage; if they are found, the cleanup cost is to be borne by them. However, very often what happens is that the cleanup cost is beyond the financial means of the party involved, or the property had changed hands and there is no trace of the original responsible party. Such situations call for legal experts to sort out the details. Of course, laws are subject to interpretation and are often challenged in the courts demanding services of other legal professionals. Environmental problems have resulted in much litigation and have led to the development and flourishing of the new specialty of environmental law. Many legal experts believe it will be the most lucrative specialty in the years ahead. This has resulted in more and more individuals entering the fields of environmental sciences and environmental law. In one case a student, after receiving her Master's degree in urban environmental geology from the University of Missouri-Kansas City, worked for a few years in a consulting environmental engineering firm, then enrolled in law school to get her law degree. In another case, a practicing lawyer decided to enroll in the same master's program, and over a long period of time (remember, he had an ongoing legal practice) completed all the requirements and received his M.S. degree in urban environmental geology in 1992. These two examples illustrate the specialization trend in environmental disciplines.

The interdisciplinary nature of hazardous waste management requires that there should always be clear and open communication among the experts. Many times, professional scientists or engineers get so used to their own jargon that they do not realize that the term(s) they are so familiar with may have a different connotation to other professionals. For example, geologists use the term *well-graded* for a sediment that is made up of mineral grains falling in a *narrow* size range; geotechnical engineers use the same term for a soil that is made up of particles falling over a *wide* size range. Beach sands represent well-graded sediment to a geologist but are considered poorly or uniformly graded soil by the geotechnical engineer. Obviously, the danger and confusion inherent in use of such terms without realizing that the other expert may have a different meaning cannot be over emphasized. The important point to remember is that use of such terms must be avoided—facts should be stated as clearly and unambiguously as possible. Working with people from different backgrounds also calls for good interpersonal skills. Failure to abide by these simple rules could result in serious problems.

STUDY QUESTIONS

1. What is the relationship between industrialization and environmental impact?
2. What is the latency effect? Why do some health problems take so long to manifest?
3. What is geology? Discuss the role of geologists in contemporary society.
4. What is a hazardous waste? A sample of sand that got contaminated with trichloroethylene (TCE) during a lab test was stored in an airtight jar with proper labels for recovery of the TCE. Would you consider the contaminated sand sample a hazardous waste? Why or why not?
5. Is environmental geology the same as engineering geology? Why or why not?
6. What is the National Priority List? Does it include hazardous waste sites owned by the U.S. Department of Defense?
7. What are the historical aspects of our environmental awareness?
8. Why should hazardous waste management be considered an interdisciplinary science? Explain with examples.

REFERENCES CITED

Carson, R.L., 1962, Silent Spring: Boston, Houghton Mifflin Company, 368 p.

Environmental Protection, 1993, Sites/Facilities Awaiting Cleanup: Environmental Protection, v. 4, n. 8, p. 8.

Federal Register, April 25, 1995, Superintendent of Documents, Government Printing Office No. 60FR, p. 20330, Washington, D.C.

Keller, E.A., 1992, Environmental Geology (6th ed.): New York, Macmillan Publishing Co., 521 p.

Levey, M., 1966, Medieval Arabic Toxicology, *The Book on Poisons* of Ibn Washiya and its relation to early Indian and Greek texts: Transactions, American Philosophical Society, v. 56, part 7, 130 p.

RCRA/Superfund Hotline, 1993: (800) 424-9346, Crystal City, VA.

Time Magazine, January 2, 1989: New York, Time, Inc., p. 30.

Turner, A.K., and D.M. Coffman, 1973, Geology for planning: a review of environmental geology: Quarterly of the Colorado School of Mines, v. 68, n. 3, 127 p.

U.S. Environmental Protection Agency, 1995, CERCLIS Site/Event Listing: EPA List 8E Report for Microcomputers: Government Printing Office Document No. SUB5358, Feb. 1995.

SUPPLEMENTAL READING

Brown, L.R., et al., 1994, State of the World: New York, Norton and Co., 265 p. (This is an annual publication that covers important topics relating to the Earth's environment.)

Basic Geologic Concepts

CONCEPT 1

Geology plays a central role in our lives, though we do not generally recognize it. From the time an individual is born until the person's death, he or she consumes large quantities of the Earth's resources. It is not only food and water that constitute the resources, but also the various mineral products and energy resources necessary to sustain our lives. Table 2–1 shows the variety of resources that each of us consumes daily in our lives.

Consumption of energy and mineral resources is a good yardstick to measure the affluence and state of industrialization of a nation. In general, the higher the consumption of resources, the higher the level of industrialization and the living standard of a country. This relationship between energy consumption and level of industrialization is clearly displayed in Figure 2–1. Industrialization has caused many of the environmental problems confronting these nations.

CONCEPT 2

As in other disciplines of science, geology can be grouped into classical and applied branches; applied geology is more important in hazardous waste management. The subdisciplines of geology that have direct bearing on hazardous waste management include geomorphology, historical geology, hydrogeology, environmental geology, engineering geology, and, to a lesser extent, economic geology.

Geomorphology is the branch of geology that deals with the Earth's processes that have been responsible for the development and modification of the landscape. Geomorphic factors are very important in all jobs involving siting of a waste disposal facility or design of a remediation plan, and assume critical

TABLE 2–1 Some Mineral Products Used in Our Daily Life (adapted from U.S. Geological Survey, 1975)

Materials	Sources
Building materials	Sand, gravel, stone, brick (clay), cement, steel, aluminum, asphalt, glass
Plumbing and wiring	Iron, steel, copper, brass, lead, cement, asbestos,* glass, tile, plastic
Insulating materials	Rockwool, fiberglass, gypsum (as plaster and wallboard)
Paint and wallpaper	Mineral pigments and fillers (iron, zinc, titanium); talc
Floor tiles and other plastics	Petroleum products; mineral fillers and pigments
Appliances	Iron, copper, and many rare metals
Furniture	Synthetic fibers made from coal and petroleum products; steel springs; mineral varnish
Clothing	Mineral fertilizers for growing natural fibers; synthetic fibers made from coal and petroleum products
Food	Grown with mineral fertilizers and processed and packaged by machines made of metals
Drugs and cosmetics	Mineral chemicals
Other items	Windows, screens, lightbulbs, porcelain fixtures, china, utensils, jewelry—all made from mineral products

*Asbestos was commonly used for insulation and heat resistance; it was banned in 1982 by the EPA because of its carcinogenic nature.

importance in understanding the contaminant migration in groundwater in unconsolidated aquifers.

Historical geology deals with history of the Earth since the time of its formation, about 4.6 billion years ago, until the present. However, in a pragmatic sense the most recent time of the Earth's history, called the Quaternary Period, is of main interest in the hazardous waste field. The most recent episode of climatic changes that occurred during the past 1.6 million years, and the resulting sculpture of land and formation of new earth materials, is most significant in studies relating to siting of a waste disposal facility or remediation of a hazardous waste site. The case history of the Conservation Chemical Company's hazardous waste site in Kansas City, Missouri (see Chapter 11), is a good illustration of the influence of geologic history of the area on designing a remediation plan for a hazardous waste site.

Engineering geology is an applied branch of geology. It deals with the application of geologic principles in siting, design, construction, and maintenance of engineering structures. Some practitioners of engineering geology tend to consider this field to be the same as environmental geology. Recently, Keaton and Hempen (1993) tried to equate the two branches and have advocated the view that environmental geology is the same as engineering geology. In reality, though, there is a need to draw a line—however fine—between the two specialties.

Environmental geology, like engineering geology, is an applied branch of geology. It relates to the application of geology in solving problems encountered

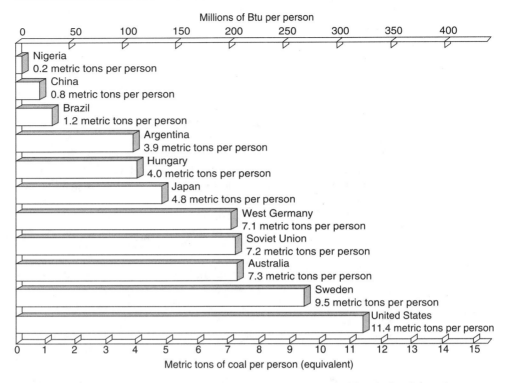

FIGURE 2–1   Relationship between energy consumption and level of industrialization. (From ReVelle and ReVelle: *The Global Environment* © 1992, Jones and Bartlett Publishers, reprinted with permission)

in our physical environment. Environmental geology deals with the study, evaluation, and mitigation of hazardous earth processes, waste management, cleanup of contaminated sites, environmental aspects of mineral- and energy-resources exploitation, landuse planning, environmental health, and problems related to global temperature change, ozone depletion, and acid rain. Of these, waste management, because of the great dimension of the problem, has placed a heavy demand on geologists trained to work in this area. As a result, some educational institutions have developed degree programs in environmental geology. However, as of 1994 there were only eight colleges/universities that offered a program leading to a bachelor's degree in environmental geology. Only one university offered a master's degree in urban environmental geology (Hasan, 1994). Currently, many geology and earth sciences departments in the United States are planning to add an environmental geology option to their degree programs.

CONCEPT 3

There is a periodicity in the natural processes; most of them follow a cycle. The geologic cycle is responsible for the formation, occurrence, and distribution of

FIGURE 2–2 The rock cycle

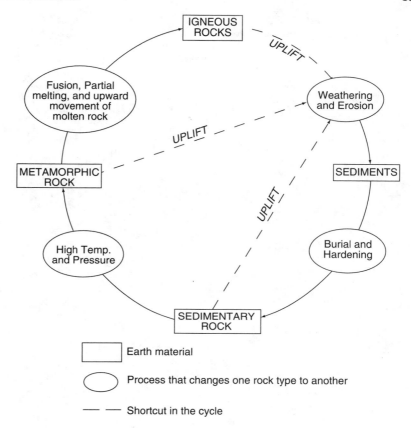

the Earth's materials, landforms, and climate. Some of the important geologic cycles are

- Rock cycle
- Hydrologic (water) cycle
- Tectonic cycle
- Geochemical cycle
- Climatic cycle

Rock Cycle

The rock cycle is the name given to one of the Earth's cycles that results in the creation, destruction, and maintenance of the Earth's materials. It is a good example of recycling of materials in nature. All three major classes of rocks—igneous, sedimentary, and metamorphic—and their numerous varieties owe their origin to the rock cycle. In addition, various minerals, including those that are used to extract metals, non-metals, coal, petroleum, and nuclear fuel materials, are all products of the rock cycle. Different types of sediments, including such common materials as sand and gravel, represent a particular stage in the rock cycle. Figure 2–2 shows various paths in the rock cycle and the resulting products.

FIGURE 2–3   Generalized hydrologic cycle

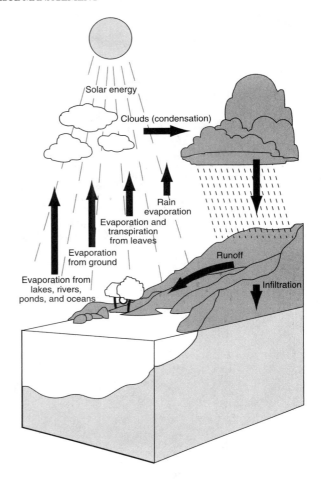

Hydrologic (Water) Cycle

Hydrologic cycle relates to the occurrence and movement of water through the lithosphere, biosphere, and atmosphere (Figure 2–3). The total quantity of water that occurs at or near the surface of the Earth, within a geographic boundary, is a function of four components, expressed by the equation

$$P = R + I + E$$

where P = precipitation, R = runoff, I = infiltration, and E = evapotranspiration.

This equation is known as the Universal Hydrologic Equation and is used to estimate the water budget of an area. For example, in the 48 conterminous states in the United States the average daily precipitation is 15,897 liters (4200 billion gallons), of which runoff accounts for 5117 billion liters (1352 billion gallons) and infiltration for 231 billion liters (61 billion gallons). Putting these numbers in the universal hydrologic equation, one can easily determine the amount of evapotranspiration, which is 10,549 billion liters (2787 billion gallons)/day.

$$E = P - (R + I)$$
$$= 15,897 - (5117 + 231) = 10,549 \text{ billion liters/day}$$

Tectonic Cycle

Tectonics is related to the study of the occurrence and nature of forces inside the Earth and the resulting changes it produces in the lithosphere. The tectonic cycle is the major Earth process that is responsible for the formation of large scale features of the Earth, such as mountain ranges, oceans, and continents. The tectonic cycle also controls the occurrence of earthquakes, volcanic activities, and the rock cycle. Tectonic forces cause rocks to deform and to develop certain structures. Common geologic processes displayed in rocks, such as tilting, folding, faulting, and jointing, result from tectonics.

Prior to 1960 not much was known about the dynamics of the lithosphere below the ocean floors. Earth scientists have developed a clear understanding of the nature of the lithosphere and the type of movements associated with it mainly from a long and ongoing exploration of ocean floors across the globe. Meticulous analyses and synthesis of a huge volume of data led to the formulation of the *theory of plate tectonics* in the late 1960s. The plate tectonics theory has been regarded as the most important concept in Earth sciences, and has unified a diverse group of ideas. The theory successfully explains: the distribution of oceans and continents; the present positions and similarity in the margins of the continents; the disposition of mountain ranges; past climatic conditions; and the occurrence of various types of faunal, mineral, and rock assemblages in continents that are now separated by oceans. It also explains why some locations, such as California in the United States and parts of the Middle East and Japan, are far more susceptible to earthquake activities than other places on the Earth. The occurrence of volcanoes, both present and past, is also satisfactorily explained by the theory of plate tectonics. Although a detailed discussion of the theory is beyond the scope of this book, some of the main features of plate tectonics must be presented to allow for full understanding of the relationship of geology to hazardous waste management.

Summary of the Main Points of Plate Tectonics Theory

We know that the Earth's plates include the lithosphere, which itself floats over the upper mantle. The energy required to drive these gigantic plates comes from the deeper parts of the Earth's interior. In order to gain a proper understanding of the Earth's tectonics, we need to know the internal makeup of the Earth.

Earth's internal structure. Earth is not a homogeneous solid from its surface to the center, a distance of 6731 km (4183 mi). It is made up of layers having different physical and chemical characteristics (Figure 2–4). The outermost layer of the Earth, the crust, is a hard layer of solid materials, 7–70 km (4.35–43.5 mi) thick. Below the crust is a layer of solid rock, called mantle, up to 2885 km (1793 mi) thick. The mantle is not uniform, but comprises an upper layer of solid and rigid, brittle rocks, 70–225 km (43.5–140 mi) thick (thicker under the continents and thinner under the oceans), called the lithosphere. Below the lithosphere, the mantle is made up of another layer of hot and weaker rocks that behave like a ductile solid, i.e., they are capable of flow when stressed. This layer, approximately 475–630 km (295–392 mi) thick, is called the asthenosphere (Figure 2–5). Below the mantle is the outer core, a 2270-km (1411 mi) thick layer of molten metallic materials. The outer core overlies the inner core, a solid mass of materials rich in

FIGURE 2–4   Internal structure of the Earth

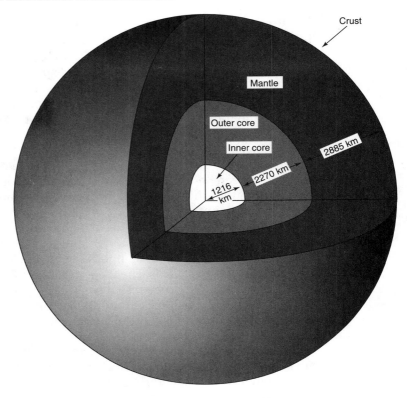

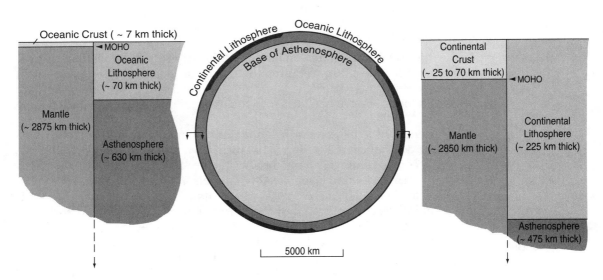

FIGURE 2–5   Relative thicknesses of crust and lithosphere under continents and oceans (adopted from
Davis, 1984)

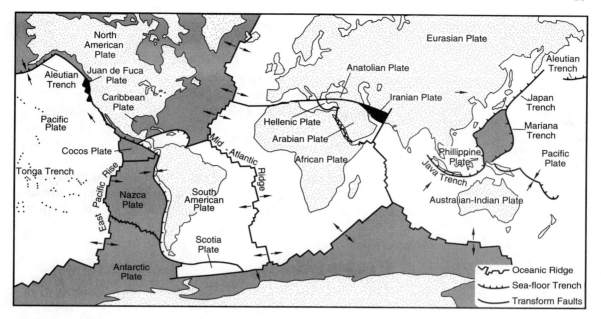

FIGURE 2–6   Plates of the Earth

iron and nickel, having high density and a radius of 1216 km (756 mi). The composition of the inner and outer cores is the same; the main difference is in the physical state of the matter. It is in the solid form in the inner core because of the extremely high pressures (more than 3000 kbars or 300 GPa; cf: 1 atmos = 1.01×10^{-3} GPa) that prevent elements like iron from melting despite the prevalent high temperature of more than 2000°C. On the other hand, the pressure and temperature in the outer core region are balanced such that iron exists in a molten state.

Plates of the Earth. The lithosphere is not a continuous solid shell that envelopes the mantle (egg shell), but is broken up into a number of large and small segments (a cracked walnut shell). These segments of the hard, brittle lithosphere are called plates. The term *plate* is used for solid bodies in which the third dimension (thickness) is very small compared to the other two dimensions (length and width). Looking at Figure 2–6, it is apparent that the plates vary in width from several hundred to several thousand kilometers, but their average thickness is very small. There are six large and eight smaller plates that have been mapped across the globe. The boundaries of the plates, called *plate margins,* show different kinds of motions and produce different features. The plate boundaries have been drawn on the basis of the study of sediment and core samples from the ocean floors and records of paleomagnetism, earthquakes, and volcanic activities.

The process of exploration of the world's oceans is not yet complete and as new data are being analyzed plate boundaries are being refined and small, new plates are being delineated. We do, however, have a fairly accurate picture of the overall distribution of the plates of the Earth.

Plate margins. Three types of plate margins have been identified. These are related to the nature of the forces acting at the plate margins. Depending on the stresses (force per unit area) acting on the plate margins, we have

1. *Divergent plate margins.* Forces are tensile (pulling apart); the adjacent plates move away from each other along a fracture in the lithosphere.
2. *Convergent plate margins.* Stresses are compressive (pushing toward each other); adjacent plates either collide with each other, producing a *collision zone*, or one of the plates sinks beneath the other, producing a *subduction zone*.
3. *Transform fault plate margins.* Shearing stresses (sideways motion) operate here; plates slide past each other, producing grinding and abrading actions at the edges of the lithosphere. Figure 2–7 shows the three types of plate margins.

Plate motion. Earth scientists have found that the plates of the Earth, along with their mountains, oceans, and continents, are in a continuous state of motion. They slip and slide over the softer asthenosphere much as a contact lens slides over the wearer's eyeball. The rate of movement of the plates is variable and has been found to range from 1.3 cm/yr (0.5 in/yr) to 18.3 cm/yr (7.2 in/yr)

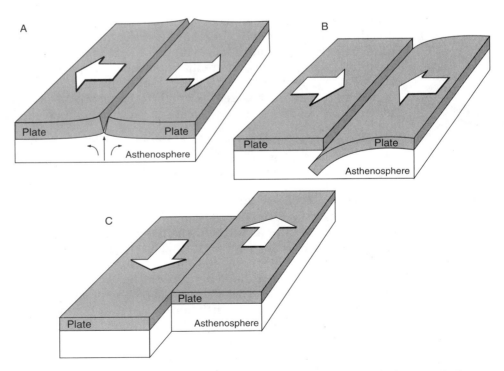

FIGURE 2–7   Various plate margins: (A) Divergent; (B) Convergent (subducting); (C) Transform fault (from Tarbuck and Lutgens, 1993)

(Figure 2–8), based on actual measurements using orbiting satellites and electronic distance measuring (EDM) devices.

Earth's features resulting from plate motion. Depending on the nature of the stresses acting on adjacent plate boundaries and whether the adjacent plates are part of the (1) continental lithosphere, (2) oceanic lithosphere, or (3) continental-oceanic lithosphere, different types of features are produced. Table 2–2 summarizes these processes and the resulting features.

Significance of plate movement. Plate movement affects everything on the Earth; most of the changes are subtle and occur over long periods of time (thousands to millions of years). It is due to the slow but sure movement of the plates that new deposits of minerals and petroleum are formed; the occurrence of earthquakes and volcanoes is also a direct consequence of plate movement. Good examples of how plate movements affect our daily lives come from California, where large earthquakes are common, and the June 1992 eruption of Mt. Pinatubu in the Philippines, which resulted in a massive evacuation of people and closure of the U.S. Clark Air Force Base.

Over periods of millions of years, plate movement may modify the flow pattern of oceanic waters, heralding climatic changes. This may result in some areas receiving higher precipitation than in the past, and others becoming drier. The Pacific plate is one of the largest, and the El Niño phenomenon that causes unusual climatic phenomena across the Pacific Ocean could possibly be linked to movements of the Pacific plate.

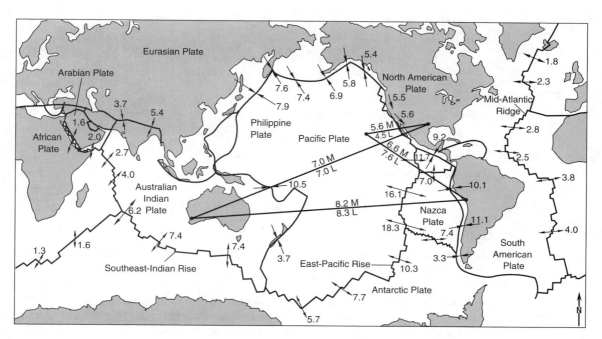

FIGURE 2–8 Present-day plate velocities in cm/yr; L = laser ranging; M = estimate from magnetic measurements (from Skinner and Porter: *The Dynamic Earth: An Introduction to Physical Geology,* 2nd ed., ©1992, John Wiley & Sons, Inc., reprinted with permission)

TABLE 2–2 Features Associated with Different Plate Margins (from Skinner and Porter: *The Dynamic Earth: An Introduction to Physical Geology,* 2nd ed., ©1992, John Wiley & Sons, Inc., reprinted with permission)

Crust on Each Plate	Feature	Kind of Margin		
		Divergent	**Convergent**	**Transform Fault**
Oceanic-Oceanic	Topography	Oceanic ridge with central rift valley	Seafloor trench	Ridges and valley created by oceanic crust
	Earthquake	All foci less than 100 km deep	Foci from 0 to 700 km deep	Foci as deep as 100 km
	Volcanism	Basaltic pillow lavas	Andesitic volcanoes in an arc of island parallel to trench	Volcanism rare; basaltic along "leaky" faults
	Example	Mid-Atlantic Ridge	Tonga-Kermadec Trench; Aleutian Trench	Kane Fracture
Oceanic-Continental	Topography Earthquake Volcanism	——— ——— ———	Seafloor trench Foci from 0 to 700 km deep Andesitic volcanoes in mountain range parallel to trench	——— ——— ———
	Example	(No examples)	Western Coast of South America	(No examples)
Continental-Continental	Topography	Rift valley	Young mountain range	Fault zone that displaces surface features
	Earthquake	All foci less than 100 km deep	Foci as deep as 300 km over a broad region	Foci as deep as 100 km throughout a broad region
	Volcanism	Basaltic and rhyolitic volcanoes	No volcanism; intense metamorphism and intrusion of granitic plutons	No volcanism
	Example	African Rift Valley	Himalaya, Alps	San Andreas Fault

Geochemical Cycle

The geochemical cycle is responsible for movement and interaction of chemical elements and compounds through various paths in the lithosphere, hydrosphere, biosphere, and atmosphere. It is nature's way of causing concentration of chemical elements and compounds at one location or their deficiency (removal) at other locations. Pollution of soil, water, and air is largely a result of human activities. Pollution represents an excess of materials that may be harmful to life and the environment. Improper waste management and release of toxic substances into the environment have caused serious environmental pollution.

Climatic Cycle

Another cycle that relates to changes in the climate over periods of (geologic) time is the climatic cycle. It is well known from the records left in ancient rocks and sediments that there have been major episodes of climatic change over a given region. For example, deposits of glacial origin are found in places like the tropics, where the present climate is warm and humid and not conducive to formation and growth of glaciers. Yet some 300 million years ago, during the Mississippian and Pennsylvanian Periods (the Carboniferous Period), many places in Africa, South and Central India, South America, and Australia were blanketed by a thick ice sheet. Subsequent climatic changes resulted in melting and disappearance of the ice sheet, but the glacial deposits were preserved in the rock records, which reveal the presence of past ice ages in the region.

Earth scientists have recognized at least four major episodes of warming and cooling of the Earth that occurred during the Quaternary Period—the most recent time in earth's history. During the Pleistocene Epoch (began 1.6 Ma), Earth's climate alternately ranged from cold (glacial) to warm (interglacial). The duration of these glacial and interglacial times was variable. During the past one million years, each glacial–interglacial cycle was about 100,000 years long; earlier cycles were about 40,000 years long. Many scientists believe that some of the climatic changes that are being experienced now are related to these warming-cooling cycles of the Earth's climate. Figure 2–9 shows climatic changes since the Pleistocene, based on the study of oxygen isotopes from deep-sea sediments.

Changes in climate are also brought about by volcanic eruptions, greenhouse effects, and disturbance in wind patterns in oceans (such as El Niño events). The climatic changes caused by these processes are rather short-lived, a few to several years. But even this short period may be sufficient to trigger earth processes and activities, such as landslides and river flooding, that pose serious threat to human lives and property and may cause substantial damage.

Anthropomorphic Effects

Despite the fact that humans have been the last to appear on the Earth, they have—in a relatively short period—altered the environment in a manner unprecedented in the Earth's history. The present threat from hazardous wastes, with its adverse impact on environmental quality, is entirely the result of human activities. Such human-induced actions are termed anthropomorphic and must be recognized in all phases of hazardous waste management. The domain of human interaction with the environment, the one that is subjected to maximum abuse, is being termed the *homosphere*.

CONCEPT 4

In order to comprehend the full magnitude of geologic cycles and processes, it is essential to expand our understanding of time and space. As humans, we tend to reckon events in terms of our lifetime: years, decades, and centuries. However, geologic time is enormous and our frame of reference must change from days, years, and decades to thousands, millions, and even billions of years. The following simple mathematical exercise illustrates the enormity of geologic time.

FIGURE 2–9   Duration of the glacial–interglacial cycles during the last 2 m.y., based on oxygen-isotope variations in deep-sea cores (from Skinner and Porter: *The Dynamic Earth: An Introduction to Physical Geology,* 2nd ed., ©1992, John Wiley & Sons, Inc., reprinted with permission)

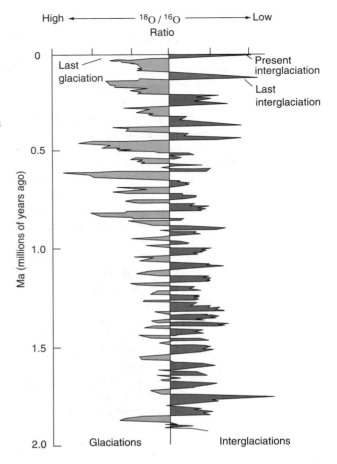

It is known that the Earth was formed about 4.6 billion years ago. If we compress this period of 4.6 billion years into one human year of 365 days, then

Each *month* of Earth time = 383.33 m.y. of human time

Each *day* of Earth time = 12.8 m.y. of human time

Each *hour* of Earth time = 0.53 m.y. of human time

Each *minute* of Earth time = 8874 yr of human time

Each *second* of Earth time = 148 yr of human time

Most geologic processes take hundreds of thousands to millions of years.

In terms of size and space, geologic features cover a wide range. For example, a fault, a common geologic structural feature, could range from microscopic—a fraction of a millimeter—discernible only under high-power magnification, to extremely large—several hundreds of kilometers long, such as the San Andreas Fault in California. Despite this variation in scale, geologic processes are continuous and ongoing. The Earth is constantly undergoing changes, but the rate of change is highly variable; from sudden and dramatic, such as eruption of a volcano or occurrence of an earthquake, to extremely slow and imperceptible, such as progressive downslope movement of a mass of earth material and the gradual breakdown of large rock mass into its individual constituent mineral grains. The important point is: The Earth is a dynamic system undergoing changes all the time.

CONCEPT 5

There are certain geologic processes that pose hazards to people, and in many cases human activities have caused an increase in the magnitude and frequency of these hazards. Examples of Earth processes that are hazardous include coastal processes, earthquakes, landslides including subsidence, river flooding, and volcanic activity. An important aspect of such processes is that, except for coastal erosion, the associated hazards are noncontinuous, meaning that they occur only periodically. For example, most of the time when water flows in a stream it is confined within its channel, but once in a while the available water is more than the channel capacity. When this happens water spills over the bank, causing flooding. There are a number of natural and human-induced factors that cause an increase in the quantity of water flowing through a river or stream. In fact, except for volcanic actions and hurricanes, human activities are known to cause the most significant increase in both the magnitude and frequency of the hazardous processes.

Another important point about hazardous Earth processes is that they are a part of the broad geologic cycle and have occurred throughout most of the Earth's history. They have occurred in the past, they are occurring now, and they will continue to occur in future. *We do not have any way to stop them from happening.* A river will flood when its channel capacity is exceeded, an earthquake will strike when the stresses along the lithospheric plates exceed the strength of rocks that make up the plates, and materials from slopes will move downward when the driving forces exceed the resisting forces. Environmental geologists, who study these processes, try to find ways to *minimize* the risk associated with such hazardous processes rather than trying to *eliminate* them.

It is very important that the occurrence and significance of hazardous geologic processes be properly recognized and carefully evaluated in any plan involving the location of a hazardous waste facility or the design of a remediation plan. Delineation of flood plains and identification of areas susceptible to landslides, earth movements, and earthquakes are some of the considerations that go into site selection for hazardous waste disposal or treatment facilities.

STUDY QUESTIONS

1. What 10 mineral and energy resources are used to sustain our lifestyle?
2. Define geomorphology and historical geology. What is their importance in hazardous waste site remediation?
3. What is the difference between engineering geology and environmental geology?
4. What are the four major subcycles that comprise the geologic cycle?
5. What is the difference between sediment and sedimentary rock?
6. Draw a sketch showing the various zones in the Earth's interior. Why is the outer core liquid and the inner solid?
7. What are the various plate boundaries? What kind of stresses are associated with each type? How has the rate of movement of the plates been ascertained? What is the range of this rate?

8. What evidence do we have to conclude that the Earth's climate was different in the past?
9. What is meant by anthropomorphism? How is environmental pollution related to this? Give examples.
10. How old is the Earth? Why is the Pleistocene Period so significant in hazardous waste management?
11. Give five examples of hazardous earth processes. Is it possible to stop the occurrence of these hazardous processes? Why or why not?

REFERENCES CITED

Davis, G.H., 1984, Structural geology of rocks and regions: New York, John Wiley and Sons, Inc., 492 p.

Hasan, S.E., 1994, Practice and training in environmental geology: Boulder, CO, Geological Society of America, Abstracts with Programs, vol. 26, no. 7, p. A389

Keaton, J.R., and G.L. Hempen, 1993, Environmental geology is engineering geology: AEG News, Association of Engineering Geologists, vol. 36, no. 1, p. 36–39.

Keller, E.A., 1992, Environmental geology (6th edition): New York, Macmillan Publishing Company, 521 p.

ReVelle, P., and C. ReVelle, 1992, The global environment: securing a sustainable future: Boston, Jones and Bartlett Publishers, Inc., 480 p.

Skinner, B.J., and S.C. Porter, 1992, The dynamic earth: an introduction to physical geology: New York, John Wiley and Sons, Inc., 570 p.

Tarbuck, E.J., and F.K. Lutgens, 1993, The Earth: an introduction to physical geology: New York, Macmillan Publishing Company, 654 p.

U.S. Geological Survey, 1975, Mineral resources perspectives: Washington, D.C., Professional Paper No. 940, p. 1.

SUPPLEMENTAL READING

Keller, E.A., 1992, Environmental geology (6th edition): New York, Macmillan Publishing Company, 521 p.

Skinner, B.J., and S.C. Porter, 1992, The dynamic earth: an introduction to physical geology: New York, John Wiley and Sons, Inc., 570 p.

Definition and Sources of Hazardous Waste

REGULATORY FRAMEWORK

In Chapter 1 we looked at a simple definition of hazardous waste. However, the definition that any waste that has the potential to harm human health or the environment is a hazardous waste is not adequate from a legal standpoint. Therefore, in order to comply with laws dealing with hazardous wastes, the U.S. Congress, in the Resource Conservation and Recovery Act (RCRA) of 1976, defined hazardous waste as *a waste, or combination of wastes, which because of its quantity, concentration, or physical, chemical, or infectious characteristics may (1) cause or significantly contribute to an increase in mortality or an increase in serious irreversible or incapacitating reversible illness, or (2) pose a substantial present or potential hazard to human health or the environment when improperly treated, stored, transported, or disposed of.*

EPA Definition

The U.S. Environmental Protection Agency (EPA) considers a waste to be hazardous if (a) it possesses certain characteristics (ignitibility, corrosivity, reactivity, or toxicity), or (b) it is on a list of specific wastes that are determined by the EPA to be hazardous. The former are called *characteristic wastes* and the latter *listed wastes*. RCRA regulations, found in 40 Code of Federal Regulations (CFR), Part 261 (40 CFR, 1991), give the listing of hazardous wastes and also describe hazardous waste characteristics and specify test methods to determine whether or not a waste is hazardous.

According to the EPA, a solid waste* will be considered hazardous if it meets *any* of the following four criteria:

1. It is a listed waste, i.e., the waste material is listed as a hazardous waste in 40 CFR, Part 261. Such wastes are designated by an alphabetic prefix (F, K, P, or U) followed by a three-digit number, and are considered hazardous regardless of concentration. F series wastes are generated by a variety of industrial processes (nonspecific sources) and are broken down into solvent wastes (F001–F005), electroplating wastes (F006–F012, F017), and dioxin wastes (F020–F023, F026–F028). Table A–1 (Appendix A) gives a list of F wastes.

2. It is a *characteristic* waste, meaning that it exhibits any of the following characteristics: ignitibility, corrosivity, reactivity, and toxicity (ICRT).

3. It is a mixture containing both hazardous and nonhazardous waste. Exception: The mixture is specifically excluded or no longer exhibits any of the characteristics of a hazardous waste.

4. It is not specifically excluded from regulation as a hazardous waste (a listing of *excluded* wastes is given in Table 3–1.

The second category of listed wastes includes those generated by specific sources. These are designated by the letter K and come from various industrial materials and processes: metal processing, wood preservation, petroleum products, acids and caustics, pesticides or related chemicals; dyes, paints, printing inks, thinners, solvents, or cleaning fluids; explosives, and the like. Table A-2 (Appendix A) contains a list of K wastes.

The third category of listed waste includes commercial chemical products. These are designated by the letter P or U. The P wastes are acutely hazardous (contain chemicals that are fatal to humans in small doses) and are subject to more stringent requirements for empty containers and quantity limits (1 kg for acute hazardous waste; 1000 kg for nonacute hazardous waste). All P series wastes, F020–F023, and, F026–F028 are acutely hazardous; U series wastes are not acutely hazardous.

In addition to the above, any by-products coming from treatment of any hazardous waste should also be considered hazardous, unless they have been specifically excluded as such.

Characteristic Waste

The EPA has described in detail the terms *ignitibility, corrosivity, reactivity,* and *toxicity.* It has also set the specifications and test procedures to determine these characteristics. The idea is that using these procedures and specifications, it will be easy for waste generators to ascertain if their waste is hazardous. The law puts the responsibility on the waste generators to determine whether their wastes are hazardous or not.

Available test procedures are accurate enough to enable determination of the four characteristics. However, test protocols and data interpretation for measuring

---

*The term *solid* waste is a misnomer because it does not refer to the physical state of the waste. In the EPA's meaning of the term, it includes both solid and liquid wastes and their mixtures, with or without gases.

TABLE 3–1   Excluded Hazardous Wastes (adapted from 40 CFR, Part 261, 1990)

1. Domestic sewage
2. Any mixture of domestic sewage and any other waste that passes through a sewer system to a publicly owned treatment works (POTW)
3. Irrigation return flows
4. Source, special nuclear, or by-product material as defined by the Atomic Energy Act (AEA)
5. Materials subjected to in situ mining techniques that are not removed from the ground during extraction
6. Certain pulping liquors used in the production of paper
7. Spent sulfuric acid used to produce virgin sulfuric acid
8. All household wastes and resource recovery facilities that burn only household waste (Hotel, motel, septic sewage, and campground wastes are all considered household waste.)
9. Materials returned to the soil as fertilizers, such as manure and crops
10. Mining overburden returned to the mine site
11. Fly ash waste, bottom ash waste, slag waste, and flue gas emission control waste generated primarily from the combustion of coal or other fossil fuels (the "utility waste exemption")
12. Drilling fluids, produced waters, and other wastes associated with the exploration, development, or production of crude oil, natural gas, or geothermal energy
13. Specific wastes from the tannery industry containing primarily trivalent chromium instead of hexavalent chromium
14. Solid waste from the extraction and beneficiation of ores and minerals, including phosphate rock overburden from uranium mining (the "mining waste exclusion," or the Bevill Amendment)
15. Cement kiln dust
16. Discarded wood that fails only the Characteristic Toxicity Test (a test to determine if a waste exhibits a hazardous characteristic) for arsenic as a result of being treated with arsenical compounds

carcinogenicity, mutagenicity, phytotoxicity, and potential for bioaccumulation are either poorly developed or are too complex and require high levels of expertise. Because of these limitations, the EPA has not included these characteristics in its current definition of hazardous waste. It is likely that with the availability of simpler test procedures and equipment, combined with the desired level of confidence, the EPA may, in the future, include these characteristics with the four now commonly used to characterize a waste as hazardous.

The characteristics of ignitibility, corrosivity, reactivity, and toxicity are discussed in the following sections.

Ignitibility

Ignitibility relates to the potential of waste material to cause a fire during storage, disposal, or transport. Used organic solvents and waste oils are examples of ignitable materials; if they are part of a waste stream, the waste will be classified as hazardous.

If a representative sample of a waste exhibits *any* of the following properties, it will be ruled hazardous:

1. It is a liquid, other than an aqueous solution containing less than 24 percent alcohol by volume, and has flash point of less than 60°C (140°F), as determined by a Pensky-Martens Closed Cup Tester, using the test method specified in ASTM Standard D-93-90 (ASTM, 1992), or by a Setaflash Closed Cup Tester, using the test method specified in ASTM Standard D-3278-78 (ASTM, 1993).

2. It is not a liquid and is capable, under standard temperature and pressure, of causing fire through friction, absorption of moisture, or spontaneous chemical changes, and when ignited burns so vigorously and persistently that it creates a hazard.

3. It is an ignitible compressed gas as defined in 49 CFR 173.300 and as determined by ASTM Test D-323.

4. It is an oxidizer as defined in 49 CFR 173.151.

A solid waste that exhibits the characteristic of ignitibility has the EPA Hazardous Waste Number D001.

Corrosivity

Hydrogen ion concentration (pH) is a good measure of corrosivity. Wastes containing materials of very high or very low pH can produce dangerous reactions with other materials in the waste. Acidic wastes (from many industrial processes) and *pickle liquor* (from steelmaking) are examples of corrosive wastes. Many hazardous wastes contain materials that corrode steel drums, causing the hazardous waste to leak into the environment.

A solid waste should be considered corrosive if a representative sample from the waste has either of the following properties:

1. It is aqueous and has a pH of 2 or less, or greater than or equal to 12.5 (Figure 3–1), as determined by a pH meter, using either an EPA test method or an equivalent test method.

2. It is a liquid and corrodes steel (SAE 1020) at a rate greater than 6.35 mm (0.25 in.) per year at a test temperature of 55°C (130°F), as determined by the test methods specified in National Association of Corrosion Engineers (NACE) Standard TM0169-76 (1991), and standardized in *Test Methods for Evaluating Solid Waste, Physical/Chemical Methods*, commonly known as SW-846 (EPA, 1986), or an equivalent approved test method.

A solid waste that exhibits corrosivity has the EPA Hazardous Waste Number D002.

Reactivity

Some constituents of a waste may be unstable and may have the potential to cause explosion at any stage of the waste management cycle. Used cyanide solvents and water from TNT operations are examples of reactive wastes.

FIGURE 3–1 pH range of corrosivity

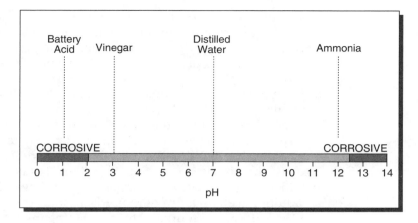

A solid waste exhibits the characteristic of reactivity if a representative sample of the waste has *any* of the following properties:

1. It is normally unstable and readily undergoes violent change without detonating.
2. It reacts violently with water.
3. It forms potentially explosive mixtures with water.
4. When mixed with water, it generates toxic gases, vapors, or fumes in a quantity sufficient to present a danger to human health or the environment.
5. It is a cyanide- or sulfide-bearing waste which, when exposed to pH conditions between 2 and 12.5, can generate toxic gases, vapors, or fumes in a quantity sufficient to present a danger to human health or the environment.
6. It is capable of detonation or explosive reaction if it is subjected to a strong initiating source or is heated under confinement.
7. It is readily capable of detonation or explosive decomposition or reaction at standard temperature and pressure.
8. It is a forbidden explosive as defined in 49 CFR 173.51, a Class A explosive as defined in 49 CFR 173.53, or a Class B explosive as defined in 49 CFR 173.88.

A solid waste that exhibits the characteristic of reactivity has the EPA Hazardous Waste Number D003.

Toxicity

Toxicity is the ability of a substance to cause death, injury, or impairment to an organism that comes in contact with it. Ingestion, inhalation, and dermal contact are common modes of contact. The EPA developed the toxicity characteristic test to identify wastes that are likely to leach hazardous constituents into groundwater from improperly managed facilities. The leach test simulates natural leaching action that may occur in a landfill. A solid waste will be considered toxic if, using the standard test methods, the extract from a representative sample of the waste contains any of the contaminants listed in Table 3–2 at a concentration equal to or greater than the respective value given in the table.

The Extraction Procedure (EP) leach test was used after 1980 to determine if a waste was likely to leach certain metals or pesticides into groundwater.

In March 1990, the EPA issued a new Toxicity Characteristics rule, the Toxicity Characteristic Leaching Procedure (TCLP) (40 CFR, 1990). This test is a modification of the 1980 standard, the EP leach test. TCLP is designed to test for an additional 25 organic compounds, besides the 15 metals and pesticides in the 1980 EP test. Essentially the new listing includes new chemicals that are determined to be toxic. It is likely that, as more test results become available, the list will be expanded. The additional 25 constituents included under the 1990 TCLP test are listed in Table 3–3.

If the waste is in liquid form, it is not required to be subjected to leaching. The liquid is considered an extract and analyzed for contaminants accordingly.

Flow charts developed by the EPA can be used to determine whether or not a solid waste is hazardous (Figures 3–2, 3–3). Details of the test procedures, apparatus used, and methods of calculation are given in 40 CFR, 261, Appendix II and III.

TABLE 3–2   Chemical Constituents Included in the EP Rule (adapted from 40 CFR, 261.30, 1990)

Chemical	Maximum Allowable Concentration (mg/L)
Arsenic	5.0
Barium	100.0
Cadmium	1.0
Chromium	5.0
Lead	5.0
Mercury	0.2
Selenium	1.0
Silver	5.0
Endrin	0.02
Lindane	0.4
Methoxychlor	10.0
Toxaphene	0.5
2,4-Dichlorophenoxyacetic acid	10.0
2,4-Dinitrotoluene	0.13
2,4,5-Trichlorophenoxypropionic acid (Silvex)	1.0

TABLE 3–3 New Constituents Included in the 1990 TCLP Test (adapted from 40 CFR, 261.30, 1990)

Chemical	Maximum Allowable Concentration (mg/L)
Benzene	0.5
Carbon Tetrachloride	0.5
Chlordane	0.03
Chlorobenzene	100.0
Chloroform	6.0
m-Cresol	200.0
o-Cresol	200.0
p-Cresol	200.0
Cresol	200.0
1,4-Dichlorobenzene	7.5
1,2-Dichloroethane	0.5
1,1-Dichloroethylene	0.7
2,4-Dinitrotoluene	0.13
Heptachlor (and its hydroxide)	0.008
Hexachloro-1,3-butadiene	0.5
Hexachlorobenzene	0.13
Hexachloroetahne	3.0
Methyl Ethyl Ketone	200.0
Nitrobenzene	2.0
Pentachlorophenol	100.0
Pyridine	5.0
Tetrachloroethylene	0.7
Trichloroethylene	0.5
2,4,5-Trichlorophenol	400.0
2,4,6-Trichlorophenol	2.0
Vinyl Chloride	0.2

SOURCES AND GENERATORS OF HAZARDOUS WASTES

In the process of producing goods and services, we also generate wastes, and in many cases these wastes are hazardous. Everything from hair dressing to production of drugs, from sign painting to steelmaking, involves hazardous substances. The level of industrialization of a nation and the consumer demand result in processes that generate hazardous wastes. The major industries that generate hazardous wastes in developed countries include

- *Petrochemical industry:* Phenols, metals, acids, caustics, and organic compounds
- *Metal industry:* Heavy metals, fluorides, cyanide, acids, alkali, solvents, and phenols
- *Leather industry:* Heavy metals and sulfides

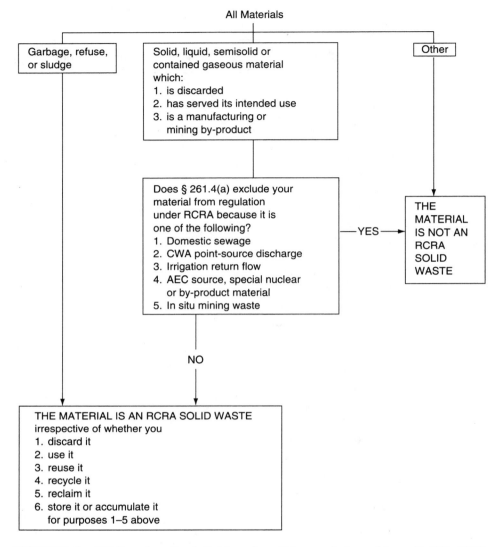

FIGURE 3–2   EPA flowchart for definition of a solid waste (adopted from 40 CFR, 1990)

The following is a partial list of other industries generating hazardous wastes:

- Primary metals smelting and refining
- Paints and allied products
- Organic chemicals, pesticides, and explosives
- Electrical and electronics
- Electroplating and metal finishing
- Rubber
- Batteries
- Pharmaceutical

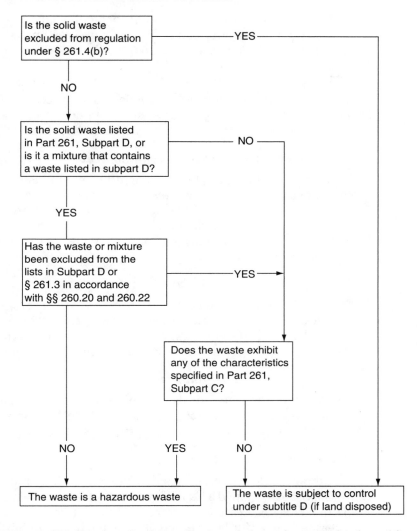

FIGURE 3–3   EPA flowchart for determination of a hazardous waste (adapted from 10 CFR, 1990)

- Textiles, dyeing, and finishing
- Petroleum refining
- Special machinery
- Leather tanning and finishing
- Plastics
- Waste oil re-refining

In addition, military operations also generate large quantities of hazardous wastes that come from explosives, naval paints, plating, and related processes. Defense activities of the past have resulted in a large number of hazardous waste sites that are awaiting cleanup. According to the 1991 Annual Defense Environmental Restoration Program Report, the 40-year Cold War left behind

14,401 toxic waste sites at 1579 military facilities in the United States (Kansas City Star, 1991). Cost of cleanup of these sites is estimated to range between $20 billion and $40 billion, according to the EPA, and may take decades.

EPA Categorization of Hazardous Waste Sources

In terms of categorization of hazardous waste sources, the EPA has grouped hazardous wastes as: (a) hazardous wastes from nonspecific sources, (b) hazardous wastes from specific sources, and (c) commercial chemical products.

Hazardous Wastes from Non-Specific Sources

Generic wastes produced by manufacturing and industrial processes are included in the list of hazardous wastes from nonspecific sources. Spent halogenated solvents, bottom sludge from electroplating operations, wastes from various chemicals manufacturing, and the like are examples of hazardous wastes from nonspecific sources. Table A–1 (Appendix A) gives a list of such wastes; these are designated by the prefix F.

Hazardous Wastes from Specific Sources

This list includes sludge, still bottoms, wastewaters, spent catalysts, and residues. These are generated by specific industrial processes, such as wood preserving, petroleum refining, and organic chemical manufacturing. Table A–2 (Appendix A) gives a list of such wastes; these are designated by the prefix K.

Commercial Chemical Products

This list comprises chemicals like chloroform, creosote, acids, pesticides, and numerous other commercial chemicals. Table A–3 (Appendix A) lists all such wastes; these are designated by the prefixes P and U.

Hazardous Constituents

In addition to the materials listed in Tables A–1, A–2, and A–3 (Appendix A), the EPA has also developed a list of hazardous constituents which it considers capable of producing toxic, carcinogenic, mutagenic, or teratogenic effects on humans and other life forms. This list is based on research results published in reputable journals and identified by the EPA's Carcinogenic Assessment Group (CAG). It is included in Table A–4 (Appendix A). The EPA has also compiled a list of hazardous constituents that it uses to identify a waste as hazardous. This list (Table A–5 in Appendix A) is also used for de-listing wastes (from hazardous to nonhazardous).

Industry Categorization of Hazardous Waste Sources

The hazardous waste management industry recognizes five generic categories of wastes:

1. Metals/metal finishing
2. Paints/solvents/coatings
3. Organic
4. Petroleum
5. Inorganic

In 1991—the year for which the latest data are available—there were 23,426 generators who collectively produced 306 million tons (MT) of hazardous waste in the United States (EPA, 1994). The quantity generated in 1989 was 198 MT, which apparently means that the quantity of hazardous waste produced has increased. This is not the case; the ambiguity is due to the inclusion of 25 new constituents in the 1990 TCLP rule discussed previously. This means that 137 MT of waste that could have passed the older EP test (pre-1990) was included for 1991. Therefore, for comparison purposes this quantity should be excluded from the 1991 figure, which means that the quantity produced in 1991 was actually 29 MT less than in 1989.

Table 3–4 shows the number of generators and the quantity of hazardous waste produced in 1991 in the 50 states and the territories of the United States. Texas was the largest producer of hazardous waste (104 MT, or 34 percent of the national total), followed by Michigan (32 MT), Louisiana (31 MT), and New Jersey (29 MT). Together these four states accounted for 64 percent of the total quantity produced.

Table 3–4 also shows that New York had the largest number of hazardous waste generators (2627, or 11.2 percent), followed by California (2116), New Jersey (1661), and Ohio (1542). Collectively, these four states accounted for 34 percent of all generators in the country.

TABLE 3–4   Quantity of Hazardous Waste Generated and Number of Hazardous Waste Generators in the United States in 1991 (adopted from EPA, 1994)

State	Hazardous Waste Quantity			Hazardous Waste Generators		
	Rank	Tons Generated	Percentage	Rank	Number of Generators	Percentage
Alabama	27	559,823	0.2	23	2177	1.2
Alaska	42	24,141	0.0	45	58	0.2
Arizona	34	158,279	0.1	26	249	1.1
Arkansas	23	748,018	0.2	34	149	0.6
California	8	12,925,393	4.2	2	2116	9.0
Colorado	30	478,343	0.2	35	146	0.6
Connecticut	15	2,062,163	0.7	15	483	2.1
Delaware	43	20,531	0.0	43	63	0.3
District of Columbia	53	975	0.0	52	11	0.0
Florida	28	508,839	0.2	13	399	1.7
Georgia	22	757,885	0.2	18	399	1.7
Guam	55	346	0.0	53	8	0.0
Hawaii	51	2032	0.0	48	35	0.1
Idaho	12	4,350,064	1.4	47	40	0.2
Illinois	7	13,086,020	4.3	7	1229	5.2
Indiana	19	1,633,861	0.5	11	671	2.9
Iowa	37	126,218	0.0	32	155	0.7
Kansas	13	3,215,044	1.1	29	177	0.8
Kentucky	29	487,622	0.2	16	465	2.0

TABLE 3–4   *(continued)*

State	Hazardous Waste Quantity			Hazardous Waste Generators		
	Rank	Tons Generated	Percentage	Rank	Number of Generators	Percentage
Louisiana	3	31,486,169	10.3	22	309	1.3
Maine	46	11,657	0.0	32	155	0.7
Maryland	39	75,911	0.0	17	430	1.8
Massachusetts	32	274,985	0.1	14	552	2.4
Michigan	2	31,862,518	10.4	9	755	3.2
Minnesota	11	5,662,647	1.9	24	276	1.2
Mississippi	9	8,050,831	2.6	28	178	0.8
Missouri	24	686,651	0.2	20	389	1.7
Montana	47	11,177	0.0	46	54	0.2
Nebraska	40	35,705	0.0	40	86	0.4
Nevada	48	9951	0.0	41	71	0.3
New Hampshire	44	17,309	0.0	30	166	0.7
New Jersey	4	29,490,704	9.6	3	1661	7.1
New Mexico	35	155,943	0.1	42	68	0.3
New York	5	18,036,041	5.9	1	2627	11.2
North Carolina	31	281,849	0.1	13	582	2.5
North Dakota	25	685,256	0.2	51	16	0.1
Ohio	16	1,809,547	0.6	4	1542	6.6
Oklahoma	20	933,230	0.3	31	161	0.7
Oregon	36	132,297	0.0	27	191	0.8
Pennsylvania	18	1,692,608	0.6	6	1264	5.4
Puerto Rico	14	3,120,686	1.0	39	97	0.4
Rhode Island	45	14,653	0.0	37	107	0.5
South Carolina	26	604,456	0.2	21	337	1.4
South Dakota	52	979	0.0	50	21	0.1
Tennessee	17	1,697,402	0.6	10	683	2.9
Texas	1	104,079,270	34.0	5	1394	6.0
Trust Territories	49	2835	0.0	54	3	0.0
Utah	21	900,643	0.3	38	99	0.4
Vermont	41	35,565	0.0	43	63	0.3
Virgin Islands	54	811	0.0	55	1	0.0
Virginia	38	96,169	0.0	25	264	1.1
Washington	6	14,726,588	4.8	8	939	4.0
West Virginia	10	7,619,802	2.5	36	120	0.5
Wisconsin	33	253,308	0.1	12	607	2.6
Wyoming	50	2127	0.0		4928	0.1
Total		305,708,881	100.0		23,426	100.0

Categorization of Generators Based on Waste Quantity

The law defines a generator as any person whose act or process produces hazardous waste, or whose act first causes a hazardous waste to become subject to regulation. This places the liability not only on the first producer of the waste, but also on those who get involved in its subsequent handling and/or removal. If the owner of a raw material storage tank hires a contractor to remove waste from the tank, then both the tank cleaner and the tank owner are considered generators according to the EPA definition of generator in 40 CFR, 262.10.

Before the passage of the Hazardous and Solid Waste Amendments (HSWA) of 1984, there were only two categories of hazardous waste generators: those generating <1000 kg/month and those generating 1000 kg/month or more. HSWA reorganized the categories of generators and developed a three-tier classification of generators. If the total quantity of hazardous waste produced in a calendar month is <100 kg for nonacutely hazardous waste or <1 kg of an acutely hazardous waste, the generator is categorized as a *small-quantity generator* [(SQG); earlier called *conditionally exempt small quantity generator* (CESQG)]; a *medium-quantity generator* (MQG) if the quantity produced is between 100–1000 kg/month of nonacutely hazardous waste; and a *large-quantity generator* (LQG) if the quantity produced is 1000 kg or more of nonacutely hazardous waste or >1 kg/month of acutely hazardous waste. Special reporting, licensing, training, and other requirements have been specified by the EPA. These are summarized in Table 3–5.

TABLE 3–5 Requirements for Hazardous Waste Generators (adapted from 40 CFR, 1990)

	Requirements		
	Small-Quantity Generators	**Medium-Quantity Generators**	**Large-Quantity Generators**
Quantity limits	<100 kg/mo	100–1000 kg/mo	1000 kg/mo or greater
Management of waste	State-approved or RCRA-permitted facility	RCRA-permitted facility	RCRA-permitted facility
Manifest	Not required*	Required	Required
Biennial report	Not required	Not required	Required
Personnel training	Not required	Basic training required	Required
EPA ID number	Not required*	Required	Required
On-site storage limits	May accumulate up to 999 kg	May accumulate up to 6000 kg for up to 180 days, or 270 days if waste is to be transported over 200 miles	May accumulate any quantity up to 90 days

*Although not legally required under RCRA, many transporters will not handle hazardous waste without these items.

Categorization of generators is solely based on the quantity of hazardous waste produced per calendar month. This means that it is possible for a generator to change categories each month. For example, a generator who normally produces <100 kg of hazardous waste each month had 1200 kg of waste produced in a certain month, say June. This happened because of an unusually large order for his product. According to the EPA, the generator will be considered an LQG for June only. After that month, if the quantity of hazardous waste produced is <100 kg per month, the generator will revert to the earlier classification of SQG.

STUDY QUESTIONS

1. How does the EPA define hazardous waste? How can a generator determine if the waste produced is hazardous?
2. What are the characteristics of a hazardous waste? Define each, and explain how they can be determined.
3. List ten industries that generate hazardous waste. Are the wastes produced at defense establishments hazardous? Does the EPA have the responsibility for cleanup of such hazardous waste sites?
4. Distinguish between hazardous waste from specific and nonspecific sources; give three examples of each. What prefix is used by the EPA to designate such wastes?
5. What is a hazardous constituent? Which group within the EPA determines inclusion of a substance on its List of Hazardous Constituents? How does the group make this determination?
6. What are the five major categories of hazardous waste generators identified by the hazardous waste management industry?
7. What is the difference between acutely hazardous and nonacutely hazardous wastes?
8. What are the three categories of hazardous waste generators established by the EPA in relation to the quantity of waste generated? What is the basis of this classification? If a manufacturer produces 980 kg of nonacutely hazardous waste but >1 kg of acutely hazardous waste in one calendar month, will it be considered an MQG or an LQG?
9. What are the legal requirements for the SQG, MQG, and LQG as it relates to the quantity of waste produced, manifest requirement, and biennial reporting? Prepare a table to show the relationships.
10. Can a generator change categories? Explain.
11. Why is the total amount (306 MT) of hazardous waste generated in 1991 actually less than that for 1989 (198 MT)?

REFERENCES CITED

40 Code of Federal Regulations, Part 261, July 1, 1990, Environmental Protection Agency: Washington, D.C., U.S. Government Printing Office, p. 20–122.

49 Code of Federal Regulations, Part 173, October 1, 1989, Department of Transportation Regulations: Washington, D.C., U.S. Government Press, p. 558, 659–660.

American Society for Testing and Materials, 1992, 1992 Annual Book of ASTM Standards: Philadelphia, American Society for Testing and Materials, Section 4, vol. 04.09, p. 24–31.

American Society for Testing and Materials, 1993, 1993 Annual Book of ASTM Standards: Philadelphia, American Society for Testing and Materials, Section 6, vol. 06.01, p. 432–438.

Kansas City Star, 1991, Military's "toxic legacy" is assailed: Capital Cities/ABC, Inc., March 13, p. A–5.

National Association of Corrosion Engineers, 1991, Standard Test Method: Laboratory Corrosion Testing of Metals for the Process Industries, Standard TM0169–76: NACE Book of Standards, 9 p.

Resource Conservation and Recovery Act of 1976, 42nd U.S. Congress 6901 et seq., Public Law 94-580 and its amendment of Nov. 1978, Public Law 95-609.

U.S. Environmental Protection Agency, 1993, The Biennial RCRA Hazardous Waste Report: Washington, D.C., U.S. Government Printing Office, 351 p.

U.S. Environmental Protection Agency, 1994, Executive Summary: The Biennial RCRA Hazardous Waste Report (Based on 1991 Data), Report No. EPA/530-S-94-039: Washington, D.C., U.S. Government Printing Office, 6 p.

U.S. Environmental Protection Agency, 1986, Test Methods for Evaluating Solid Wastes, (3rd ed.), Vols. 1A, 1B, and 1C, Laboratory Manual, Physical/Chemical Methods, Report No. SW-846: Washington, D.C., Office of Solid Waste and Emergency Response, 4 chapters plus appendices, not sequentially paginated.

SUPPLEMENTAL READING

Wagner, T.P. 1991, The Hazardous Waste Q & A: New York: Van Nostrand Reinhold, 404 p.

Environmental Law in the United States

INTRODUCTION

Among the industrialized nations of the world, the United States can claim the distinction of being the first to recognize the severity of environmental problems and to adopt measures, as a nation, to protect and maintain the quality of the environment. As discussed in Chapter 1, effects of decades of negligence and environmental mismanagement began to show up in the early 1960s. Several infamous episodes—Love Canal being the foremost—raised the consciousness of the nation and drew attention to the need for action to control environmental degradation. These efforts culminated in the passage of the National Environmental Policy Act (NEPA) in 1969.

NEPA may be considered landmark legislation, because with its passage the United States committed itself to environmental protection. NEPA's passage led to formation of the presidential council on National Environmental Quality (NEQ) and the Environmental Protection Agency (EPA). Since 1969, a number of important legislative actions have been carried out, including the Resource Conservation and Recovery Act (RCRA) in 1976 and the Comprehensive Environmental Response, Compensation, and Liability Act (CERCLA) in 1980. The latter is popularly known as the Superfund. Passage of these and related acts created an enormous and unprecedented task for the primary regulatory agency, the EPA: To translate the intent of the laws into practical and workable guidelines—a task that took a long time to accomplish. For instance, regulations for RCRA took nearly four years to complete, yet were far from perfect. Loopholes were found, which were later closed by amending the laws. RCRA was amended in 1984, and the new law came to be known as the Hazardous and Solid Wastes Amendments (HSWA); CERCLA was amended in 1986, and the new version is known as the Superfund Amendments and Reauthorization Act (SARA). Both represent culmination of a long effort, during which a great deal was learned from the shortcomings of the original acts and from the ensuing mistakes. Nonetheless, the amendments led to the removal of uncertainties and brought about the realization that containment and disposal should not be the preferred option for managing hazardous waste—alternative technologies relating to waste reduction or material substitution should be considered first. The irreducible waste produced must then be properly treated prior to disposal.

The important thing to realize is that in the 1970s the United States embarked upon a massive effort for environmental protection, which was without precedent in the world. It was only natural that mistakes or omissions would be made. But the commitment for a better environmental quality helped in modifying, redefining, and amending the laws to the point that we now—after ten years—have a hazardous waste management system that is second to none. This has been aptly stated by Fortuna (1989, p. 1.7): "Together these landmark statutes mark the end of the beginning of the national program by establishing the beginning of the end of unrestricted land disposal. They represent an enduring commitment to creation of a hazardous waste management system in the United States that we can at last look on with pride and a sense of certainty rather than look back on with chagrin."

In this chapter we will review various environmental laws that have been enacted in the United States. It should be noted that at the federal level the EPA is the agency that regulates the laws. However, the 50 states also have their own regulations dealing with management of hazardous waste and environmental pollution. Some states have stricter requirements for compliance. As a rule, federal laws take precedence over state laws except when the latter have more stringent requirements.

IMPORTANT LAWS

The following is a list of major laws relating to the environment that have been enacted in the past:

- Rivers and Harbors Act of 1899
- Atomic Energy Act of 1954
- Solid Waste Disposal Act of 1965
- National Environmental Policy Act (NEPA) of 1969
- Resource Recovery Act of 1970
- Occupational Safety and Health Act of 1970
- Clean Air Act (CAA) of 1970; amended 1990
- Federal Water Pollution Control Act (FWPCA) of 1972
- Marine Protection, Research, and Sanctuaries Act (MPRSA) of 1972
- Federal Insecticide, Fungicide, and Rodenticide Act (FIFRA) of 1972
- Safe Drinking Water Act (SDWA) of 1974; amended 1986
- Resource Conservation and Recovery Act (RCRA) of 1976
- Toxic Substances Control Act (TSCA) of 1976

- Comprehensive Environmental Response, Compensation, and Liability Act (CERCLA) of 1980; commonly called Superfund
- Hazardous and Solid Wastes Amendments (HSWA) of 1984
- Superfund Amendments and Reauthorization Act (SARA) of 1986

Rivers and Harbors Act

The Rivers and Harbors Act, passed in 1899, is also referred to as the Refuse Act. This act stated that it is unlawful to throw, discharge, or deposit any type of refuse from any source, except that flowing from streets and sewers, into any *navigable water* or its tributaries, except with the permission of the Secretary of the Army. This law may be considered the first attempt to restrict water pollution; however, the emphasis at that time was not on the environmental quality of waters but on their navigational capabilities. In 1970 the act was resurrected to effectively begin the process of elimination of indiscriminate dumping of waste materials into streams and rivers.

Atomic Energy Act

The Atomic Energy Act (1954) was developed to regulate nuclear materials and to manage nuclear wastes. The Atomic Energy Commission (AEC) was established to license the processing and use of nuclear materials in the defense and public sectors.

Through an act known as the Energy Reorganization Act (1974), the AEC was divided into two agencies: (1) The Department of Energy (DOE), and (2) The Nuclear Regulatory Commission (NRC). The DOE was made responsible for development of nuclear energy and production of nuclear devices in the defense establishments; the NRC was assigned licensing and oversight authority for civilian nuclear energy production (nuclear power plants).

Nuclear efforts since the 1940s, both for civilian and defense use of nuclear energy, led to the development of several uranium mines and processing plants across the United States. The finely ground residue produced by the processing of uranium ore (tailings) contains low concentrations of radioactive materials, including thorium and radium. Both produce radon gas, which is known to cause lung cancer. The majority of these tailings are found in the Western United States (Figure 4–1). A special act, the Uranium Mill Tailings Radiation Control Act, was passed in 1978. The act set the regulations for control and stabilization of uranium and thorium mill tailings to protect public health and the environment. It made the DOE responsible for cleanup of the 24 inactive tailing piles; 90 percent of the cost is to be paid by the federal government, with 10 percent shared by the respective states. While the DOE is currently trying to remediate such sites, some (Satchell, 1989) believe it will not be possible to completely clean up some of these sites.

The Low-Level Radioactive Waste Policy Act was passed in 1980 and amended in 1985. A definition was provided for low-level radioactive waste (LLRW), and the NRC was no longer charged with the responsibility for its management. The act gave this responsibility to the respective states where LLRW was generated, and encouraged neighboring states to form LLRW Compacts and select sites for its disposal within these compacts. The compacts were given the authority to refuse LLRW from outside their boundaries. Because of the delay in selecting

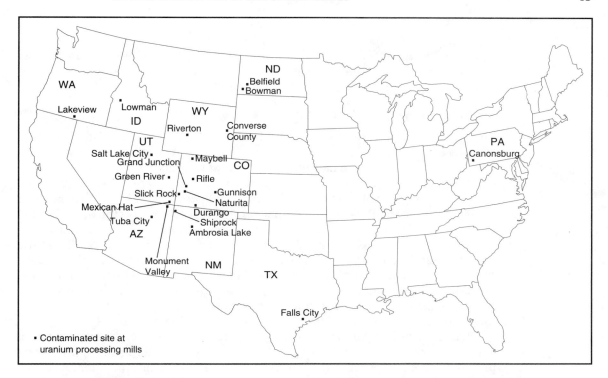

FIGURE 4–1   Distribution of uranium processing mills and associated contaminated site in the United States.

sites for disposal of LLRW within the various compacts, the Low-Level Radioactive Waste Policy Amendments Act was enacted in 1985. This act extended the deadlines for disposal facility siting to 1993, and set up specific dates for progress toward construction of new disposal facilities.

The Nuclear Waste Policy Act (1982) requires that high-level radioactive (nuclear) wastes coming from nuclear power plants and defense establishments be disposed of in geologic repositories. Criteria for siting, design, operation, and monitoring of these facilities have been established under the act, which also includes standards for protection of public health and the environment. The DOE will be responsible for operation of the repository, and the NRC will be responsible for its licensing. Presently a site in Nevada, the Yucca Mountain Site, is being investigated to determine its suitability as a permanent disposal facility. If the site meets all criteria and is approved by Congress, the United States may have its first site for disposal of high-level nuclear waste by the year 2010.

Solid Waste Disposal Act

The Solid Waste Disposal Act (SWDA) of 1965 aimed at regulation of municipal solid waste and improvement of solid waste disposal technology. The purpose was to protect human health and the environment and to reduce the volume of waste. The law provided funds for management of solid waste at the state level.

Resource Recovery Act

The Solid Waste Disposal Act was amended in 1970 and came to be known as the Resource Recovery Act. The act encouraged waste reduction and resource recovery, and created a system of national disposal sites for hazardous wastes. This legislation was the forerunner of the RCRA (1976).

National Environmental Policy Act

The National Environmental Policy Act (NEPA) of 1969 (signed into law January 1, 1970, by Richard Nixon) may be considered landmark legislation because it represents an expression of the nation's commitment to protection and maintenance of environmental quality. Several agencies and departments were created as a result of the passage of this act. The Council on Environmental Quality (CEQ) was formed to assist the President in preparing an annual report on the quality of the environment and to make recommendations on national policies for improving the environment. The CEQ analyzes conditions and trends in the quality of the environment and presents the information in the form of a report. This report, titled *Environmental Quality,* is submitted to the President, who then sends it to the Congress. These reports are a very good source of information on various aspects of the country's environment. The first report was published in 1970; as of this writing, the 24th Annual Report, for 1993, was being prepared.

The Environmental Protection Agency (EPA) was created by a Presidential action, subsequent to the passage of the NEPA. Also, the requirement for the preparation of an Environmental Impact Statement (EIS) became mandatory for all projects that are funded, even in part, by the federal government. Many states followed suit, and currently all 50 states require an EIS before the project can be approved for construction. The EIS addresses the environmental impacts of the proposed activity; unavoidable adverse impacts, alternatives to the proposed activity, and irretrievable commitments of resources. An EIS in effect forces the developer or owner of a project to fully evaluate environmental consequences of the project before the first clod of soil is turned for construction.

Occupational Safety and Health Act

The Occupational Safety and Health Act of 1970 is the most important law dealing with the protection of health and safety of employees in the workplace. The Occupational Safety and Health Administration (OSHA) was created to enforce the regulations. OSHA, though not an environmental agency per se, has developed guidelines, in consultation with the EPA, to regulate health and safety of personnel involved in hazardous waste investigation and cleanup operations. One of the provisions of the act is the requirement for maintaining up-to-date health records for all employees who are exposed to chemical substances.

Clean Air Act

The Clean Air Act (CAA), enacted in 1963 and amended in 1970, 1977, and 1990, aims at controlling sources of air pollution and improving air quality. The CAA and its amendments govern emission of pollutants into the atmosphere from industrial and commercial activities. Air-quality standards and requirements to control pollutant release have been set forth in the laws. The maximum levels of

TABLE 4–1 Health Threats Associated with Major Air Pollutants (adopted from EPA, 1989b)

Pollutant	Health Concerns
Carbon monoxide	Cardiovascular, nervous, and pulmonary systems
Hydrocarbons	Various compound-specific health hazards
Lead	Retardation and brain damage in children
Nitrogen dioxide	Respiratory illness and lung damage
Ozone	Respiratory tract
Particulate matters	Eye and throat irritation
Sulfur dioxide	Respiratory tract

six pollutants that affect air quality and pose significant threat to human health were established pursuant to the 1970 CAA amendments (a seventh pollutant, airborne lead, was added to the list in 1987). Health effects of these pollutants are given in Table 4–1.

The 1970 amendments also contain provisions for control of five airborne hazardous chemicals that are known to be dangerous to human health even at low concentrations. These chemicals and their health threats are shown in Table 4–2.

The 1977 CAA amendments resulted in establishment of the National Ambient Air Quality Standards (NAAQS), which established two standards for air quality: *primary* and *secondary*. Primary standards represent concentration levels of the six pollutants at which a particular pollutant becomes dangerous to human health and may produce damaging effects on structures and plants. Secondary standards relate to the concentration level at which the same pollutant could cause a degradation of the overall environmental quality.

The 1990 CAA amendments gave greater enforcement powers to the EPA, enabling the agency to use new civil enforcement (fines up to $25,000 per day for each violation) and enhanced criminal proceedings ($25,000 to $1,000,000 fines and up to 15 year prison terms) against the violator. If the person or company has a prior conviction for the same violation, the maximum penalties for any subsequent conviction will be doubled. Under the laws, individuals as well as companies can be found liable for criminal penalties. The 1990 amendments also authorized the EPA to pay a reward (up to $10,000), to any person who provides

TABLE 4–2 Hazardous Chemicals and Their Health Threat (adopted from EPA, 1989b)

Hazardous Air Pollutants	Health Threats
Asbestos	Lung cancer
Benzene	Multisystem cancer
Beryllium	Lung disease
Mercury	Brain, renal, and gastrointestinal diseases
Vinyl chloride	Lung and liver cancer

information or services that will lead to a criminal conviction or payment of a civil penalty by violators. This is known as the *whistle-blower/bounty provision.*

In order to determine compliance with the laws, air quality in the United States is continuously monitored and the information is maintained in a database called the Toxics Release Inventory (TRI). Table 4–3 shows the improvement in air quality following enforcement of the CAA.

Federal Water Pollution Control Act

The Federal Water Pollution Control Act (FWPCA) of 1972 is also known as the Clean Water Act. The legislation was enacted with the main purpose of restoring and maintaining the chemical, physical, and biological integrity of the waters of the United States. The act also gave authority to the EPA and states to take appropriate measures to control water pollution. Before the passage of the act, many of the nation's streams and lakes had been polluted to excessive levels. One of the goals of the Clean Water Act was to restore the nation's waters to fishable and swimmable conditions. Under the act, the EPA sets limits on the quantities of pollutants that may be discharged into surface waters by industries and municipalities, who are required to obtain either from the EPA or the state authorities, a permit for discharge. This permit is part of the National Pollution Discharge Elimination System (NPDES). The act was amended in 1977 and 1981.

The Clean Water Act classifies wastes as (a) toxic, (b) conventional, and (c) nonconventional. Water containing toxic materials must be treated before being released into water bodies (stricter standards have been imposed on this category under the RCRA). Conventional pollutants include those found in municipal wastes; for such water existing technology has to be employed to control the waste before it is released into water bodies. Unconventional pollutants should be removed using the Best Available Technology (BAT). The Clean Water Act has resulted in an overall improvement in the water quality of the nation's rivers and lakes, many of which have been returned to fishable and swimmable conditions.

Marine Protection, Research, and Sanctuaries Act

The Marine Protection, Research, and Sanctuaries Act (MPRSA) of 1972, also known as the Ocean Dumping Act, relates to the protection of the marine environment. Specifically the act prevents or limits dumping of wastes that would adversely affect human health or the marine environment into the oceans. All

TABLE 4–3 Emissions of Hazardous Substances in the Atmosphere (EPA, 1989a, 1993)

Year	Non-Point-Source Emission (billion pounds)	Point-Source Emission (billion pounds)
1987	0.8	1.8
1991	0.6	1.3

wastes to be dumped at sea must have a permit. The following categories of wastes are prohibited from ocean dumping:

- High-level radioactive wastes
- Biological, chemical, or radiological warfare materials
- Incompletely characterized materials
- Persistent inert synthetic or natural materials that float or remain suspended in the ocean and are capable of causing interference with fishing, navigation, or other use of the ocean

Federal Insecticide, Fungicide, and Rodenticide Act

The Federal Insecticide, Fungicide, and Rodenticide Act (FIFRA) of 1972 and its 1978 amendments set two broad regulations: (1) on storage and disposal of pesticides, and (2) on requirements for accurate and informative labeling of products. Pesticides are classified as *general use* or *restricted use*. Restricted-use pesticides require user certification. Restriction on the use of a certain pesticide can be imposed in areas where there has been a case of groundwater contamination or where a potential for such contamination exists. FIFRA requires premarket clearance of pesticides to prevent potential hazard to people and the environment. The EPA has been collecting information on various pesticides, and those suspected of causing serious problems have been withdrawn from the market; EDB (ethylene dibromide) and DBCP (dibromochloropropane) are examples. The EPA also requires extensive field monitoring of some pesticides by the manufacturers.

Safe Drinking Water Act

The Safe Drinking Water Act (SDWA), first passed in 1974, has been amended twice—in 1977 and 1986. This act and its 1977 amendments set minimum standards for safe drinking water in the United States.

The SDWA also aims at protection of "sole source" aquifers and other aquifers from contamination resulting from the underground injection of waste. All aquifers, including current or potential drinking-water aquifers, with a total dissolved solids (TDS) concentration of less than 10,000 mg/L (cf freshwater: 0–1000 mg/L) are to be protected from possible contamination from deep-well injection of liquid waste. The act was further amended in 1986 (the Gonzales Amendment), authorizing the EPA to designate aquifers that are specially valuable because they are the only source of drinking water in an area. The amendment directed the states to take measures to protect the surface area around public water supply wells from potential contamination from hazardous wastes, pesticides, and leaking underground storage tanks (USTs).

Underground injection of hazardous wastes and other materials is also regulated under the SDWA. Injection wells are classified as class I, class II, or class III wells. Class I wells include wells used for disposal of hazardous waste into isolated strata below the current or potential drinking-water aquifer. Class II wells include those used for recirculation of oilfield brines, and class III wells are those that are used in solution mining of minerals. The SDWA requires the EPA to establish drinking-water standards for protection of public health. A large number of chemicals and related substances are regulated, and the maximum contamination level

(MCL) for each has been established. Table 4–4 gives the MCLs for various chemicals and other regulated substances in drinking water.

Resource Conservation and Recovery Act

The Solid Waste Disposal Act of 1965 and the Resource Recovery Act of 1970 were amended in 1976 under the Resource Conservation and Recovery Act (RCRA). The act authorized the EPA to regulate all aspects of waste management, including its generation, storage, transportation, treatment, and disposal. Because of the all-inclusive nature of the act, it came to be known as the *cradle-to-grave* concept. Although the term "solid waste" is used in the law, all hazardous wastes in the form of solids, liquids, sludges, and containerized gases are regulated under the RCRA. The act has established criteria for defining hazardous waste and has set specific tests to determine whether a given waste should be classified as hazardous.

The RCRA has three major components: Subtitles C, D, and I. Subtitle C regulates hazardous waste; Subtitle D regulates solid (nonhazardous) wastes; and Subtitle I regulates underground storage tanks (USTs) that hold petroleum and other hazardous substances.

The RCRA aims at regulating hazardous wastes generated since the passage of the law (1976) and into the future, thereby excluding the hazardous wastes generated prior to 1976. The law mandates a permit-and-manifest system for all generators, transporters, and treatment, storage and disposal (TSD) facility owners/operators. It requires new land disposal facilities to have a groundwater monitoring system, and requires old facilities to be retrofitted for groundwater monitoring. Hazardous waste landfills and lagoons are required to have a double-liner system for leachate management. Land disposal of liquid hazardous wastes and certain other hazardous wastes is prohibited.

Comprehensive Environmental Response, Compensation, and Liability Act

The Comprehensive Environmental Response, Compensation, and Liability Act (CERCLA) of 1980 is commonly known as Superfund. Its primary goal is to clean up hazardous waste sites that were in existence prior to the passage of RCRA, i.e., the historical sites. It also contains a provision for emergency response to hazardous materials spills and cleanup. The law includes a provision known as *joint and several liability*, which permits the EPA to recover full cost of cleanup from *any* of the responsible parties, even if the party was responsible for only part of the waste. The law requires the EPA to establish a National Priority List (NPL) of hazardous waste sites to be targeted for remedial action. As of May 1994, there were 1232 sites on the NPL. This includes over 100 sites at federal facilities.

Superfund Amendments and Reauthorization Act

The Superfund Amendments and Reauthorization Act of 1980 is commonly referred to as SARA. The first reauthorization came in 1986; the second reauthorization was being reviewed in the U.S. Congress as of June 1994. SARA addresses some important concepts regarding financial liability for haz-

TABLE 4–4 Maximum Contaminant Level (MCL) for Various Chemicals in Drinking Water, Set by the EPA

Chemical	MCL (µg/L)	Chemical	MCL (µg/L)
Synthetic Organic Chemicals			
Acrylamide	Treatment technique	Epichlorohydrin	Treatment technique
Adipates (di(ethylhexyl)adipate)	500	Ethylbenzene	700
		Ethylene dibromide (EDB)	0.05
Alachlor	2	Glyphosate	700
Aldicarb	3	Heptachlor	0.4
Aldicarb sulfoxide	4	Heptachlor epoxide	0.2
Aldicarb sulfone	2	Hexachlorobenzene	1
Atrazine	3	Hexachlorocyclopentadiene (HEX)	50
Benzene	5	Indenopyrene	0.4
Benzo[a]anthracene	0.1	Lindane	0.2
Benzo[a]pyrene	0.2	Methoxychlor	40
Benzo[b]fluoranthene	0.2	Methylene chloride	5
Benzo[k]fluoranthene	0.2	Monochlorobenzene	100
Butylbenzyl phthalate	100	Oxamyl (vydate)	200
Carbofuran	40	PCBs as decachlorobiphenol	0.5
Carbontetrachloride	5	Pentachlorophenol	1
Chlorodane	2	Picloram	500
Chrysene	0.2	Simaze	4
Dalapon	200	Styrene	100
Dibenz[a,h]anthracene	0.3	2,3,7,8-TCDD (dioxin)	3×10^{-8}
Dibromochloropropane (DBCP)	0.2	Tetrachloroethylene	5
o-Dichlorobenzene	600	1,2,4-Trichlorobenzene	70
p-Dichlorobenzene	75	1,1,2-Trichloroethane	5
1,2-Dichloroethane	5	Trichloroethylene (TCE)	5
1,1-Dichloroethylene	7	1,1,1-Trichloroethane	200
cis-1,2-Dichloroethylene	70	Toluene	1000
trans-1,2-Dichloroethylene	100	Toxaphene	3
1,2-Dichloropropane	5	2-(2,4,5-Trichlorophenoxy)-propionic acid (2,4,5-TP, or Silvex)	50
2,4-Dichlorophenoxyacetic acid (2,4-D)	70		
Di(ethylhexyl)phthalate	6	Vinyl Chloride	2
Diguat	20	Xylenes (total)	10,000
Dinoseb	7		
Endothall	100		
Endrin	2		

TABLE 4–4   *(continued)*

Chemical	MCL (µg/L)	Chemical	MCL (µg/L)
Inorganic Chemicals		*Radionuclides*	
Antimony	6	Radium 226	20 pCi/L
Arsenic	50	Radium 228	20 pCi/L
Asbestos (fibers per liter)	7×10^6	Radon 222	300 pCi/L
Barium	2000	Uranium	20 µg/L (30 pCi/L)
Beryllium	4		
Cadmium	5	Beta and Photon emitters (excluding radium 228)	4 mrem ede/yr
Chromium	100		
Copper	1300	Adjusted gross alpha emitters (excluding radium 226, uranium, and radon 222)	15 pCi/L
Cyanide	200		
Fluoride	4000		
Lead	15		
Mercury	2		
Nickel	100		
Nitrate (as N)	10,000		
Nitrite (as N)	1000		
Selenium	50		
Sulfate	4×10^5–5×10^5		
Thallium	5		

Note: A pCi (picocurie) is a measure of the rate of radioactive disintegration; mrem ede/yr is a measure of the dose of radiation received by either the whole body or a single organ.

ardous waste cleanup. The *strict liability* provision means that a person could still be liable even though that person was not responsible for the problem. Accordingly, a current owner of a piece of real estate may be held liable for cleanup costs even if the contamination was the result of an earlier owner's action. Another provision of the law makes the liability *retroactive,* meaning that the past owner of a site may be charged with current liability even if that individual had complied with all regulations existing at the time of the ownership of the real estate.

The liability burden under the law is indeed very serious, with the result that environmental audits are becoming very common and are a standard prerequisite for transfer of commercial, and even residential, properties. This works to the advantage of both the lender and the potential owner of the property by alerting them to the potential liability associated with a contaminated site. An environmental audit does not guarantee removal of liability for cleanup, but the EPA may look more favorably at cases where an audit was performed. In such situations, the EPA may remove the liability on the grounds that the buyer used *due diligence* to avoid the liability.

Hazardous and Solid Waste Amendments Act

The Hazardous and Solid Waste Amendments Act (HSWA) of 1984 broadened the scope of RCRA and includes provisions to protect the quality of groundwater. This led to the restriction of land disposal of hazardous waste by landfilling or surface impoundments. It established the requirement for double liners, leachate management systems, and groundwater monitoring programs at disposal sites. Technical standards for landfill design, leak detection systems, and underground storage of petroleum products and related hazardous substances were developed in pursuance of the act.

The HSWA also set new requirements for management and treatment of small quantities of hazardous waste, such as those generated by automotive shops or dry-cleaning businesses. It also established new regulations for underground tanks that store petroleum or other chemicals.

Underground Storage Tanks Act

Subtitle I was added to the RCRA in 1984. This law, known as USTA, deals with problems related to leaking underground tanks used for storage of petroleum and other hazardous substances. The law aims at preventing leaks and spills, and at their detection and remediation when they occur. It also ensures that the owners and operators of UST facilities will have the financial capability to correct any problems arising from leaking tanks. From a historical point of view, the law has set different requirements for "new" and "existing" owners and operators of USTs: Any UST installed after December 1986 is considered new, and those installed earlier are considered existing.

As of July 1994, there were 1.2 million USTs in the United States, of which an estimated 262,829 were leaking. Cleanup of 101,309 USTs was completed by July 1994, while cleanup action was initiated for 202,866 USTs (RCRA/Superfund/UST Hotline, 1994).

Toxic Substance Control Act

The Toxic Substance Control Act (TSCA) of 1976 was aimed at regulating chemicals. To do so, the *TSCA Registry* was established; any new chemical manufactured or imported has to be registered. Information on the environmental fate of a chemical and associated health effects has to be furnished. Requirements for periodic reporting on production or importation and any investigations related to alleged health effects have been established under the TSCA.

STUDY QUESTIONS

1. Which act led to the creation of the council on National Environmental Quality and the EPA?
2. Who was made responsible for managing the LLRW following the passage of the Low-Level Radioactive Waste Policy Act? Explain the concept of LLRW Compacts.
3. When is a permanent facility for the disposal of HLRW likely to be built in the United States? Where will it be located?

4. How is the Primary Air Quality Standard different from the Secondary Air Quality Standard? What is the whistle-blower/bounty provision in the law?
5. What is the cradle-to-grave concept? Explain with appropriate examples.
6. What are the three major laws that relate to waste management? Discuss the one that regulates solid waste, hazardous waste, and USTs.
7. What are the two special provisions in SARA that seem very harsh? Explain these by using suitable examples. What is due diligence?

REFERENCES CITED

RCRA/Superfund/UST Hotline, 1994: (800)424-9346, Crystal City, VA.

Fortuna, R.C., 1989, Hazardous-Waste Treatment Comes of Age, *in* Freeman, H.M. (editor), 1989, Standard Handbook of Hazardous Waste Treatment and Disposal: New York, McGraw-Hill, Inc., p. 1.7.

U.S. Environmental Protection Agency, 1989a, The Toxics-Release Inventory: A National Perspective: Washington, D.C., Office of Pesticides and Toxic Substances, Report No. EPA 560/4-89-005, p. 109.

U.S. Environmental Protection Agency, 1989b, National Air Pollutant Emissions Estimates: 1940–1987: Washington, D.C., Office of Pesticides and Toxic Substances, Report No. EPA 450/4-88-002, 73 p.

U.S. Environmental Protection Agency, 1993, 1991 Toxics-Release Inventory: A National Perspective: Washington, D.C., Office of Pesticides and Toxic Substances, Report No. EPA 745-R-93-003, 364 p.

Satchell, M., 1989, Uncle Sam's Toxic Folly: U.S. News and World Report, 27 March, p. 20–22.

SUPPLEMENTAL READING

Government Institutes, 1993, Environmental Law Handbook (12th ed.): Rockville, MD, Government Institutes, 550 p.

5

Environmental Legislation in Other Countries

INTRODUCTION

Global concern for preservation of environmental quality has resulted in the passage of laws aimed at proper management of hazardous wastes in many countries of the world. Growing world population and the thrust for industrialization are the two main factors that adversely impact our environment. While a majority of the developed (industrialized) nations have adopted adequate legislation to regulate all aspects of hazardous waste management, many developing nations either do not have any laws to control such wastes or have very rudimentary legislation that is often poorly enforced.

It is not possible to include all countries of the world in this discussion. Instead, nine countries have been selected. These countries were carefully chosen to include both the developed (industrialized) and developing countries. The following discussion provides an overview of environmental laws dealing with hazardous waste management in various countries.

AUSTRALIA

Australia is a developed nation with a vast land area and sparse population. Its history of environmental law can be traced back to 1855, with its enactment of the Yarra Pollution Prevention Act. At that time waters of the Yarra River were being polluted from dumping of waste materials by local industries, located by the river, in Melbourne. The law was intended to stop industries that produced obnoxious and odorous compounds from locating on the river. The final legislation allowed existing factories to stay but prevented further expansion of these factories or locating of new factories by the Yarra River. There were, however, problems in enforcement of the law.

In 1957, the first contemporary pollution control legislation, the Victorian Clean Air Act, was enacted. This act was aimed at controlling air pollution from industrial emissions.

In 1974, Australia passed the Environment Protection Act. This act ensures that environmental considerations receive equal weight in the decision-making process along with other traditional factors, such as the social and economic advantages of a development. Unlike the United States, Australia has made the preparation of an EIS voluntary.

Australia has a Federal Department of Health and Environment that sets broad guidelines and acts in an advisory capacity. Another federal agency that serves in an advisory capacity, offering advice to individual states in formulating their environmental goals, is the Department of Science and the Environment. Actual formulation of specific laws and their enforcement are left to the provincial (state) or the territorial government.

Australia is divided into six provinces/territories. Each has its own laws to regulate hazardous waste. The following agencies are responsible for hazardous waste management in various provinces or territories:

- Waste Management Commission of South Australia
- Waste Management Authority of New South Wales
- EPA of Victoria and Western Australia
- Chemical Hazards and Emergency Management Unit of Queensland
- Department of Environmental Planning, Province of Tasmania

Environmental Laws in the Province of South Australia

South Australia is one of the six provinces/territories in Australia. It is located in the south-central part of the country, has an area of about 100,000 km^2 (1/10 that of the United States), and a population of 1.5 million. It has a variety of industries, including a large timber industry and automobile and allied industries. We will look at the province of South Australia as an example of environmental legislation in Australia.

South Australia has a Waste Management Authority that was created in 1979. It has the responsibility to oversee all phases of hazardous waste management, and is empowered to impose fines of Australian \$20,000 (U.S. \$17,000) and jail sentences for noncompliance.

Waste Management Act

The Waste Management Act was passed in 1987 to regulate hazardous waste. This act repeals the South Australia Waste Management Act of 1979 (Parliament of South Australia, 1987). A Waste Management Commission (WMC) was created to oversee and control hazardous waste subsequent to the passage of the Waste Management Act. The act requires

- Licensing for any industry that produces hazardous waste (maintains a listing of hazardous substances, similar to the U.S. EPA's list)
- Waste depot licensing (a waste depot is any facility involved in storage, treatment, or disposal of hazardous wastes)

- Any carrier of hazardous waste has to be licensed by the WMC
- A four-copy manifest system to track the hazardous waste (the original has to be returned to the WMC within seven days after arrival of hazardous waste at the waste depot)

Chlorinated Wastes

South Australia banned chlorinated wastes and PCBs in 1988. Because it does not have a high-temperature incineration facility, PCB wastes were sent to the United Kingdom for disposal.

South Australia has a flourishing timber industry in the southern part of the province. Large quantities of chrome, copper, and arsenate (CCA) residues are produced during wood-preservation operations. These are placed in 205-liter drums and shipped to plants in Adelaide for chemical treatment. The treated CCA is then subjected to an extraction procedure toxicity test (very similar to the U.S. EPA's toxicity test). If the treated residue passes the test, it is buried in a licensed landfill.

CANADA

Environment Canada, the federal department of the environment, was created subsequent to passage of the Government Organization Act of 1970. The federal Environmental Protection Service (EPS) is a unit within Environment Canada. The EPS is responsible for the enforcement of environmental legislation on federally owned lands, in interprovincial regions, and in international jurisdictions. The ten provincial governments in Canada have been authorized by the federal government to enact appropriate legislation to manage the environment. Some of the important environmental laws are discussed in the following sections.

Canadian Environmental Protection Act

Many environmental protection statutes regulating air and water quality were in existence in Canada, and various provinces (states) had their own laws to control wastes. In June 1988, however, the comprehensive Canadian Environmental Protection Act (CEPA) came into force. It is considered to be the flagship of federal environmental laws and is the first major federal statute in over a decade. The CEPA has two main purposes:

1. Consolidation of existing environmental protection statutes into a single, stronger piece of legislation; the existing legislation included
   - The Clean Air Act
   - The Canadian Water Act
   - The Environmental Contaminants Act
   - The Ocean Dumping Control Act
2. To ensure uniform national standards for management of toxic substances

The CEPA is made up of nine parts. Part 2 of the CEPA deals with toxic substances. It also uses the cradle-to-grave concept for management of toxic substances. It authorizes formulation of regulations to control the entire life cycle of toxic substances—from the initial import by a manufacturer through its handling and use, to final disposal or release into the environment.

The CEPA requires preparation of a list of toxic substances that are manufactured in or imported into Canada in a quantity of 100 kg/yr or more. This list, known as the Domestic Substances List (DSL), is used to determine if the substance is already existing in the country or is "new." In addition, there is also a Non-Domestic Substances List (NDSL) that includes toxic substances that are in use worldwide but not in Canada. This list is based on the U.S. EPA's 1985 list developed under the Toxic Substances Control Act (TSCA) of 1976. Any substance that is not on the existing U.S. EPA's list of hazardous substances (list F, D, K, U, etc.) is a new substance. All Canadian manufacturers or importers of an NDSL substance must provide information to the government about its characteristics, including potential toxic effects. A DSL and an NDSL were published in January 1991.

Substances that are either toxic or are capable of becoming toxic are listed in a separate list, the Priority Substances List (PSL). The fact that a substance has been placed on the PSL does not mean that it has to be regulated; it only provides for the *possibility* of regulation. The CEPA allows any person to file in writing a request that a substance be added to the PSL. Once a substance on the PSL is determined to be toxic, the report of the determination must be made public, and a decision on whether or not the substance will be regulated should be made.

The CEPA also empowers government agencies to conduct inspections without a warrant, or to obtain warrants for search and seizure. It allows for fines of up to $1 million per day and imprisonment of up to five years for parties convicted of a violation of the law. In extreme circumstances, life imprisonment and unlimited fines may result from a conviction arising out of criminal negligence causing death.

The CEPA also regulates the transportation of hazardous waste (Regulation 309). This regulation imposes an information system to track hazardous wastes by requiring the use of a manifest system. Shippers and carriers of dangerous goods are required to notify authorities in the event of a spill. Part IX of the CEPA details requirements for notification, cleanup responsibilities, and liability arising from a spill.

Transportation of Dangerous Goods Act

Transportation of hazardous waste is regulated by the Transportation of Dangerous Goods Act (TDGA) of the Canadian federal government. The TDGA has set up criteria by which a product or substance may be classified as a hazardous waste. Minimum concentrations of certain chemicals have been used to establish the standards. Any time hazardous waste is to be transported within the country or imported into the country, a notification to this effect has to be made to the authorities 60 days before such transport.

Transportation of hazardous wastes between Canada and the United States is controlled by the Basle Convention (1986). Both the United States and Canada signed an agreement called the Canada-U.S.A. Agreement on the Transboundary Movement of Hazardous Waste. This agreement has essentially permitted free transport of hazardous wastes between the United States and Canada, with provision that prior notification is made and approval of the importing country is obtained. The purpose of this agreement is to provide proper management of nearly 100,000 metric tons of hazardous wastes that cross the border annually.

Workplace Hazardous Materials Information System

Another national law, which became effective in October 1988, relates to occupational health and safety of workers dealing with hazardous materials. The Workplace Hazardous Materials Information System (WHMIS) is intended to provide information to the workers about hazardous materials used in the workplace. It aims at reducing the number of injuries and illnesses resulting from the handling of hazardous materials in the workplace. Under the WHMIS, there are three ways in which information on hazardous materials can be provided:

1. Labels on the containers of hazardous materials
2. Material Safety Data Sheet (MSDS) to supplement the label with detailed hazards and precautionary information
3. Worker education program

COMMONWEALTH OF THE FORMER SOVIET UNION

Stories of gross neglect of the environment and mismanagement of nuclear waste in the U.S.S.R. became known to the world after the breakup of the Soviet Union in December 1991. The former Soviet Union is sitting on the biggest pile of radioactive waste in the world. There are 640 radiated sites, which were expected to be cleaned up by 1995 (Newsweek, 1992). How many of these will actually be remediated, and to what levels, remains to be seen.

Despite the need to address overwhelming environmental problems, promulgation of environmental law in the former Soviet Union occurred only in the late 1980s. The U.S.S.R., in October 1989, passed a law that led to the creation of the first environmental agency in the country. This agency was known as the State Commission for Protection of Nature (SCPN). An All-Union Prosecution Office was also created, which, for the first time, gave citizens the right to challenge environmental decisions made by the state officials. In June 1990, the SCPN passed a regulation requiring an EIS for *all* construction projects.

Interstate Ecological Agreement

In February 1991, some of the republics in the Soviet Union developed the Interstate Ecologic Agreement (IEA) and urged legislatures in the republics to adopt it. The IEA assigned high priority to remediation of sites that cut across the republics' boundaries. These included the Chernobyl remediation project, as well as various seas and lakes. The agreement also set forth some basic guidelines that included citizens' right to a clean environment, common policy for use of natural resources, general rules for setting environmental standards, environmental incentives, and creation of an interrepublic environmental fund and an environmental information bank.

The above policies were included in the December 1991 treaty that resulted in the creation of the Commonwealth of Independent States. In February 1992, the Commonwealth States signed several other agreements on environmental policy and also agreed to joint financing of environmental projects.

Law on Protection of the Environment

Russia has moved aggressively in taking measures to protect the environment. The Law on Protection of the Environment (LPE) was passed in December 1991. This law gave additional powers to the Russian Department of Ecology. Under the provisions of the LPE, protection of the environment was given priority in all activities involving both the government and the private sector. Anyone failing to meet environmental standards will be assessed a special tax. At the same time, those engaged in environmentally sound activities are to receive economic incentives for pollution prevention.

Under the LPE, an EIS is required for all new projects. The EIS must be made available to the public for comment. Import of hazardous waste into Russia was prohibited under the new law. The enforcement of LPE is to be carried out by the Arbitration Court and Civil Courts. Individuals or firms are held responsible for cleanup and restoration of contaminated sites, if found responsible, regardless of fault (this resembles CERCLA's liability concept).

Ukraine, Estonia, and other republics have adopted similar laws. The biggest problem is lack of technology for effective management of contamination; innovative technologies for remediation of hazardous waste are not available. It is believed that the new economic incentives will act as a catalyst for introduction of sound technologies.

INDIA

India is a good example of a developing country where rapid industrial growth is taking place, boosting the economy and improving the living standard of its 920 million people. Environmental laws are formulated and enacted by the Central (federal) government; state governments do not have their own laws. Environmental laws were practically nonexistent until the 1980s. The December 1984 accident at the Union Carbide plant in Bhopal, which released the deadly methyl isocyanate gas that caused the death of over 2000 people, was a major impetus for the new environmental laws in India.

Two of the earliest legislative acts are the Water (Prevention and Control of Pollution) Act of 1974 and the Air (Prevention and Control of Pollution) Act of 1981. Both laws were designed to control pollution of waters and air from industrial activities. However, enforcement of these laws was very lax.

The Environment (Protection) Act

The Environment Act represents the first comprehensive legislation directed toward maintenance and protection of the environment. As a participant at the United Nations Conference on the Human Environment held in Stockholm, Sweden, in June 1972, India had committed itself to implement the decisions made at the conference. However, the passage of the law did not occur until 1986.

The act provides a definition for *the environment, environmental pollutant, environmental pollution, handling, occupier,* and *prescribed* and *hazardous substances.* Unlike most other countries, the word *environment,* besides including water, air, land, and humans and other life forms, also includes property (Government of India, 1987). *Hazardous substance* has been defined as "any substance or preparation which, by reason of its chemical or physico-chemical properties or handling, is liable to cause harm to human beings, other living creatures, plants, microorganisms, property, or the environment." The act authorized the Ministry of Environment and Forests (MEF) to initiate action to close down any firm that fails to comply with the act's restrictions on effluent discharge; it also permits individuals to initiate legal action against anyone violating the act. The act establishes formation of the Central Pollution Control Board to enforce the laws, and provides for civil and criminal actions against violators of the law. A fine of up to Rupees 100,000 (U.S. $3192), with or without imprisonment for up to five years, can be imposed. If the violation continues, an additional daily fine of Rupees 5000 (U.S. $160) may be imposed. These fines apply to both private-sector businesses and government-owned operations; the responsible party, in both cases, is the owner or the head of the organization.

In 1989, the government issued rules for the management and handling of hazardous waste. Hazardous substances were categorized as toxic chemicals, flammable chemicals, and explosives. All operations involving the use of hazardous substances were to follow proper rules for their handling, storage, and disposal. However, sites for disposal have not been identified by the state governments as of this writing. Legislation passed in January 1991 requires public-liability insurance coverage to pay for claims arising out of accidents involving hazardous substances. Environmental audits and the preparation of an EIS are required for specified industries.

Despite the regulatory changes and more stringent actions by the courts, compliance with the law has been sluggish. Some of the reasons: understaffed Pollution Control Board, lack of comprehensive industrial zoning policy, financial constraints faced by companies, and lack of cost-effective technology.

JAPAN

Unlike other Asian countries, Japan addresses the environmental problem comprehensively. This is done through various laws that have been enacted to control air emissions, discharges into water, solid and hazardous wastes, noise, and chemicals. Both the national and local governments are responsible for management of the environment. Specifically, the national government prescribes baseline standards which are implemented by local authorities, who have the power to impose more stringent requirements.

Noncompliance with the environmental requirements in Japan may result in imprisonment and/or fines. Litigations are rather uncommon among the Japanese, who dislike litigation and prefer to use negotiation and persuasion for compliance with laws.

Unlike the United States, various aspects of regulating the environment are not vested with one agency; the responsibility for setting standards and regulating industries is distributed among various agencies. Japan has an Environmental Agency (EA), in addition to several other national agencies that deal with environmental problems. The major ones are

- Ministry of Trade and Industry (MTI)
- Ministry of Health and Welfare (MHW)
- Ministry of Transportation (MOT)
- Ministry of Science and Technology (MST)

The EA is responsible for environmental policies and planning. It sets standards for air quality, water quality, and noise. It is also charged with setting standards for disposal of hazardous wastes and pesticides. The MTI is closely involved in setting these standards, and makes sure that the regulations are not so stringent as to impede economic growth.

Waste collection and handling is regulated by the MHW. The MOT handles oil discharges from the ships and transportation of hazardous substances, and the MST regulates radioactive wastes.

In 1990, Japan initiated a manifest system for tracking hazardous wastes.

Air Pollution Control Act

The Air Pollution Control Act (APCA) was enacted in 1968. The Environmental Agency has prescribed national ambient air quality standards for sulfur dioxide, nitrogen oxide, carbon monoxide, and suspended particulate matter. Local and state governments may impose more stringent requirements.

Water Pollution Laws

Several laws address water pollution and waste discharges, including

- The Sewerage Law of 1958
- The Marine Pollution Control Law of 1970
- The Water Pollution Control Law of 1970
- The Clean Lakes Law of 1984

The EA sets the guidelines for effluent discharges from specific industries.

Waste Management Law

Solid and hazardous wastes are regulated under the Waste Management Law of 1970. The EA prescribes standards for disposal; the MHW regulates other aspects of hazardous and solid waste management. Wastes are divided into two general categories:

1. General (domestic) wastes
2. Industrial wastes, which include
   - General
   - Industrial
   - Hazardous

General (solid) waste may be disposed of along with domestic wastes. Industrial wastes comprise wastes that belong to one of the 19 classes. Hazardous wastes are those that are harmful to humans or the environment. A waste is considered hazardous waste if it contains: mercury and mercury compounds; cadmium and cadmium compounds; lead and lead compounds; hexavalent chromium compounds; arsenic and arsenic compounds; cyanide compounds; organo-phosphate compounds; and PCBs (Passman, 1986).

Different landfills are designated for disposal of various categories of wastes, e.g., open dumps for glass and ceramics, and landfills with liners and leachate collection systems for hazardous wastes. Landfills are extensively used for disposal of domestic wastes. Incineration of industrial waste is allowed after obtaining government's approval. Industrial wastes have to be sampled and analyzed prior to transportation or disposal.

Chemical Control Law

Chemicals are regulated under the Chemical Control Law of 1973. Several agencies are involved in the evaluation of existing and new chemicals. This law is comparable to the United State's Toxic Substances Control Act of 1976, and calls for the review and testing of new chemicals before they are manufactured or imported, to ensure that use of such chemicals will not adversely affect human health and the environment.

MEXICO

In recent years, Mexico has committed itself to improving and preserving the environment. Environmental protection has been given priority since 1988, and the country maintains that industrial development will not occur at environmental cost. Three articles—25, 27, and 73—in Mexico's Constitution deal with the environment.

General Law of Ecological Equilibrium and Environmental Protection

The General Law of Ecological Equilibrium and Environmental Protection (GLEEP), passed in March 1988, addresses environmental pollution, resource conservation, environmental impact, risk assessment, and ecological zoning and sanctions. It authorizes the Secretariat of Urban Development and Zoning (SEDUE*), an agency equivalent to the U.S. EPA, to develop detailed regulations for environmental protection. SEDUE is also responsible for implementation of the GLEEP. Four regulations relating to environmental impact assessment, air pollution, Mexico City air pollution, and hazardous waste have been promulgated.

State and local governments may promulgate their own regulations, but the federal regulations and standards take precedence over state and local regulations, unless the latter are more stringent than the former. As of 1992, 26 of the 32 Mexican states and the Federal District had adopted their own environmental laws (Alonzo, 1992).

Under the Environmental Impact Review Process, all new public and private construction projects, including expansion of existing facilities, which may cause ecological imbalance are required to demonstrate compliance with legal standards and to seek prior authorization from SEDUE before beginning operations.

Enforcement of the law is done through plant closure, fines, criminal penalties, and administrative arrest. Fines up to $80,000 can be imposed for environmental noncompliance. The law has a special provision, administrative arrest, which deprives a corporate officer of his or her freedom for up to 36 hours (a kind of house arrest).

Considerable attention has been given to air pollution control in Mexico City. Regulations to replace old diesel engines in buses with cleaner-burning diesel engines, requirement of catalytic convertors on all 1991 and later cars,

---

*SEDUE is the Spanish acronym.

phasing out leaded fuel, and economic incentives for unleaded fuel use have been very successful. Significant drops in lead, carbon monoxide, and sulfur dioxide in Mexico City's air have been noted. In 1991, the Pemex Oil Refinery, the largest and most-polluting oil refinery in Mexico, was closed for violating air quality standards. It has been estimated that this closure resulted in the loss of 5000 jobs and $500 million in revenue (Alonzo, 1992).

With the passage of the North America Free Trade Agreement (NAFTA), bilateral efforts between the United States and Mexico to improve the environment have been accelerated. NAFTA is a landmark international agreement in the sense that, for the first time in modern history, environmental issues were included in a treaty.

THE NETHERLANDS

The Netherlands has enforced pollution control laws since the 1970s. Its National Environmental Policy Plan is designed to reduce *all* forms of pollution by 70 to 90 percent by the year 2000 or earlier. The country has adopted extraordinary measures to control water, land, and air pollution. For example, licensing requirements have been established for discharge of wastewater into sewers, surface water bodies, or the ocean. Stringent criteria, under the Surface Water Pollution Act (1970) and the Sea Water Pollution Act (1975), have been established that must be met before wastewaters can be discharged into natural systems (Dekker et al., 1987). The Best Available Technology (BAT) is to be used for eliminating concentrations of cadmium and other hazardous substances before wastewaters are released into rivers, lakes, or the sea.

Chemical Waste Act

The Chemical Waste Act (CWA) of 1976 and its amendments are designed to control industrial wastes that are potentially harmful to human health and the environment. A list of hazardous substances was developed that categorizes chemicals into seven groups: metals (As, Ba, Cd, Cr, Co, Cu, Hg, Mo, Ni, Pb, and Se), inorganics (NH_4, F, cyanides, S, Br, and PO_4), aromatic compounds (benzene, phenols, etc.), polycyclic hydrocarbons (anthracene, nepthalene, pyrene, etc.), chlorinated hydrocarbons, pesticides, and others (pyridine, styrene, fuel oil, etc.). Safe levels for these chemicals were established to determine whether or not the soil or groundwater is contaminated (Siegrist, 1989). This list came to be known as the Dutch List, and was adopted by other European nations.

Under the CWA, generators of hazardous wastes can only send their waste to a licensed Treatment, Storage, and Disposal (TSD) facility, and are required to maintain proper records of all such activities.

TAIWAN

In 1987, the Environmental Protection Administration (EPA) was established at the federal level in Taiwan. The purpose was to control environmental degradation brought about by the country's massive effort for industrialization, without concurrent environmental control.

Waste Disposal Law

The Waste Disposal Law was first promulgated in 1974 and was amended in late 1988. The law defines various types of wastes as: general waste (same as municipal solid waste, MSW), general industrial wastes, and hazardous industrial wastes. General industrial waste includes nonhazardous wastes generated by industrial and commercial sources. Hazardous industrial wastes, on the other hand, are wastes that are toxic or dangerous and have the potential to cause harm to human health and the environment. The law established criteria for identification of hazardous substances. Hazardous industrial wastes are required to be properly stored, transported, treated, and disposed of. The responsibility for proper disposal of hazardous industrial waste lies with the generators, who can either clean and/or treat and dispose of the wastes or send the wastes to an authorized waste disposal facility. These facilities are both state- and private-owned. Hazardous industrial wastes are disposed of on land and by incineration. To track waste movement, a six-copy manifest is required. The law has established a reward system for industries who comply with the law and reduce pollution. At the same time, fines and/or suspension of license may be imposed on the violators.

Enforcement of the law is carried out at state, city, or local levels, under federal authority. Over $4 billion was expended in 1992 for environmental maintenance and restoration. This included monies spent for new treatment and disposal projects, waste-to-energy incineration plants, clean air projects, and solid waste projects.

UNITED KINGDOM

The United Kingdom has had various laws relating to the environment. These include the 1857 law to protect the public from fumes and smog coming from the Thames River, the Public Health Act of 1875 to control the spread of disease, the Prevention of Pollution Act of 1876, and the Control of Pollution Act of 1974. However, these laws were narrow in scope and were loosely defined.

It was not until 1990 that a comprehensive law, the Environmental Protection Act (EPA), was passed. It has been in force since January 1991. The law applies to England, Wales, Scotland, and parts of Northern Ireland. The United Kingdom, unlike the United States, allows co-disposal of wastes, meaning that household and industrial wastes can be disposed of in the same landfill. The main features of the EPA as it relates to hazardous waste are summarized below.

- The law classifies all wastes into
    1. Household waste (MSW or garbage).
    2. Industrial waste that comes from industries, airports, seaports, rail and road terminals, gas and oil or water supply companies, sewers, and the like.
    3. Commercial wastes, including those generated by business and by sports and recreational facilities.
    4. Special waste, called Controlled Waste (CW). CW is defined as household, industrial, and commercial or any such waste that is

capable through quantity or concentrations of causing harm to mankind or any other living organism supported by the environment. (In a general sense, CW is similar to hazardous waste.)

- The Secretary of State for the Environment (SSE) is the final authority and can prohibit or restrict the import or export of substances that pose a threat to human health or the environment. The SSE also has the authority to control and regulate industries that generate CW, and has the power to designate a process as being subject to control. The SSE does *not* have authority over control of air pollution. This is done by the London boroughs and the district councils. The SSE is responsible for setting up standards for concentrations, amounts, or characteristics of discharges into the environment.

- Enforcement of the law is through Her Majesty's Inspectorate of Pollution (HMIP). At county and local levels, enforcement is carried out by the Inspector's offices.

- A license is required to operate a TSD facility. Stiff fines, up to £20,000 (about U.S. $36,000) and/or six month jail imprisonment can be imposed for noncompliance. The idea is to create a chain of responsibility at every stage of CW movement.

- The law also controls the import or export of genetically altered organisms and hazardous biological materials. Such substances may be allowed only with prior approval of the Chief Inspector of Pollution (CIP).

  Properly characterized waste can be imported and exported with the permission of the CIP (Australia's PCB wastes were imported for destruction by incineration in the United Kingdom).

  Because of low costs, land disposal of hazardous wastes is most common in the United Kingdom. During 1986–1987, for example, 83 percent of all hazardous waste was landfilled, 8 percent disposed of in the sea, 7 percent disposed of after chemical or physical treatment, and less than 2 percent was disposed of by incineration. Relative disposal costs in the same time period were: landfilling, $3.50 to $8.00/ton of dry waste; incineration, from about $200/ton for simple industrial waste to more than $3000/ton for PCB waste (House of Commons, 1989).

- The law requires manufacturers to divulge detailed information about the process: its chemical components, equipment used, and throughput (raw materials processed in a given time). This has raised some concern, as it requires commercial publication of confidential information. A provision in the law, however, allows the manufacturer to withhold the information in the event that disclosure will seriously affect its commercial interests.

- Operators of the licensed TSD facilities are required to dispose of the CW in an environmentally safe way. The onus of proof to show that there was no better technology available to dispose of the waste rests with the operator.

- One of the provisions of the EPA calls for use of Best Available Techniques Not Entailing Excessive Cost (BATNEEC). The idea is to prevent or minimize the release of detrimental substances into the environment at a reasonable cost.

TABLE 5–1 Approaches to Environmental Management (adapted from Woodward-Clyde, 1993)

American	British	European	International
Organized medium by medium	Preventive planning and action important	Focuses on pollution prevention	Pollution prevention, eco-friendly design
Based on regulatory compliance to achieve zero release of pollutants and to lower health risks	Environmental management involves an evaluation and decision process that is part of an overall plan	Growing interest in the use of quantitative environmental analyses and plans	Environmental management fully integrated with the rest of business, leading to total quality management
Driven by liability and litigation; emphasis is on documentation	Focuses on good information system	Public reporting of environmental performance; lack of emphasis on inactive sites	Environmental tools linked to money-making strategies

FUTURE TRENDS

The overview in this chapter gives a general idea of the differences in approaches to waste management and related environmental issues in various countries of the world. The difference in scope of the laws and their enforcement among the developed and the developing nations is obvious. However, there are also differences in the approach to waste management among the developed nations. With the global emphasis shifting toward sustainable development, a new paradigm in environmental management that focuses on pollution prevention, environmental design of products, and total quality management is emerging. Table 5–1 summarizes the various approaches to environmental management.

There is a move to develop international standards for environmental management. The International Standards Organization (ISO) is working toward this goal, and it is expected that the standards will be completed by 1997. The standards comprise: (a) environmental management systems, (b) life-cycle assessment of products, and (c) eco-labeling. It is believed that once this standard is available, many other countries, who presently do not have well-defined standards or environmental regulations, will follow the ISO's lead and formulate their own environmental standards.

STUDY QUESTIONS

1. What are the similarities and differences between the hazardous waste laws in Australia and the United States?
2. How is the Domestic Substance List different from the Non-Domestic Substance List in the Canadian Environmental Protection Act (CEPA)? Which one of the two is similar to the U.S. EPA's list? What is the Priority Substance List?

3. What are the main features of the Workplace Hazardous Materials Information System of the CEPA?
4. What agencies in Japan regulate hazardous materials and hazardous wastes? What role does the Environmental Agency play in the management of waste?
5. What is the Japanese Waste Management Law?
6. What are the salient features of the Mexican General Law of Ecological Equilibrium and Environmental Protection?
7. What is the Dutch List? How is it utilized in the Netherlands?
8. Why does the United Kingdom permit co-disposal of household and industrial wastes?

REFERENCES CITED

Alonzo, A.L, 1992, Recent Development in Mexico, *in* International Environmental Law, Special Report: Rockville, MD, Government Institutes, Inc, p. 301–313.

Dekker, L., B.T. Bower, and Koudstaal, 1987, Management of Toxic Materials in an International Setting: A Case Study of Cadmium in the North Sea: Rotterdam, A.A. Balkema, 116 p.

Government of India, 1987, The Environment (Protection) Act, 1986: New Delhi, Manager Government of India Press, 10 p.

House of Commons, U.K., 1989, Toxic Waste, Second Report, Volume I, by Environment Committee: London, Her Majesty's Stationery Office, p. xxvi

Newsweek, Inc., 1992, Get Out the Geiger Counters: Newsweek Magazine, November 2, p. 64.

Parliament of South Australia, 1987, Waste Management Act, 1987: South Australia, Director and Government Printer, 19 p.

Passman, P.S., 1986, Japanese hazardous waste policy: Signalling the need for global and regional measures to control land-based sources of pollution: Virginia Journal of International Law, vol. 26, no. 4, p. 921–964.

Siegrist, R.L., 1989, International perspective for cleanup standards for contaminated land, *in* Proceedings, Third International Conference for Hazardous Waste Management, September 10–13, 1989, Pittsburgh, PA: Office of Research and Development, Environmental Protection Agency, PA/600/9-98/072, Cincinnati, OH; p. 348–359.

Woodward-Clyde Group, Inc., 1993, Corporations Turn Green to Stay in the Black: InSite, Fall 1993, p. 3–7.

SUPPLEMENTAL READING

Government Institutes, 1992, International Environmental Laws, Special Report: Rockville, MD, Government Institutes, 395 p.

C H A P T E R

6

Philosophy and Approaches to Hazardous Waste Management

INTRODUCTION

Our philosophy and approach to management of solid and hazardous waste have undergone many changes over time. These changes reflect the level of industrialization, societal attitudes, and population trends. During the first century of the Industrial Revolution (1760–1860), when the steam engine was invented and use of dynamite as a powerful explosive was discovered, we adopted the philosophy of *dilute and disperse* to handle the waste. Waste materials were either dumped on the land or released into the water and air. Generally, the low concentration of toxic chemicals and the relative purity of the environment worked to the advantage and the polluted air or water cleansed itself. This approach served the intended purpose for a period of time. Soon, however, with the growing population and the synthesis of a variety of new chemicals and their heavy use in industry, the realization dawned that the earlier philosophy would not work any more.

A new approach, *concentrate and contain,* came in vogue during the second century of the Industrial Revolution. This approach worked satisfactorily for some time; it was later realized that containers are subject to leak, thereby posing threat to the environment. During the 1970s and early 1980s a new philosophy evolved that called for *conservation and recycling*. While this approach has served us well in conserving valuable resources, it has not helped solve the problem of *reducing* the quantity and toxic nature of some hazardous wastes. In addition, any material set aside for recycling remains a waste until it is reused. For example, accumulated waste paper and glass that sits in warehouses for lack of market demand is not truly recycled—it has only been collected and stored.

The amount of hazardous waste generated has gone up over the years, and its toxicity has attained high levels. The current emphasis is on *pollution prevention,* which can be defined as any practice that results in the reduction or elimination of any pollutant prior to recycling, treatment, or disposal. Some of the chemicals, because of their known toxicity or adverse impact on the environment, have been targeted by the U.S. EPA for immediate reduction; others have been placed on the list of chemicals of concern. Table 6–1 lists such chemicals. *Integrated waste management* is the current waste management philosophy and incorporates various options available for effective management of hazardous waste.

TABLE 6–1   Chemicals of Concern to the EPA
(adopted from EPA, 1991)

Targeted for Immediate Reduction

Metals
 Cadmium
 Chromium
 Lead
 Mercury
 Nickel

Chlorinated Compounds
 Carbon tetrachloride
 Chloroform
 Dichloromethane
 Tetrachloroethylene
 Trichloroethane
 Trichloroethylene

Other Toxic Compounds
 Benzene
 Cyanide
 Toluene
 Xylene
 Methyl ethyl ketone
 Methyl isobutyl ketone

Other Chemicals of Concern
 Acrylonitrile
 Arsenic
 Carbon disulfide
 Ethylene disulfide
 Formaldehyde
 Styrene
 Vinyl chloride

Pesticides
 Alachlor
 Carbaryl
 Captan
 Chlorpyrifos
 2,4-Dichlorophenoxyacetic acid

INTEGRATED WASTE MANAGEMENT

Integrated waste management represents a comprehensive approach to pollution prevention and includes the following components:

- Source reduction—includes elimination and substitution of toxic materials *at the source*
- Recycling—includes recovery, reuse, and treatment
- Residual disposal—component of the waste stream left after recycling, which has to be disposed

Integrated waste management requires that greater emphasis be placed on waste elimination and reduction at the source than on disposal. Doing so reduces the quantity of waste material that needs disposal. A reduction in the quantity of waste will minimize the release of pollutants in the environment, resulting in pollution prevention. Of the various waste management options currently available, the following represent the preferred order, based on the potential impact on the environment from a particular method of disposal:

Pollution Prevention	**HIGHEST PRIORITY**
Recycling	
Treatment	
Disposal	**LOWEST PRIORITY**

One of the main aspects of an integrated waste management philosophy is that it forces waste generators to take a hard look at the *entire* manufacturing process that creates waste rather than only considering what comes at the end of the process—the *end-of-the-pipeline* approach. The difference between the two is that an integrated philosophy allows for elimination, substitution, and reduction of hazardous materials at various stages of the manufacturing process, while the end-of-the-pipeline approach is devoid of these options and offers no choice other than to accept and manage the waste that is generated as a result of the process. Integrated waste management involves: (a) identifying which steps of the process generate hazardous waste and, (b) exploring ways to eliminate or minimize the waste. Simple housekeeping measures, such as segregating wastes so they can be more easily reused, sometimes result in surprisingly large waste reductions. Other options include modifying manufacturing processes, using different raw materials, and replacing hazardous products with safer substitutes.

INCENTIVES FOR WASTE REDUCTION

Federal law requires all generators of hazardous waste to implement methods to reduce or eliminate hazardous wastes. The Hazardous and Solid Waste Amendments (HSWA) passed in 1984 state: "The Congress hereby declares it to be the national policy of the United States that, wherever feasible, the generation of hazardous waste is to be reduced or eliminated as expeditiously as possible" (U.S. Congress, 1984, p. 3224).

In addition to the legal requirements, there are many reasons why a manufacturer should seek to minimize waste generation. These incentives could be grouped under: (a) economic, (b) regulatory, (c) liability, and (d) public relations.

Economic Incentives

Economic incentives include

- Tax breaks
- Savings on cost of land disposal
- Avoiding expensive alternative treatment
- Savings in cost of raw materials
- Manufacturing cost savings

Generators of hazardous wastes who send their waste off site for treatment, storage, or disposal are levied a tax of $5–$100/ton. If the waste is recycled, the fee is waived.

In the United States, cost of landfilling hazardous waste rose to more than $250/ton—a sixteen-fold increase since the 1970s. The cost of incineration is between $500 and $1500/ton. Indeed, the cost of waste management is a major item of expenditure for many companies. For example, DuPont, the largest chemical manufacturing company in the United States, was spending over $100 million annually for waste management in the late 1980s.

Regulatory Incentives

RCRA and other waste management acts require the generators of hazardous wastes to establish certification and reporting programs. By adopting a waste minimization policy, the generator fulfills these legal requirements.

Another legal requirement is certification by the generator, on the hazardous waste manifest, that a waste minimization program is in place. Other regulatory measures include

- Biennial waste minimization program reporting
- Stricter permitting requirements for waste handling and treatment
- Land disposal restrictions and bans

In 1990, the EPA imposed significant restrictions on land disposal of certain wastes, including sludge, solvents, toxic chemicals, and wastewater, without prior treatment to bring the contaminant level to acceptable standards. The 1993 Land Disposal Restriction bans land disposal of hazardous waste without pretreatment. Table 6–2 gives examples of concentration limits for selected chemicals in wastewaters and nonwastewaters.

Liability

Potential reduction in the generator's liability for environmental problems at both on-site and off-site treatment, storage, and disposal facilities is another powerful incentive for pollution prevention. There is also a potential reduction in liability for worker safety.

Public Image and Environmental Concern

Administrative incentives for pollution prevention result in a better public image of the company. These incentives are also well accepted by employees, resulting in increased productivity. Finally, these incentives enable the company to project a positive expression of its concern for the environment.

TABLE 6–2   Concentration Limits for Selected Chemicals in Wastewater and Nonwastewater (40 CFR, 1992)

Constituent	Wastewaters (mg/L)	Nonwastewaters (mg/kg)
Acetone (F039)	0.28	160
Chloroform (K009–K010)	0.10	6.0
Total Cyanides (F011–F012)	1.9	110
Cyclohexane (U057)	0.36	NA
Dimethyl phthalate (U102)	0.54	28
Mercury (K071)	0.030	NA
Toluene (U220)	0.080	28
Xylenes (U239)	0.32	28

Manufacturers of products are actively seeking ways that the product or its components, after reaching the end of their life, could be recycled into other products or materials so they do not require disposal. This has given rise to a new concept known as the *cradle-to-cradle* concept (as opposed to the cradle-to-grave concept). More environmental friendly materials and products are being used in industries, with the underlying emphasis on design for the environment. This undoubtedly will be a major trend in the future.

The above incentives have encouraged many companies to implement a waste minimization program. Invariably, such efforts entail considerable savings in manufacturing and disposal costs. Table 6–3 gives a summary of successful hazardous waste reduction efforts.

WASTE MINIMIZATION TECHNIQUES

Various techniques of waste minimization are shown in Figure 6–1. Source reduction includes any activity that reduces or eliminates the generation of hazardous waste at the source, usually within a process. Source reduction can be accomplished by product changes and source control; the latter includes changes in the input (raw) materials, changes or modifications in the existing technology, and good operating practices.

Recycling involves use, reuse, or reclamation of any material. A material is used or reused if it is either: (a) used as an ingredient to make a product, or (b) employed in a particular process as an effective substitute for an existing product. A material is reclaimed if it is processed to recover a useful product or if it is regenerated. Recovery of lead from spent batteries and silver from photographic film processing are examples of reclamation.

A viable waste minimization program should place greater emphasis on source reduction than on recycling. This hierarchy of effort is desirable from the environmental standpoint, because source reduction techniques avoid generation of hazardous waste, thereby reducing the quantity of hazardous waste generated (pollution prevention). Recycling, on the other hand, explores the possibilities of

TABLE 6–3   Selected Successful Industrial Waste Reduction Efforts (from Postel, 1988, Controlling toxic chemicals, *in* Brown et al., *State of the World:* ©1988, Worldwatch Institute, Washington, D.C., reprinted with permission)

Company/ Location	Products	Strategy and Effect
Astra Södertälje, Sweden	Pharmaceuticals	Improved in-plant recycling and substitution of water for solvents reduced toxic wastes by one-half.
Borden Chemicals California, U.S.A.	Resins; adhesives	Modification of rinsing and other operating procedures cut organic chemical in wastewater by 93 percent; sludge disposal costs reduced by $49,000 per year.
Cleo Wrap Tennessee, U.S.A.	Gift wrapping paper	Substitution of water-based ink for solvent-based ink virtually eliminated hazardous waste, saving $35,000 per year.
Duphar Amsterdam, The Netherlands	Pesticides	New manufacturing process reduced toxic waste per unit of one chemical produced from 20 kg to 1 kg.
Duphar Barranquilla, Colombia	Pesticides	New chemical-recovery system for fungicide manufacturing reclaims materials valued at $50,000 annually; waste discharges cut by 95 percent.
Du Pont Valencia, Venezuela	Paints; finishes	New solvent-recovery unit eliminated disposal of solvent wastes, saving $200,000 per year.
3M Minnesota, U.S.A.	Varied	12-year pollution prevention effort has halved waste generation, yielding total savings of $300 million.
Pioneer Metal Finishing New Jersey, U.S.A.	Electroplated metal	Redesigned treatment system reduced water use by 96 percent and sludge production by 20 percent; annual net savings of $52,500; investment paid back in three years.

utilizing hazardous materials for a beneficial purpose; it does not reduce the volume of waste.

Source Reduction

Source reduction includes good operating practices, technology changes, and material or product changes. These are discussed in the following sections.

Good Operating Practices

Good operating practices are related to human aspects of manufacturing operations. Any procedural, administrative, or institutional measures that a company

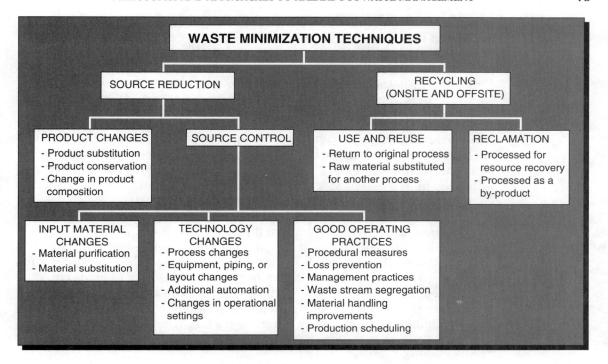

FIGURE 6–1   Waste minimization techniques (adopted from U.S. EPA, 1988)

may adopt to reduce waste would constitute good operating practices. These can often be implemented with little cost, and therefore have a high return on investment. These practices can be implemented in all areas of a plant, including raw materials and product storage, production, and maintenance operations.

Technology Changes

Sometimes owners of manufacturing plants take for granted that the particular process used in manufacturing their product optimizes the use of raw materials, energy, and personnel to yield the most desirable quality and quantity of the product. This attitude precludes any consideration of introducing change in the process technology. However, taking a careful look at the process and considering equipment and process modifications can lead to a significant reduction in hazardous wastes. Technology changes include changes in production process, changes in equipment used and layout, use of automation, and changes in process operating conditions, such as flow rates, temperatures, pressures, and residence times. Some of these changes can be accomplished in a few days at low costs; others may involve long times and large capital investments.

Example. A manufacturer of fabricated metal products employed an alkaline bath to clean nickel and titanium wires prior to their use in the final product. In 1986, the company started using a mechanical abrasive system in which the wires were cleaned by passing them, under pressure, through silk and carbide pads. The system worked, but required passing the wires through the system twice to ensure complete cleaning. In 1987, the company bought a second abrasive unit and installed it in series with the first unit. The new system produced clean wires, shortened the cleaning time, and completely eliminated use of a chemical cleaning bath. The hazardous materials were altogether eliminated, thus permanently eliminating disposal cost of the hazardous waste solution.

Input Material Changes

Input material changes accomplish waste minimization by eliminating or reducing hazardous materials that enter the production process. By changing the input material (feedstock), it is possible to avoid hazardous waste generation within the production process. Input material changes can be accomplished by (a) material purification, and (b) material substitution.

Example. An electronic manufacturing facility of a large diversified corporation originally cleaned printed circuit boards with chemical solvents. The company found that by switching from a solvent-based cleaning system to a water-based system, the same operating conditions and workloads could be maintained. The water-based system was found to be six times more efficient in cleaning the circuit boards. This change resulted in a lower product rejection rate and, more importantly, eliminated a hazardous waste.

Product Changes

A manufacturer may eliminate or reduce hazardous wastes by substituting a nonhazardous material for a hazardous one, through product conservation, or by changing the product composition.

Example. Solvent-based paints are being rapidly phased out in the paint industry. Water-based paints have eliminated the use of toxic or flammable chemicals. Water-based paints do not require chemical solvents to clean paintbrushes and applicators. Emissions of VOCs from solvent-based paints have also been eliminated.

Product Conservation

This involves careful management of inventory of hazardous material. Chemicals that have short shelf-lives should not be stocked beyond their expected shelf-life. Either smaller quantities of such hazardous materials should be stocked, or they should be replaced by similar materials that have longer shelf-lives; both will ensure complete utilization of such materials, avoiding treating the otherwise unused materials as hazardous wastes. The American Chemical Society (1985) recommends purchasing many small containers rather than one large container for the most cost-efficient and safest way to acquire laboratory chemicals for use in universities' teaching and research laboratories.

Change in Product Packaging

Sometimes the simple process of changing the packaging from metal cans to paper containers can eliminate hazardous waste. Dow Chemical Company used to manufacture a wettable powder pesticide and sell it in 2-lb metal cans. The empty cans, if not properly decontaminated before disposal, constituted a hazardous waste. The company now sells the pesticide in a 4-oz water-soluble package that dissolves when the product is mixed with water before use (U.S. Congress, Office of Technology Assessment, 1986).

Recycling

Recycling (use or reuse) can be done at the site where waste is generated or it can be done offsite. The recycled material can either be used in the originating process as a substitute or as an input material, or in another process as an input material.

Example. A printer of newspaper advertisements in California bought an ink-recycling unit to produce black newspaper ink from its waste inks. The unit mixes the different colors of waste inks with fresh black ink and black toner to create the black ink. This ink is then filtered to remove flakes of dried ink, after which it is ready for use in place of fresh black ink. The reuse has eliminated the need for the company to ship the waste ink offsite for disposal and has entailed significant cost savings. The cost of the recycling unit was paid off in nine months, based on the savings on the purchase of fresh black ink and the cost of shipping and disposal of the waste ink.

Reclamation

Reclamation is the recovery of a valuable material from a hazardous waste. It is different from use and reuse techniques because the recovered material is not used in the facility where it was generated; rather, it is sold to another company.

Example. A photo processing company uses an electrolytic deposition cell to recover silver from the rinse water from film processing equipment. The silver is sold to a small recycler. By removing silver from wastewater, the wastewater is no longer considered hazardous waste and is discharged into the sewer system without additional pretreatment by the company. The cost of the cell was paid off in less than two years with the value of the silver recovered. The company also collects used film and sells it to the same recycler. The recycler burns the film and collects the silver from the residual ash. By removing silver from the ash, the ash is rendered nonhazardous.

WASTE EXCHANGES

Reclamation of useful materials from waste streams has gained popularity during the past 10–15 years. A waste exchange is a matchmaking operation based on the idea that one manufacturer's waste may be another manufacturer's feedstock. Waste exchanges are private- or government-funded organizations that can help bring together generators of hazardous waste and companies that can use the waste as feedstock or substitute material in their operations.

There are two types of waste exchanges: *information exchanges* (passive exchanges) and *material exchanges* (active exchanges). Information exchanges are clearinghouses for information on wastes generated by one manufacturer that may be needed by another as feedstock. They put generators in touch with waste users for the purpose of recycling waste materials back into manufacturing processes. Material exchanges take actual physical possession of hazardous waste and may initiate or actively participate in the transfer of wastes from the generator to the user.

Solvents, organics, acids, and alkalies are the most commonly recycled hazardous materials. Most transactions involve relatively "pure" wastes that can be used directly with minimal processing. Example entries from a waste exchange catalog are shown below.

> *Formaldehyde Surplus.* Formaldehyde solution. Potential Use: embalming fluid. Type 1: Contains 25% formaldehyde with 10% glycerine, 10% alcohols (ethanol, isopropanol, methanol) and distilled water by wet weight. Type 2: Contains 25% formaldehyde with 25%–35% alcohols (ethanol, isopropanol, methanol) and distilled water by wet weight. 165,000 gallons in 15-gallon drums/plastic carboys in steel drums. One time. Independent analysis (specifications) available.

> *Paraffin Wax.* Paraffin wax from clean-out of chewing gum base mixers. Fully refined. Potential use: firelogs, crayons, etc. Contains traces of gum base and calcium carbonate. 80,000 lbs. in 50-gallon drums. Quantities continuous. Thereafter 40,000 lbs/quarter.

> *1,1,1-trichloroethane.* 1,1,1-trichloroethane from asphalt extractions. Contains 90% 1,1,1-trichloroethane with 10% asphalt and 1% oil. 220 gallons in drums available. Quantities vary. Thereafter 220 gallons/year. Sample available.

Appendix G gives a list of hazardous waste exchanges in the United States, Canada, and Taiwan. Appendix H lists hazardous waste information exchanges in the United States.

STUDY QUESTIONS

1. Historically, how has waste management philosophy evolved? Discuss the implications of this evolution.
2. What is integrated waste management? What are its main components? How is integrated waste management different from the earlier philosophies?
3. Why should a generator of hazardous waste consider waste minimization? What incentives are available? How may these incentives help this business?
4. What are the various techniques of waste minimization? Give appropriate examples for each.
5. What is a waste exchange? How is an active waste exchange different from a passive exchange? Explain.

REFERENCES CITED

40 Code of Federal Regulations, 1992, § 268.43, Protection of Environment: Washington, D.C., U.S. Government Printing Office, p. 817–839.

American Chemical Society, 1985, Less is Better: Laboratory Chemical Management for Waste Reduction: Washington, D.C., Bulletin published by the American Chemical Society, 16 p.

Goldman, B.A., J.A. Hulme, and C. Johnson, 1986, Hazardous Waste Management: Council on Environmental Priorities: Washington, D.C., Island Press, 316 p.

Postel, S., 1988, Controlling Toxic Chemicals, *in* Brown et al. (eds.), State of the World, 1988: New York, W.W. Norton and Co., p. 118–136.

U.S. Congress, Office of Technology Assessment, 1986, Serious Reduction of Hazardous Waste: Washington, D.C., Superintendent of Documents, Government Printing Office, p. 83.

U.S. Congress, 1984, The Hazardous and Solid Waste Amendments of 1984: United States Statutes at Large, vol. 98, Part 3, p. 98STAT.3224, Washington, D.C., U.S. Government Printing Office, 3652 p + indices.

U.S. Environmental Protection Agency, 1988, Waste Minimization Opportunity Assessment Manual: Cincinnati, OH, Hazardous Waste Engineering Research Laboratory, Report No. EPA/625/7-88/003, 25 p. plus appendices.

U.S. Environmental Protection Agency, 1990, Waste Minimization: Environmental Quality with Economic Benefits: Report No. EPA/530-SW-90-044, 2nd ed., 34 p.

U.S. Environmental Protection Agency, 1991, EPA Clean Products Research Program: Cincinnati, OH, EPA Office of Research and Development, Risk Reduction Engineering Laboratory, EPA Report No. EPA/600/D-91/029, 11 p.

7

Physical, Geotechnical, and Geochemical Properties of Earth Materials

INTRODUCTION

In the hazardous waste management field, the media that most commonly become contaminated are water, soil, and rock. A clear understanding of the origin, geologic nature, and occurrence of these materials is therefore essential for proper management of hazardous waste, particularly the aspects relating to evaluation of contamination of the media and subsequent remediation. This chapter presents a discussion of various properties of rocks and soils that are important in hazardous waste site evaluation, site selection, and remediation. (Water is discussed separately in Chapter 8.) Minerals, though of great importance in nations' economies, are not of special relevance in hazardous waste management (with the exception of clay minerals) and are not discussed in detail.

Soils, rocks, and minerals together constitute what, in geology, are known as earth materials. These are naturally occurring substances found practically everywhere. Earth materials are used for a variety of purposes, including producing food and building structures. Soils are especially significant for all life forms, including humans. As long as they are "clean," they serve to support life forms. However, upon pollution the same life-supporting earth materials may become life threatening.

ROCKS AS EARTH MATERIAL

Rocks are part of the lithosphere and represent usually hard, compact, and firm materials formed in distinct geologic environments. When solid rocks form as a result of cooling and crystallization of minerals from a molten material, they are known as *igneous* rocks. Sediments deposited in oceans, after undergoing physical, chemical, and biochemical changes, produce compacted, indurated, and layered rocks that are called *sedimentary* rocks. Any preexisting rock acted upon by the high temperatures and pressures that prevail in the earth's interior changes into a new set of rocks called *metamorphic* rocks.

Fresh, unaltered rocks, in general, possess high strength. Upon weathering, however, their strength may diminish to the point that they are no better than sediments. It is therefore extremely important to make a distinction between rock and soil—a seemingly difficult task complicated by the degree of weathering and presence of discontinuities. *Discontinuity* is a general term for planes of weakness in rock. Most of the common planar structures found in rocks are also planes of discontinuities. Bedding planes, foliation planes, joints, fractures, faults, shear zones, and lineations are common examples of discontinuities found in rocks. In nature we find fresh, unaltered rocks with few discontinuities representing one extreme, and rocks—with many discontinuities—that have been completely weathered into loose mineral grains and behave like soil as the other extreme. Every conceivable intermediate variety occurs in nature (Figure 7–1).

Origin and Classification of Rocks

Petrology is the branch of geology that deals with the origin, occurrence, composition, classification, and distribution of rocks. Even a partial discussion of fundamental concepts of petrology is beyond the scope of this book. Nevertheless, a brief discussion of the origin, classification, and geotechnical properties of major rock types is included.

Origin of Rocks

The origin of rocks is intimately related to the broad tectonic cycle. As discussed in Chapter 2, the rock cycle specifically is responsible for the creation and destruction of solid earth materials.

The three main families of rocks that comprise the lithosphere are igneous, sedimentary, and metamorphic. Igneous rocks are the most abundant—65 percent

FIGURE 7–1   Degree of weathering and material strength

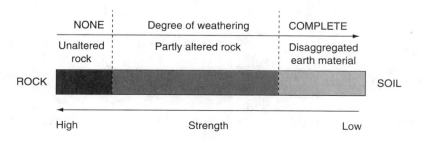

by volume of the Earth's crust; metamorphic rocks—27 percent; followed by sedimentary rocks—8 percent.

Igneous rocks. When the molten earth material, *magma,* cools and solidifies at some depth below the Earth's surface, the resulting rocks are called *intrusive* igneous rocks. When the molten material comes to or above the surface of the Earth, it is known as *lava;* the rocks formed by cooling and solidification of lava are called *extrusive* igneous (or *volcanic*) rocks. Because the same magma is the source of both intrusive and extrusive rocks, the mineral composition of the two is essentially the same. The difference is in the texture of the rocks; intrusive igneous rocks generally have larger mineral crystals, 1–10 mm and greater, which can be readily discerned with an unaided eye. Extrusive igneous rocks are extremely fine grained, generally less than 1 mm, and the individual minerals cannot be recognized without a hand lens. Granite, diorite, and gabbro are common examples of intrusive igneous rocks; their extrusive equivalents are rhyolite, andesite, and basalt.

Sedimentary rocks. The loose mineral grains, and occasionally rock fragments, after being transported by wind, ice, and water, are ultimately deposited on the floors of oceans, where they are buried under the column of water and the continuous incoming load of sediments. Pressure of the overburden material, including water, aided by chemical and biochemical reactions that occur in oceans, initiate the process of conversion of sediments into hard layers of rocks, called sedimentary rocks. Local conditions of water currents, microorganisms, salinity, temperature, and pressure control the formation of various types of sedimentary rocks. For example, highly saline waters in enclosed basins, in a dry climatic environment, will result in the formation of bedded rock salt, gypsum, and potash. These are collectively known as *evaporite* deposits. Evaporites can also form in lakes where the water is saline; sodium chloride ($NaCl$), sodium carbonate (Na_2CO_3), sodium sulfate (Na_2SO_4), and borax ($Na_2B_4O_7.10H_2O$) are common salts that are produced as a result of evaporation of saline lake waters in dry climatic environments.

Recrystallization and cementation of mineral and rock particles, called clasts, result in the formation of *clastic* sedimentary rocks. Sandstone, conglomerate, breccia, siltstone, and shale are examples of clastic sedimentary rocks characterized by different sizes and shapes of clasts that make up the rocks.

Sediments produced as a result of physiological activities of organisms and from animal and plant remains, after being subjected to physical and biochemical changes, are converted into *biogenic* sedimentary rocks. Limestone, coal, and oil shales are examples of common biogenic sedimentary rocks.

Metamorphic rocks. When the preexisting sedimentary, igneous, or even metamorphic rocks are subjected to the high temperature, high-pressure environment inside the Earth, changes in mineral assemblage and texture of the pre-existing rocks take place. The new rocks formed as a result of these changes are called metamorphic rocks. All changes occur in solid state, meaning there is no melting of the preexisting rocks. The range of temperatures and pressures at which metamorphic processes operate is generally +200°C (392°F) to 600–800°C

(1112°–1472°F) and 300 MPa to about 1200 MPa (3000–12,000 atmosphere). The upper range of temperature for metamorphism is a function of the amount of water present in the rocks. A wet rock will melt at a lower temperature than when it is dry, i.e., water lowers the melting point. High pressure also aids in lowering the melting point. Therefore, in metamorphic processes, presence of water at higher pressures will bring about changes in preexisting rocks at a lower temperature than would have occurred had the rock been dry and at lower pressures.

Metamorphic rocks are classified into two main groups: *foliated* metamorphic rocks and *non-foliated* metamorphic rocks. Foliated metamorphic rocks display a preferred alignment (general parallelism) of mineral grains that imparts a layered effect to the rock. Non-foliated metamorphic rocks do not exhibit such layering and are called *massive*. Marble and quartzite are common examples of non-foliated metamorphic rocks; slates and schists are common foliated metamorphic rocks.

Significance of Weathering and Discontinuity

The physical and engineering properties of rocks are greatly influenced by weathering and the discontinuities present in them. A strong rock, like granite, possessing high strength, when subjected to a high degree of weathering will be reduced to a loose aggregate of minerals comprised of the original and unaltered quartz and highly decomposed silicate minerals that altered into clay minerals. These weathered materials possess strength that may be a few orders of magnitude lower than that of the unweathered rock. Weathering, in general, also causes an increase in the porosity and permeability of rocks. This aspect is of critical importance in hazardous waste management. It is, therefore, very important that complete evaluation of rocks be done to assess the degree of weathering before any remediation plan is formulated or before construction of a disposal facility begins.

Lack of a universal and quantitative weathering index makes it very difficult to distinguish the various intermediate products that fall in between rock and soil, the two extremes. In some situations a sandstone may be weathered to the point that all cementing materials holding the mineral grains together have been leached away, leaving behind a mass of loose grains (sediments). These sediments are completely different from the original rock in the sense that they are unconsolidated, have lower strength, and are disaggregated. On the other extreme, one may find a layer of clay that has been cemented by mineralized water to form a layer of hard, rock-like mass, known as *claypan*. Because of this complexity, a highly weathered rock may be considered a soil material, while a hard, cemented soil layer may be taken to be a rock material. Classifications based on these qualitative factors often result in confusion and litigation at construction projects. For example, a foundation material classified as "earth" (soil) for excavation (because of the absence of any bedrock) may turn out to be a layer of hard claypan or caliche possessing high strength, which requires drilling and blasting for excavation. On the other hand, a foundation material considered rock (because of nearby outcrops and lack of any sediment cover in the area) may turn out to be a highly weathered, soft material, that can be excavated by common earthmoving equipment. With the cost of rock excavation being 4–10 times greater than that for soil excavation, the problems created by such classifications are obvious. It is therefore very important to classify the earth materials as soil or rock on the basis of their strength, and not on their geologic origin. A convenient classification for engineering contract purposes classifies soil excavation as one that can be done by using conventional earthmoving equipment; rock cannot be excavated in a similar manner and requires drilling and blasting. In another case, the distinction between rock and soil has been drawn on the basis of the unconfined compressive strength (q_u) of the earth material. The material is classified as rock if its q_u is greater than 1379 kN/m² (200 psi), and soil if it is less than 1379 kN/m².

Numerous studies have been done at various locations across the world to evaluate the degree of weathering of rocks. Based on these studies, several schemes for rating the rocks in terms of their degree of weathering have been developed. However, all have the drawback of being site-specific and complex. No classification has yet been developed that can be considered reliable, relatively easy to use, and universal in application.

Generally, rock weathering is evaluated by studying the degree of alteration of minerals under a hand lens or a microscope. Granites have been commonly used to illustrate various levels of weathering. It has been found that higher levels of chemical weathering produce progressively higher levels of alterations of minerals in granites and similar rocks. For example, Kiersch and Treasher (1955) used degree of alteration of feldspar and other minerals in a representative sample of rock to classify granodiorite into four categories:

Degree of Weathering	Mineral Alteration Features
Fresh rock (none)	All minerals unaltered
Slightly weathered	Feldspars slightly bleached and fissured
Moderately weathered	Feldspars more intensely bleached and fissured; bleaching of biotite; slight rounding of quartz grains; limonite appears as specks and coatings
Highly weathered	Feldspars highly bleached and fractured; quartz grains highly rounded; limonite commonly occurs as accessory mineral; rock easily scratched with a steel nail

Degree of weathering has also been studied in situ. A qualitative classification to describe the degree of weathering, developed by British engineering geologists, is given in Table 7–1.

Williamson (1984), in attempting to devise a unified rock classification system, developed a five-point scale to categorize various levels of rock weathering, based on field observations. Despite its simplicity the system has not been widely used. Table 7–2 gives weathering categories and associated features of the rock.

Geotechnical Properties of Rocks

General Discussion

Geotechnical properties of rocks are highly variable. Various factors, such as degree of weathering, nature of discontinuities, mineralogy, geologic occurrence (near a large fault zone, regions of repeated structural deformation, proximity to active plate margins), and local climatic and biologic conditions, modify rock

TABLE 7–1 Intact Rock Weathering Classification (adopted from London Geological Society Engineering Group Working Party, 1977)

Weathering Class	Description	Grade
Fresh	No visible sign of rock material weathering	IA
Faintly weathered	Discoloration on major discontinuity surfaces	IB
Slightly weathered	Discoloration indicates weathering of rock material and discontinuity surfaces. All rock material may be discolored by weathering and may be somewhat weaker than in its fresh condition	II
Moderately weathered	Less than half of the rock material is decomposed and/or disintegrated to a soil. Fresh or discolored rock is present either as a continuous framework or as corestones	III
Highly weathered	More than half of the rock material is decomposed and/or disintegrated to a soil. Fresh or discolored rock is present either as a discontinuous framework or as corestones	IV
Completely weathered	All rock material is decomposed and/or disintegrated to soil. The original mass structure is still largely intact	V
Residual soil	All rock material is converted to soil. The mass structure and material fabric are destroyed. There is a large volume change, but the soil has not been significantly transported.	VI

TABLE 7–2 Weathering Classification (adopted from Williamson, 1984)

Degree of Weathering	Characteristic Features	Symbol
Micro Fresh State	Viewed under 10x hand lens, the rock does not show alteration of any mineral components	MFS
Visually Fresh State	Viewed with unaided eye, rock shows uniform color, maximum unit weight, maximum strength, and the least water absorption	VFS
Stained State	Rock shows complete discoloration; specimen cannot be remolded with finger pressure	STS
Partly Decomposed State	Rock is solid when in place, but can be desegregated into gravel or larger size pieces set in soil matrix by applying finger pressure	PDS
Completely Decomposed State	By applying finger pressure the material disaggregates into gravel or larger size pieces of original rock; can be easily molded	CDS

properties and influence its engineering behavior. Therefore, values of rock properties found in handbooks should be used with extreme caution, because in most cases the influence of the above factors is not considered. The reported values represent average measurements performed on *fresh* rock specimens that do not necessarily replicate conditions at the site. Tables, such as Table 7–3, are intended to provide generalized values of geotechnical properties of rocks and should be used with caution. They should *not* be used for design purposes. However, handbooks listing geotechnical properties may be used to compare the difference in the geotechnical properties of various rock types.

In terms of their strength and performance, intrusive igneous rocks are more suitable for engineering construction. Extrusive igneous rocks, because of their susceptibility to weathering and the presence of a large number of discontinuities, are weaker and generally less suitable. At uncontrolled hazardous waste sites, such rocks are more likely to cause serious problems in contaminant transport because of higher values of hydraulic conductivity and intricate flow paths.

Engineering use and performance of sedimentary rocks are highly variable. In general, coarse-grained clastic rocks and limestone are suitable for most engineering construction, except when water retention is the main concern, as in dams and reservoirs. Limestone, because of its susceptibility to dissolution in rain water, tends to develop voids and other solution features, resulting in the *karst topography*. Sinkholes and subsurface openings are potential pathways of contaminant movement, and require careful investigation. Contaminated surface and groundwater may travel rapidly over large distances through the network of solutioned openings in limestones. *Dye tracing* is often used to determine the travel rate and paths of water through such openings.

From an engineering standpoint, metamorphic rocks, in general, are suitable for most construction projects. They have served as a disposal medium for deep well injection of hazardous waste in liquid form. Surplus chemicals from World War II were disposed of at a depth of 3.6 km (2.2 mi), in metamorphic rocks, at the Rocky Mountain Arsenal Site near Denver, Colorado, between 1962 and 1965.

Yet another important aspect of the geotechnical properties of rocks relates to the difference between *rock material* and *rock mass*. The former relates to properties observed and measured in a sample of rock (a rock specimen), while the latter is the property of the rock *in place,* which includes both rock outcrop and rock formations encountered in the subsurface. Rock material is a sample of fresh rock taken from the rock formation in place and usually is not representative of the degree of weathering and the nature of the discontinuities. As discussed previously, these features are of great importance in hazardous waste management because they control the permeability of rock mass. Therefore, it is a common practice to carry out field tests to determine the permeability of rock mass, and not to rely on results obtained by running permeability tests on rock specimens in the laboratory, to get a more reliable value for hydraulic conductivity.

Important Geotechnical Properties of Rocks

Specific gravity (*SG*), moisture content (ω), porosity (*n*), unit weight (γ), absorption, permeability, compressibility, and strength are some of the common geotechnical

TABLE 7–3 Selected Geotechnical Properties of Common Rocks (from Hunt: *Geotechnical Engineering Investigation Manual* ©1984, McGraw-Hill, Inc., reprinted with permission)

Rock Type	Texture (grain size, mm)	Discontinuities	ρ_d kg/m^3	q_u MPa	$E_{in\ situ,}$ $\times 10^4$ kg/cm^2
A. *IGNEOUS*					
Granite			2690	69–172	28–49
Diorite	Intrusive; coarse to medium (2.0–0.2)	Tight to widely spaced joints, 3 sets common; sometimes massive	2620	69–172	35–56
Gabbro			2880	103–206	49–84
Rhyolite	Extrusive; fine (0.02–0.06)	Extensive jointing, often vesicular	2590	69–172	35–56
Andesite			2660	69–172	42–63
Basalt			2850	172–275	49–90
Obsidian	Extrusive, glassy	Massive cemented ash and rock fragments; porous	2200	14–55	7–28
Tuff	Extrusive, coarse (2.0–0.6)		1600	1–7	1–7
B. *SEDIMENTARY*					
Conglomerate	Clastic; coarse (2.0–0.6); rounded grains	Bedded and cemented	2480	34–103	7–35
Breccia	Clastic; coarse (2.0–0.06); angular grains	Bedded and cemented	2530	34–103	7–35
Sandstone	Clastic; medium (0.6–0.2); rounded to subrounded grains	Bedded and cemented	2350	28–63	7–21
Siltstone	Clastic; fine (0.2–0.06); subangular to rounded grains	Bedded and cemented	1800–2400	<1–34	3–14
Shale	Clastic; very fine (<0.06), subangular to rounded grain	Thinly bedded and jointed (compaction); Bedded (cementation shale)	1600–2200	<1–34	3–14
Limestone	Biogenic; fine (0.06–0.2)	Massive to bedded and jointed; forms solution cavities	2640	34–103	14–42
Dolomite	Biogenic; fine (0.06–0.2)	Massive; recrystallized minerals	2670	48–137	26–56
C. *METAMORPHIC*					
Gneiss	Coarse to medium (0.2–2)	Banded and foliated, joints common	2700	69–137	28–56
Schist	Medium to fine (0.06–0.6)	Foliated and jointed	2670	34–103	14–35
Slate	Fine (0.06–0.2)	Slaty cleavage; platy	2690	69–137	35–56
Quartzite	Medium to fine (0.06–0.6)	Massive; jointed	2660	103–240	42–56
Marble	Medium to very fine (<0.06–0.6)	Massive; jointed	2690	82–206	49–70

properties of rocks that are important in engineering projects and in hazardous waste management.

Specific gravity. Specific gravity (SG) of a rock material is the ratio of its mass (or weight) to that of an equal volume of water. It can be expressed as

SG = Mass (or weight) of the rock in air ÷ Mass (or weight) of an equal volume of water.

Mass density (ρ_m) of a rock is related to its specific gravity by the equation

$$\rho_m = SG \times \rho_w \tag{7-1}$$

where ρ_w = the density of water (1 g/mL; 1000 kg/m^3 or 62.4 lb/ft^3).

Mass density gives the maximum value of rock density and assumes that the rock is massive with no openings (pores, microfractures). In reality, however, most rocks have some porosity, which results in a lower density, less than ρ_m.

Specific gravity of the rock, on the other hand, takes into account the openings in the rock and can be expressed as bulk specific gravity (SG_B); bulk specific gravity, saturated surface dried (SG_S); and apparent specific gravity (SG_A). These can be calculated by using the following expressions.

$$SG_B = A \div (B - C) \tag{7-2}$$
$$SG_S = B \div (B - C) \tag{7-3}$$
$$SG_A = A \div (A - C) \tag{7-4}$$

where A = the weight of rock in air (after 24-hr oven drying), B = the weight of rock in air (after 24-hr soaking in water), C = the weight of saturated rock in water.

Absorption. The amount of water that can enter the pores in a saturated rock material is called absorption. Using the above nomenclature,

$$\text{Absorption} = \{(B–A) \div A\} \times 100 \tag{7-5}$$

Absorption is an important property of rock materials used as aggregate for concrete or as foundation for highways and other construction projects. Most state highway departments have set the upper limit of absorption at 5 percent.

Porosity. The ratio of the volume of voids to the total volume of rock is its porosity (n); it is expressed as a percentage.

Permeability. The ability of rock to allow for the movement of fluids through its mass without any change in its structure or fabric is its permeability. The *coefficient of permeability* is a measure of this property.

Igneous and metamorphic rocks, because of the way they are formed, generally do not have open spaces (voids) between the solids (minerals). In other words, they lack *primary porosity*, defined as the porosity that develops in rocks during their formation. Sedimentary rocks do possess primary porosity. However, all rocks develop joints and other discontinuities that impart *secondary porosity* to an otherwise nonporous and impermeable mass. Secondary porosity relates to openings developed *after* the formation of the rock, usually by tectonic forces. These secondary openings often result in movement of water and other fluids through such rocks, giving rise to *secondary permeability*. It is because of secondary permeability that rocks such as basalt and granite can hold a sufficient quantity of groundwater to yield a small to moderate water supply. Secondary porosity and permeability should always be given due consideration while assessing fluid migration through rocks.

Rock quality designation (RQD). Rock quality designation is an extremely useful parameter that relates to the evaluation of rock mass for a variety of engineering purposes. It was developed by Deere et al. (1967). RQD is an empirical relationship that originally involved measurement of NX size (54.7 mm or 2.15 in.) core pieces obtained by drilling through rock formations. Of the total core run, the lengths of all pieces 0.10 m (4 in.) or larger are added and the sum is divided by the total core run. The result, when expressed as a percentage, is the RQD. In Figure 7–2 the total core run is 6.0 m and the sum of all core pieces 0.10 m or larger is 4.68 m; the RQD is calculated by

$$RQD = [\Sigma \text{ of all core pieces} >0.10 \text{ m} \div \text{total core run}] \times 100 \qquad (7\text{–}6)$$
$$= (4.68 \div 6.0) \times 100 = 78\%$$

Table 7–4 gives the relation between RQD and rock quality.

Although Deere et al. (1967) had recommended NX-size cores for RQD determination, Deere and Deere (1989) later found that core sizes slightly larger or smaller (NQ: 47.6 mm or 1.88 in. diameter) than NX are also acceptable. RQD has become a standard feature in rock exploration for engineering construction and is always shown on a rock drill log alongside other borehole information.

FIGURE 7–2   RQD determination from drill cores

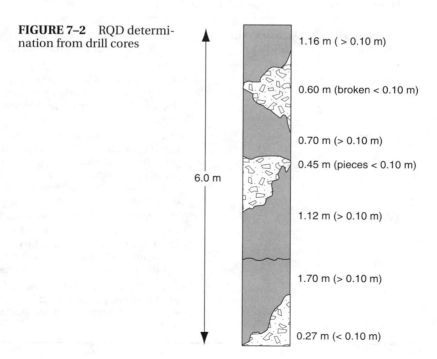

6.0 m

1.16 m (> 0.10 m)

0.60 m (broken < 0.10 m)

0.70 m (> 0.10 m)

0.45 m (pieces < 0.10 m)

1.12 m (> 0.10 m)

1.70 m (> 0.10 m)

0.27 m (< 0.10 m)

TABLE 7–4   RQD and Rock Quality

RQD (%)	Rock Quality
0–25	Very Poor
25–50	Poor
50–75	Fair
75–90	Good
90–100	Excellent

The fundamental premise of RQD is that discontinuities control the strength and mechanical behavior of rock mass. The higher the frequency of discontinuities and their aperture, the lower its strength. These aspects will also result in poor core recovery. While determining the RQD, care must be taken to see that only natural fractures (discontinuities) are considered for measuring length of the core pieces. Breaks that occur during core handling and the drilling process should not be taken into account. In general, high RQD has been found to correlate with high compressive and shear strengths and fewer discontinuities in the rock mass. Hatheway (1990), in discussing representative uses of RQD in rock engineering, indicated that low-RQD zones in a rock mass correspond to flow paths of contaminants, since the fractures will facilitate flow of groundwater.

TABLE 7–5   Selected Geotechnical Properties of Rocks and Their Significance in Hazardous Waste Management (adapted from Hatheway, 1990)

Property/ Parameter	Measured in Field / Lab		Importance in Hazardous Waste Management
Density		X	Rock formations of lower density may have a tendency to fracture if high pressures are used for deep well injection of hazardous waste in liquid form.
Porosity		X	Controls fluid flow to a great extent; an important input in contaminant transport modeling. High porosity combined with discontinuities may render an "impermeable" rock like shale a permeable formation, allowing for contaminant movement.
Permeability	X	X	Contaminated fluid transport in rocks is, to a large extent, affected by its permeability. Field permeability should be determined while assessing the nature and extent of groundwater contamination and modeling studies.
RQD		X	A relatively inexpensive and versatile indicator of important properties of rocks; can be used to estimate contaminant transport through rock formations.

The fundamental geotechnical properties of rocks discussed above must be evaluated in any hazardous waste disposal and remediation project. Table 7–5 is a summary of important geotechnical properties of rocks and their significance in hazardous waste management.

Geochemical Characteristics of Rocks

A brief discussion of the geochemical nature of rocks is included to provide a background on the relative abundance and occurrence of common elements in rocks. Rocks are the primary earth materials that, upon weathering, release various chemical elements and compounds that later become part of the environment. The geochemical properties of sediments, soils, surface water, groundwater and sea water are the manifestation of the geochemical cycle, which, in a general sense, is responsible for the dispersion and concentration of various elements and compounds in the environment. At places where chemical concentration exceeds the permissible levels, we have the problem of pollution (chemical pollution is an excess of certain chemical elements or compounds in water, air, or soil, making it harmful to any living form). For example, the Safe Drinking Water Act has established the Maximum Contaminants Levels (MCLs) for various chemicals in drinking water to safeguard public health. According to this standard, the MCL for sulfate in drinking water cannot be more than 250 mg/L. Whenever and for whatever reasons (e.g., excessive aquatic biologic population or release of sulfur gases following a volcanic eruption) the sulfate concentration goes above 250 mg/L, the water is to be considered polluted and unfit for human consumption. Uncontrolled disposal of hazardous wastes and migration of leachate from improperly designed sanitary landfills cause an increase in the ambient levels of chemical elements and compounds in water, air, and soils, causing pollution.

It should also be noted that in nature anomalous concentration of certain chemical elements and/or compounds occurs, giving rise to mineral deposits of economic importance. Deposits of metallic and nonmetallic elements, hydrocarbons, and radioactive elements represent unique aspects of the geochemical cycle that have resulted in the formation of these deposits. Depending upon their occurrence and potential for release into the environment, one could expect higher-than-average concentration of certain metals and nonmetals in sediment, soil, and water in the vicinity of such deposits.

Geochemical Nature of Igneous and Sedimentary Rocks

Studies based on a large number of chemical analyses of rocks from all over the world show that overall there is no significant difference in the chemical compositions of igneous and sedimentary rocks. This is to be expected, because igneous rocks provide the materials that, after deposition and lithification, turn into sedimentary rocks. However, some difference in the relative percentages of various elements and compounds have been observed in the two rock types. Table 7–6 gives the chemical composition of various sedimentary rocks along with that of the average igneous rock. It is evident from the table that sedimentary rocks are generally deficient in Na_2O but richer in CO_2 and SO_3. On an elemental basis, enrichment of As, U, Mo, and S has been noted in sedimentary rocks (Table 7–7).

TABLE 7–6   Chemical Composition (%) of Common Sedimentary Rocks, Compared with Average Igneous Rock (adapted from Mason and Moore, 1982)

Common Oxides	Average Igneous Rock	Average Shale	Average Sandstone	Average Limestone
SiO_2	59.14	58.10	78.33	5.19
TiO_2	1.05	0.65	0.25	0.06
Al_2O_3	15.34	15.40	4.77	0.81
Fe_2O_3	3.08	4.02	1.07	0.54
FeO	3.80	2.45	0.30	—
MgO	3.49	2.44	1.16	7.89
CaO	5.08	3.11	5.50	42.57
Na_2O	3.84	1.30	0.45	0.05
K_2O	3.13	3.24	1.31	0.33
H_2O	1.15	5.00	1.63	0.77
P_2O_5	0.30	0.17	0.08	0.04
CO_2	0.10	2.63	5.03	41.54
SO_3	—	0.64	0.07	0.05
BaO	0.06	0.05	0.05	—
C	—	0.80	—	—
Total	99.56	100.00	100.00	99.84

SOILS AS EARTH MATERIAL

Viewed from a scientific standpoint, soil is not *dirt,* although the term means different things to different people. This is yet another reason why clear and unambiguous communication must exist between various experts involved in hazardous waste management. For example, to a soil scientist—the agronomist—soil represents the top few feet of the Earth's surface that contains organic and inorganic materials and living organisms and is capable of supporting plant life. To a geologist, soil represents an end product in the weathering continuum; it is a part of the overburden. In a broad sense, soils include all loose (unconsolidated) materials, collectively referred to as sediments. These unconsolidated materials occur on the Earth's surface and mask the bedrock.

Definition and Origin of Soil

Definition

Civil engineers include all unconsolidated materials, including man-made fills, in their definition of soil. Geotechnically, soil represents a two- or three-phase system that comprises discrete solid particles (the mineral grains), both organic and inorganic, with liquids and/or gases in between. The most common liquid

TABLE 7–7   Elemental Composition in Parts Per Million of Common Sedimentary and Igneous Rocks (adapted from Mason and Moore, 1982)

Element	Shales	Sandstones*	Carbonates*	Igneous Rocks
Li	66	15	5	20
Be	3	0.X	0.X	2.8
B	100	35	20	10
F	740	270	330	625
Na	9600	3300	400	28,300
Mg	15,000	7000	47,000	20,900
Al	80,000	25,000	4200	81,300
Si	273,000	368,000	24,000	277,200
P	700	170	400	1050
S	2400	240	1200	260
Cl	180	10	150	130
K	26.600	10,700	2700	25,900
Ca	22,100	39,100	302,300	36,300
Sc	13	1	1	22
Ti	4600	1500	400	4400
V	130	20	20	135
Cr	90	35	11	100
Mn	850	X0	1100	950
Fe	47,200	9800	3800	50,000
Co	19	0.3	0.1	25
Ni	68	2	20	75
Cu	45	X	4	55
Zn	95	16	20	70
Ga	19	12	4	15
Ge	1.6	0.8	0.2	1.5
As	13	1	1	1.8
Se	0.6	0.05	0.08	0.05
Br	4	1	6.2	2.5
Rb	140	60	3	90
Sr	300	20	610	375
Y	26	15	6.4	33
Zr	160	220	19	165
Nb	11	0.0X	0.3	20
Mo	2.6	0.2	0.4	1.5
Ag	0.07	0.0X	0.0X	0.07
Cd	0.3	0.0X	0.09	0.08
In	0.1	0.0X	0.0X	0.1
Sn	6.0	0.X	0.X	2
Sb	1.5	0.0X	0.2	0.2

*X indicates order-of-magnitude estimate.

TABLE 7–7   *(continued)*

Element	Shales	Sandstones*	Carbonates*	Igneous Rocks
I	2.2	1.7	1.2	0.5
Cs	5	0.X	0.X	3
Ba	580	X0	10	425
La	24	16	6.3	30
Cc	50	30	10	60
Pr	6.1	4.0	1.5	8.2
Nd	24	15	6.2	28
Sm	5.8	3.7	1.4	6.0
Eu	1.1	0.8	0.3	1.2
Gd	5.2	3.2	1.4	5.4
Tb	0.9	0.6	0.2	0.9
Dy	4.3	2.6	1.1	3.0
Ho	1.2	1.0	0.3	1.2
Er	2.7	1.6	0.7	2.8
Tm	0.5	0.3	0.1	0.5
Yb	2.2	1.2	0.7	3.4
Lu	0.6	0.4	0.2	0.5
Hf	2.8	3.9	0.3	3
Ta	0.8	0.0X	0.0X	2
W	1.8	1.6	0.6	1.5
Hg	0.4	0.3	0.2	0.15
Tl	1.0	0.5	0.2	0.8
Bi	0.4	0.17	0.2	0.1
Pb	20	7	9	13
Th	12	1.7	1.7	9.6
U	3.7	0.45	2.2	2.7

*X indicates order-of-magnitude estimate.

in conventional geotechnical situations is water, and the most common gas is air. However, in the context of hazardous waste management, the pore fluids could be any type of organic or inorganic chemical.

For the purpose of this book, the geotechnical definition of soil is most appropriate and henceforth the term *soil* will be used in the context of this definition.

Soil Genesis

Soils represent the end product of rock weathering. When a parent material—rock or sediment—is exposed to the atmosphere for a long period of time, it is physically and chemically altered to form soil. In many cases, alteration of the original parent material is aided by biological agents, including humans.

Because the process of weathering is primarily responsible for conversion of parent material into soil, the factors that control weathering also control soil formation. The following five factors control the process of soil formation:

- Parent material
- Climate
- Topography
- Biology
- Time

It may take anywhere from a hundred years to several hundred thousand years for the formation of a particular type of soil. Soils developed from sediments, such as loess, may take as little as a hundred years to form, whereas soils developed on a crystalline bedrock, such as granite, may take more than several hundred thousand years to form.

Generally, the soil formation process is accelerated in the bottomland and depression parts of a region's topography. This is because groundwater is close to the surface in low-lying areas, and a greater amount of water is available to promote chemical decomposition of the parent material and to sustain life forms that aid in soil formation. The upland part of the landscape is generally deficient in moisture, because of accelerated erosion, faster runoff, and deeper groundwater, all of which impede soil formation.

Biologic factors include both animal and plant activities, all of which promote soil formation. Animals cause mixing of materials (burrowing animals), disintegration of parent material and nutrients recycling (plant roots), and organic-matter enrichment (decaying microorganisms).

Wet and warm climates are more conducive to soil formation than dry and cold climates. This is because the rate of chemical reactions is faster at elevated temperatures and in the presence of water. This is the reason why soil cover is nonexistent or very thin in dry and hot desert climatic environments.

Soil profile. In order to study the nature of soil in the third dimension (depth), pits and trenches are excavated. A soil profile is the view of the entire soil thickness as seen on the wall of the pit or trench. Most soils are characterized by two or more layers of different color, having different mineralogical composition. These individual layers are called the *soil horizon;* the entire thickness is called the *soil profile.* A soil profile that shows full development of the prominent horizons is called a mature soil. Figure 7–3 shows a mature soil profile developed in a warm and humid climatic environment.

Soil texture. The term *soil texture* relates to the size of mineral grains making up the soil solids. Three common textural classes are sand, silt, and clay. The range in size for each class, using the engineering soil classification, is as follows:

Sand	4.76 mm to 0.074 mm in diameter
Silt	0.074 mm to 0.002 mm
Clay	<0.002 mm (2 micron)

FIGURE 7–3 A mature soil profile

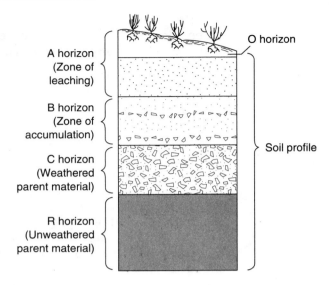

It should be noted that there are various grain-size classification schemes that are used by soil scientists, geologists, and engineers, each with its own size range for clay, silt, sand, and other particles.

The terms *sand* and *clay* are sometimes used to designate the mineralogy of soil solids in addition to the size of mineral grains. For example, a soil comprising the mineral quartz, irrespective of its grain size, is often referred to as sand or sandy soil. Similarly, the term *clay* is used for a soil dominated by clay minerals, regardless of the fact that the mineral particles may be larger or smaller than 0.002 mm in diameter.

Soil Types

All soils can be grouped into two main categories:

1. *Organic.* Rich in carbon; comprises upper layer of the soil profile

2. *Nonorganic.* Mixture of various minerals; could be further subdivided into

    a. *Residual.* In place; soil occurs at the place where it developed

    b. *Transported.* Soil removed by wind, water, ice, and gravity from its place of formation and deposited elsewhere

Table 7–8 summarizes the source material, transporting agent, and the resulting soil types.

All other factors remaining the same, different rock types will result in the formation of different types of soils. For example, sandstone will generally produce sandy soil; shales, silty and clayey soils; and crystalline rocks rich in Fe- and Mg-bearing minerals, clayey to sandy soils.

TABLE 7–8 Geologic Origin and Texture of Various Soils

Transporting Agent	Origin	Texture	Soil Type
None	Residual	Variable, depends on parent material and geologic environment	Residual
Water	Transported	Variable, generally MC, SP, GC, SP, SM, MC	Alluvial Deltaic
		ML or CL in tidal flats and marshes	Estuarine
		Clays: CH	Marine
		Beach sand: SP, SW, ML,CL,OL,MH,OH	Lacustrine
Ice	Transported	Till: highly variable; boulder to rock flour	Glacial
		Outwash: GP, SP	Glacio-fluvial
Wind	Transported	Loess; ML,CL, MH	Eolian
		Dune sands; SP	Eolian
Gravity	Transported	Variable; angular fragments	Colluvial
		ML to MH; weathers to CH (volcanic ash and cinder)	Pyroclastic

SOIL CLASSIFICATION SYSTEMS

Agricultural Classification

This classification system is also known as the 7th Approximation or Soil Taxonomy. Soils are grouped on the basis of their physical and chemical properties into Order, Suborder, Great Group, Subgroup, Family, and Series—from most general to very specific. This classification is for agricultural purposes and is of limited value in hazardous waste management. The other classification, the Unified Soil Classification System (USCS), is an engineering classification and is more relevant in hazardous waste management.

Unified Soil Classification System

This classification scheme is designed to evaluate soils in terms of their engineering performance and behavior. Using the Unified Soil Classification System (USCS), any soil may be classified into one of the 15 types. Soils are classified on the basis of

- Grain size
- Atterberg limits: the plastic limit (PL) and the liquid limit (LL)

Both tests are simple to perform, the equipment required is inexpensive, and no specialized skills are needed to run the test and interpret the results. It is for these reasons, as well as its reliability, that the USCS has been adopted world-wide. ASTM Test Designation D2487 (ASTM, 1993) provides detailed guidelines for classifying an unknown soil on the basis of data obtained from standardized tests. Table 7–9 gives the classification, along with engineering properties of the 15 soil types.

Control of Geologic Structures on Soil Types

In addition to the rock type or the sediments, geologic structures also control occurrence of soils. Folding and faulting of bedrock may result in development of different soil types adjacent to each other. Figure 7–4 shows how faults and folds—two common geologic structures—control the occurrence of different soils.

GEOTECHNICAL PROPERTIES OF SOILS

The USCS classifies soils in relation to their anticipated engineering behavior and performance. The following important geotechnical properties are routinely determined for most construction projects, including those involving character-ization of soils for hazardous waste management.

Index Properties

Because soils are made up of solid, liquid, and gaseous phases, engineering behavior of soils is heavily dependent on the relative percentages of these three

FIGURE 7–4   Control of geologic structure in soil for-mation: (A) Effect of a fault on the development of resid-ual soils; (B) Effect of a fold on the development of differ-ent types of residual soils

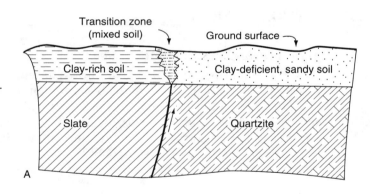

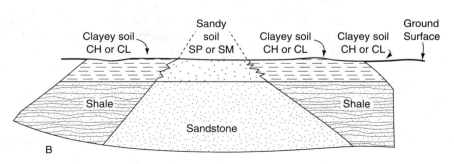

TABLE 7-9 Engineering Classification and Properties of Soils (modified from Keller, 1992)

Major Divisions				Group Symbols	Group Names	Shear Strength	Permeability	Compressibility	Use as a Construction Material
COARSE-GRAINED SOILS (Over half of material larger than 0.074 mm)	Gravels >50% retained on #4 sieve	Clean Gravels	Less than 5% fines	GW	Well graded gravel	Very high	High	Negligible	Excellent
				GP	Poorly graded gravel	High	Very high	Negligible	Good
		Dirty Gravels	More than 12% fines	GM	Silty gravel	High	Low	Negligible	Good
				GC	Clayey gravel	High–Medium	Very low	Very low	Good
	Sands >50% passes #4 sieve	Clean Sands	Less than 5% fines	SW	Well graded sand	Very high	High	Negligible	Excellent
				SP	Poorly graded sand	High	High	Very low	Fair
		Dirty Sands	More than 12% fines	SM	Silty sand	High	Low	Low	Fair
				SC	Clayey sand	High–Medium	Very low	Low	Good
FINE-GRAINED SOILS (Over half of material smaller than 0.074 mm)	Silts Non-plastic	Liquid Limit 50% or less		ML	Silt	Medium	Low	Medium	Fair
				CL	Silky Clay	Medium	Very low	High	Poor
				OL	Organic silt	Low	Low	Medium	Fair
	Clays Plastic	Liquid Limit Greater than 50%		MH	Micaceous silt	Medium-Low	Low	High	Poor
				CH	High plastic clay	Low	Very low	High	Poor
				OH	Organic clay	Low	Very low	High	Poor
Predominantly organics				PT	Peat and Muck	Very low	Low	Very high	Not suitable

97

components. Figure 7–5 illustrates the three modes of occurrence of soils based on the relative abundance/absence of the three components.

Figure 7–5(A) represents soils that comprise all three phases: solid, liquid, and gas (air). Figure 7–5(B) represents soil that has only two phases: solid (the minerals) and liquid. All submerged soils and those occurring below the groundwater table fall in this category. This is the most common occurrence of soils. Figure 7–5(C) represents another condition in which water is totally absent and the soil is made up of only two phases: solid and gas. Such soils are common in dry and hot climatic environments. Geotechnical characteristics of the same soil type under any of the above three conditions will be different. For example, the density of soil will be highest under (B), lowest under (C), and intermediate under (A); shear strength of a clayey soil will be highest under (C), lowest under (B), and intermediate under (A).

Moisture Content

Water plays a very important role in the engineering performance of soils. For this reason, determination of the moisture (water) content of the soil is routinely carried out. Moisture content, ω, is defined as the ratio of the mass (or weight) of water to that of the solids present in the soil. In Figure 7–5B, ω can be expressed as

$$\omega = M_w/M_s \times 100 \text{ (to express the result in percent)} \tag{7–7}$$

The American Society for Testing and Materials (ASTM) has standardized the test procedure for determination of soil moisture content. Details are available under ASTM Test Designation D2216 (ASTM, 1993).

Grain-Size Distribution

Soils comprise solids of varying sizes and shapes; a soil that consists of solids falling within *one* size range is extremely rare. The size of soil particles is determined from the average diameter of individual soil minerals. Simple tests, standardized by ASTM, are used to determine the grain-size distribution of soil solids. Two methods are available, depending on grain size of the soil: (a) the mechanical method, for coarse-grained, such as sandy and gravelly, soils, and

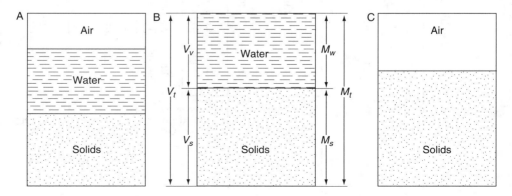

FIGURE 7–5   Three variations of soil phase

(b) the hydrometer method, for fine-grained, silty and clayey soils. Details of the test procedures, equipment needed, and data presentation are given in ASTM Test Designation D421 (sample preparation) and D422 (test procedures).

Atterberg Limits

These properties, originally proposed in 1911 by Albert Atterberg, a soil scientist from Sweden, are good indicators of soil consistency. Atterberg limits apply to fine-grained soils only. They tell us whether a soil is likely to behave as a solid, semisolid, or a viscous fluid (slurry-like mass) at a given moisture content. Of the five original limits proposed by Atterberg, two are commonly used in geotechnical engineering: the liquid limit *(LL)* and the plastic limit *(PL)*. *Plastic limit* is defined as the water content below which a soil will behave as a semisolid (nonplastic) material. *Liquid limit,* on the other hand, corresponds to the moisture content below which a soil behaves as a plastic material; conversely, liquid limit can also be defined as the moisture content above which the soil will behave like a viscous fluid (e.g., slurry). The numerical difference between the values of *LL* and the *PL* is called the *plasticity index (PI)*, determined by the equation

$$PI = LL - PL \qquad (7\text{--}8)$$

Figure 7–6 shows the two extreme states of a fine-grained soil from dry to saturated. At a given moisture content, the soil may behave like a brittle solid (extreme left of the diagram), or it may flow like a slurry (extreme right); every state in between the two extremes is possible. Thus, by knowing the natural moisture content of a soil and the values of Atterberg limits, one may estimate the physical state and engineering behavior of a soil.

Figure 7–6 also shows the relationship between the Atterberg limits and estimated strength characteristics of soil. ASTM Test Designation D4318 contains the detailed procedure and equipment requirements for determination of the liquid and plastic limits of soils.

Once the moisture content, grain–size distribution, and the Atterberg limits of a soil have been determined, it is a simple matter to classify the soil in terms of the USCS. Details of the procedure to be used for classifying the soil are given under ASTM Test Designation D2487 (ASTM, 1993). The classification also enables one to estimate other geotechnical properties of the soil. For example, dominant percentages of sand and gravel, and a low moisture content, would generally mean that a soil has high shear strength, good permeability, and negligible compressibility—all

FIGURE 7–6 Atterberg limits and shear strength of soil

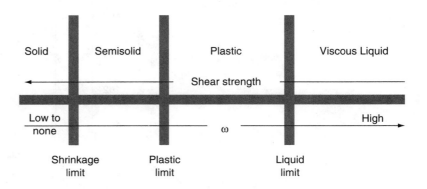

positive attributes that make such soil suitable for most geotechnical projects. On the other hand, fine-grained, clayey soil, with a high liquid limit, would imply low shear strength, extremely low permeability, and high compressibility—all undesirable properties in most geotechnical situations. However, it must be emphasized that this generalization is intended for a preliminary evaluation of engineering soils. In the ultimate analysis, soil properties must be evaluated in relation to proposed use, because an apparently undesirable charcateristic, such as very low permeability, may make the soil preferable for a construction project where liquid retention is a major concern. Such situations occur in a landfill, where leachate containment is a primary consideration. Table 7–9 summarizes important index properties of soil in relation to their anticipated engineering behavior.

Moisture-Unit Weight (Density) Relationships

This test, also known as the Proctor Compaction Test or the Compaction Test, determines the maximum unit weight (the ratio of weight to volume, γ) of a soil corresponding to a given moisture content, called the optimum moisture content, ω_{opt}. The test is based on the fact that a mass of soil, when compacted, undergoes a progressive increase in its unit weight, with increasing moisture content. However, past a certain value of the moisture content, ω_{opt}, the unit weight of the soil begins to decrease. The maximum value of the unit weight (or dry density, if the ratio of mass of dry soil and its volume is used) is called the maximum unit weight (or maximum dry density); the moisture content corresponding to the maximum unit weight is the optimum moisture content. This relationship is shown in Figure 7–7.

Compaction is the least expensive method of soil stabilization. Inadequate soil properties are routinely improved by compaction, which results in higher shear strength, lower compressibility, and higher unit weight. Two versions of the Compaction Test are used: the Standard Compaction Test, ASTM Test Designation D698; and the Modified Compaction Test, ASTM Test Designation D1557 (ASTM, 1993). Since compaction also results in decreased permeability,

FIGURE 7–7 Relationship between ω and γ_d and the values of ω_{opt} and γ_{dmax} determined from Proctor Compaction Test data

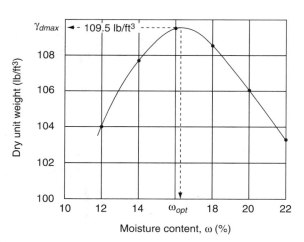

proper compaction of soil for use as liners in landfill will ensure minimum potential for leachate migration.

Compressibility

A soil mass, upon being loaded, undergoes volume reduction. Compressibility is a measure of the relative compression of a soil mass when loaded. Generally, coarse-grained soils have low compressibility, while fine-grained soils have higher compressibility. Compressibility is a time-dependent property. In coarse-grained soils, most of the compression occurs right after the load is imposed on it. In fine-grained soils, compression is a slow process; it may take from weeks to months for significant compression to occur in some soils. The Consolidation Test, ASTM Test Designation D2435 (ASTM, 1993), is used to determine the rate of change of void ratio of the soil in response to change in applied load. Figure 7–8 is a plot of the void ratio and pressure. The coefficient of compressibility, a_v, which is the slope of the curve ABC, can be calculated by the equation

$$a_v = \Delta e / \Delta p \tag{7–9}$$

FIGURE 7–8 Void ratio-pressure curve used for determination of the coefficient of compressibility, a_v

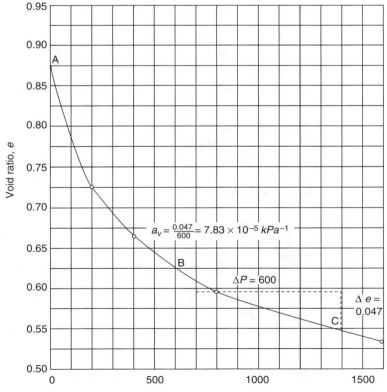

Permeability

The relative ease (or difficulty) with which a fluid can move through a soil or rock mass is a measure of its permeability. In general, coarse-grained soils and fractured and solutioned rocks have higher permeability than fine-grained clayey soils and massive rocks. Permeability of the geologic medium is very important in hazardous waste disposal, and earth materials with very low or negligible permeability are more desirable. The design of a remediation system for contaminated groundwater is based on, among other factors, the permeability of the aquifer.

In the laboratory, permeability of granular soil can be measured using the test procedures given in ASTM Test Designation ASTM D2434 (ASTM, 1993). Caution must be exercised when applying values of hydraulic conductivity obtained from laboratory tests on clayey soils to field conditions. Generally, the field conductivity values are orders of magnitude greater than those obtained from laboratory tests. This is because the effects of macropores, fine root holes, and other openings in clayey soil as it exists in the field are not present in laboratory samples.

Shear Strength

Shear strength is a measure of the ability of a soil mass to sustain the maximum shearing stresses (load) before failure occurs. Shear strength is a fundamental property used in the design of structures on soil. It is a function of the intrinsic cohesion and internal friction of a soil and the applied load. Mathematically, it is expressed as

$$S = c + \sigma \tan \phi \qquad (7\text{--}10)$$

This is known as the Mohr-Coulomb equation, where S = shear strength of the soil; c = soil cohesion; σ = applied stress (load); and ϕ = angle of internal friction of the soil.

In general, coarse-grained soils have higher shear strength and fine-grained soils have lower shear strength. ASTM Test Designations D2166 (uniaxial shear strength), D2850 (triaxial shear strength), and D3080 (direct shear test) (ASTM, 1993) describe tests used to determine shear strength of fine- and coarse-grained soils in the laboratory.

GEOCHEMICAL CHARACTERISTICS OF SOILS

Natural soil represents a microscopic and macroscopic environment characterized by physical, chemical, and biological dynamics. The three phases of a soil system—solid, liquid, and gas—aided by biologic processes and products, and the ambient temperature, pressure, and moisture conditions, complete the natural chemical laboratory where a variety of reactions take place. Soil scientists have studied soil chemistry in detail; the discussion that follows is based on the work done by soil chemists.

Mineralogy and Chemical Makeup of Soils

Minerals that make up the soil are naturally occurring chemical compounds. These minerals, derived from rocks and sediments, occur in various states of decomposition, depending on how far the weathering process has progressed toward completion. Also, the nature of the minerals comprising a soil mass is dependent on the rock type(s) and the sediment(s) over which the soil profile has developed. Despite the seemingly endless possibilities arising from the above combinations, common minerals that make up soils are, surprisingly, not many. This is due to the relative abundance of nine elements that comprise 99 percent of the crust of the Earth. Table 7–10 shows the relative abundance of chemical elements in the Earth's crust and in the soils.

From the above, it is clear that of the 103 elements described in the Periodic Table (see front end paper copy), only eight, or 7.77 percent, account for 99.34 percent of the total weight of the crustal rocks, while the remaining 95 (92.23 percent) comprise only 0.66 percent of the weight of the crust. This remarkable fact points to the unique order of things in nature. It is also clear that elemental composition of soils follows that of the rocks, with the exception of carbon and nitrogen, which occur in minute quantities in rocks (0.048 and 0.025 percent, respectively) but are relatively abundant in soils (2.5 and 0.2 percent). Common minerals found in soils are listed in Table 7–11.

The first six minerals in the list belong to the *silicate* group and are called primary minerals because they are derived from the parent material and are not formed as a result of chemical changes accompanying the weathering process. The strong covalent bond in silicate minerals gives an Si/O molar ratio of 0.25 (for olivine) to 0.50 (for quartz and feldspar). These minerals thus have the lowest solubilities in water and weak acids. A high Si/O molar ratio also correlates with weathering susceptibility of the primary silicates. Weathering resistance of minerals is related to their original order of crystallization from a cooling magma, known as the Bowen Reaction Series. As a rule, the mineral to crystallize first from a melt is also the one that is least resistant to weathering. Figure 7–9 shows the order of crystallization of minerals and their relative susceptibility to weathering.

TABLE 7–10 Abundance of Chemical Elements in the Crustal Rocks and Soils (adapted from Keller, 1992, and Sposito, 1989)

Element (symbol)	Weight (%) in Rocks	Weight (%) in Soils
Aluminum (Al)	8.23	7.2
Calcium (Ca)	4.14	2.4
Iron (Fe)	5.63	2.6
Magnesium (Mg)	2.33	0.9
Oxygen (O)	46.40	49.0
Potassium (K)	2.09	1.5
Silicon (Si)	28.15	31.0
Sodium (Na)	2.36	1.2
Total	99.33	95.8

TABLE 7–11   Common Minerals in Soils (adapted from Sposito, 1989)

Mineral	Chemical Formula	Importance
1. Quartz	SiO_2	Abundant in sandy and silty soils
2. Feldspar	$(Na,K)AlO_2[SiO_2]_3$; also $CaAl_2O_4[SiO_2]_2$	Abundant in soil that is not leached extensively
3. Mica	$K_2Al_2O_5[Si_2O_5]_3Al_4(OH)_4$; also $K_2Al_2O_5[Si_2O_5]_3(Mg,Fe)_6(OH)_4$	Source of K in most temperate-zone soils
4. Amphibole	$(Ca,Na,K)_{2,3}(Mg,Fe,Al)_5(OH)_2[Si,Al]_4O_{11}]_2$	Easily weathers to clay minerals and oxides
5. Pyroxene	$(Ca,Mg,Fe,Ti,Al)(Si,Al)O_3$	Easily weathers
6. Olivine	$(Mg,Fe)_2SiO_4$	Easily weathers
7. Epidote	$Ca_2(Al,Fe)_3(OH)Si_3O_{12}$	⎫ Highly resistant to chemical
8. Tourmaline	$NaMg_3Al_6B_3Si_6O_{27}(OH,F)_4$	⎬ weathering; used as
9. Zircon	$ZrSiO_4$	⎭ "index mineral" in
10. Rutile	TiO_2	pedologic studies
11. Kaolinite	$Al_4Si_4O_{10}(OH)_8$	11–14 are abundant in clayey soils as products of weathering; source of exchangeable cations in soils
12. Smectite*	$M_{0.6}(Al,Fe,Mg)_{4-6}(Si,Al)_8O_{20}(OH)_4$	
13. Vermiculite	$(Mg,Fe,Al)_3(Al,Si)_4O_{10}(OH)_2 \cdot 4H_2O$	
14. Chlorite	$(Al,Fe,Mg)_{4-6}(Al,Si,Fe)_4O_{10}(OH,O)_8$	
15. Allophane	$Si_3Al_4O_{12} \cdot nH_2O$	Abundant in soils derived from volcanic ash deposits
16. Imogolite	$Al_2SiO_3(OH)_4$	
17. Gibbsite	$Al(OH)_3$	Abundant in leached soils
18. Goethite	$FeO(OH)$	Most abundant Fe oxide
19. Hematite	Fe_2O_3	Abundant in warm regions
20. Ferrihydrite	$Fe_4(OH)_{12}$ to $Fe_5O_3(OH)_9$	Abundant in organic horizons
21. Birnessite	$Na_4Mn_{14}O_{27} \cdot 9H_2O$	Most abundant Mn oxide
22. Calcite	$CaCO_3$	Most abundant carbonate
23. Gypsum	$CaSO_4 \cdot 2H_2O$	Abundant in arid regions

*Vermiculite–smectite form a series with the following general formula: $M_x(Si,Al)_8(Al,Fe,Mg)_4O_{20}(OH)_4$; M = interlayer cation.

Minerals listed from 7 to 10 in Table 7–11 are grouped under *heavy minerals.* These minerals, also silicates, with the exception of rutile (an oxide), generally have higher density (>3.0) than the primary silicates (<3.0; Exception: olivine and pyroxene). Heavy minerals are resistant to weathering. For this reason they are used to estimate the degree of weathering in soils—the percentage of heavy minerals is higher in soils and sediments that have undergone a higher degree of weathering.

Minerals listed from 11 to 23 in Table 7–11 are called *secondary minerals* because they are formed as a result of weathering processes. Clay minerals, kaolinite, smectite, vermiculite, and chlorite (11–14) are of special importance in hazardous waste management. Clay minerals are characterized by very

FIGURE 7–9 Bowen reaction series showing order of crystallization of minerals and their weathering resistance

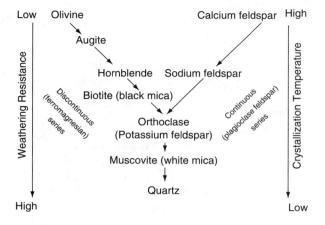

small particle size, <0.002 mm (2μ), poorly ordered atomic structure, high surface area (up to 800 m²/g for montmorillonite), and high cation exchange capacity (CEC). The latter two properties are very important in selecting a proper remediation method for contaminated soils. A large surface area allows for greater rate of chemical/microbial reaction, and the CEC may be used to advantage in removing certain toxic metals from the soil. For example, ions of Al, Ca, Mg, Na, and K can be readily exchanged by the addition of suitable chemicals to the soil.

Organic Matter in Soils

Soils support a wide variety of life forms that include both plants (flora) and animals (fauna). All flora flourish in soil, while fauna have a relatively small representation, mainly in the form of soil microorganisms. Larger animals, bigger than rodents, are rarely found in soil. It is the combined action of flora, fauna, moisture, and temperature that gives the unique characteristics to a soil mass at a given location. In addition to the minerals listed in Table 7–10, organic matter also comprises the solid phase of the soil system. These organic solids generally account for less than 10 percent of soil solids, but in peat or muck they constitute over 50 percent of soil solids.

Biochemical processes dominate the breakdown of organic matter; physical and chemical processes are less important. Proteins, carbohydrates, fats, resins, and waxes comprise important organic substances that accumulate on the ground. These substances vary greatly in their susceptibility to decomposition and alteration; the general order of their resistance to decomposition is shown below.

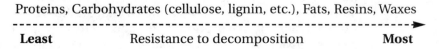

Proteins, Carbohydrates (cellulose, lignin, etc.), Fats, Resins, Waxes

Least   Resistance to decomposition   **Most**

Mineralization is the natural process that causes conversion of elements in organic substances into an inorganic form; the reverse of this process, *immo-*

bilization, causes conversion of inorganic ions or compounds into organic form.

$$\text{Organic form} \dashrightarrow \text{Inorganic form}$$
$$\text{Mineralization}$$

$$\text{Inorganic form} \dashrightarrow \text{Organic form}$$
$$\text{Immobilization}$$

Humus represents a mixture of highly altered or resynthesized products of organic matter decay. It is rich in proteinaceous materials and lacks in sugar, starch, amino acids, and other easily decomposable components of organic matters. The importance of humus is discussed in a later section.

The organic matter content (OMC) of soil is important in determining its fertility. It is also significant in assessing the availability of the right biochemical environment before planning bioremediation for cleanup of a hazardous waste site.

Organic matter in soil may be grouped into two categories: (a) humic substances, and (b) non-humic substances. Humic substances represent solid products of microbial transformation of organic matter. Humic substances generally have dark color and complex chemical structures. Except for humic and fulvic acids, not much is known about the chemical structure of other humic substances. Non-humic substances comprise all solid organic matter which is relatively undecayed—so that the nature of the original source material, such as, plant materials, lignins, and proteins, can be readily recognized.

Humic substances are divided into three categories, based on their solubility characteristics: (a) humic acids, which are insoluble in acid but soluble in basic solutions, (b) fulvic acids, which are soluble in both acidic and basic solutions, and (c) humin, which is insoluble in both types of solutions.

Despite the lack of knowledge about the structural makeup of humic substances, their elemental content and other properties are well known. For example, humic acids can be represented by the generalized chemical formula $C_{187}H_{186}O_{89}N_9S$, and fulvic acid by $C_{138}H_{182}O_{95}N_5S_2$. Both the fulvic and the humic acids contain the carboxyl (COOH) and the phenolic hydroxyl (OH) molecules.

In a generalized sense, fulvic acids have the lowest amount of C and N; humins contain the highest. Fulvic acids have the highest content of O; humin the lowest. In terms of chemical structure, humin has the largest organic molecules; fulvic acid the smallest. Humic acid has characteristics that are intermediate to those of humin and fulvic acid (Table 7–12).

Since C, N, O, H, and S are essential elements to support microorganisms, humic substances play a very critical role in bioremediation of hazardous wastes. It is known that the functional groups (the complexes—carboxyl, carbonyl, amino, imidazole, phenolic OH, alcoholic OH, and sulfhydryl) control surface reactivity, adsorption, and bonding to solids, and act as a promoter of chemical reactions. Therefore, knowledge of the complexes helps determine the best remediation method. It is known that aerobic microorganisms that utilize oxygen to oxidize organic matter do not function below pE 5 (pE is a measure of soil oxidizability, as opposed to pH, which relates to reducibility). Denitrifying

TABLE 7–12 Average Elemental Compostion of Soil Humic Substances (from Steelink, 1985)

Element	Humin* (%)	Humic Acids (%)	Fulvic Acids (%)
Carbon	54.2–58.2	53.8–58.7	40.7–50.6
Hydrogen	4.7–7.0	3.2–6.2	3.8–7.0
Oxygen	28.8–37.1	32.8–38.3	39.7–49.8
Nitrogen	2.6–6.0	0.8–4.3	0.9–3.3
Sulfur	—	0.1–1.5	0.1–3.6

*Data on humin from Hatcher et al., 1985.

bacteria thrive in the pE range of +10 to 0; sulfate-reducing bacteria do not grow at pE values of over +2. The dashed area in Figure 7–10 indicates the boundary within which the microorganisms that promote redox reactions thrive.

Soil Colloids

Colloids are finely divided particles of solids that are very sparingly soluble in water. The size range of colloids is between 1μ (0.001 mm) and 10μ (0.01 mm). Colloidal state refers to a two-phase system in which one or more substances occur dispersed through the other. The main characteristic of colloids is that the solids do not dissolve in water to form solution but remain *suspended* in the liquid. The substances occurring in a colloidal state are chemically inert, in the sense that they do not react with each other. Common examples of colloidal state include milk, fog, starch, blood, and soil. In soils, since the clay size refers to particles 2μ or smaller and silts >2μ to 0.074 mm in diameter, texturally both clays

FIGURE 7–10 pE–pH diagram showing the regions where microorganisms may occur (dashed area); shaded area shows actual occurrence in soils; solid dots represent experimental data (adopted from Sposito, 1989)

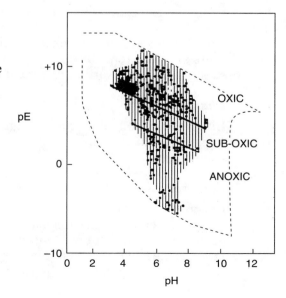

and fine silt may occur as colloidal solids in soils. Chemically these particles may comprise both inorganic and organic materials. In soils, colloidal materials commonly comprise both the inorganic and organic solids. The inorganic solids include clay minerals, such as kaolinite, vermiculite, smectite, and chlorite, and the metal hydroxides—gibbsite, goethite, limonite, and the amorphous mineral allophane ($Al_2O_3 .2SiO_2H_{2O}$). Organic solids in colloids include soil humus.

Colloidal suspensions in which no measurable gravitational settling occurs within a 2- to 24-hr period are called *stable*. If settling occurs in a few hours' time, they are called *unstable*. Stability of colloids is closely related to chemical transport in soils. Stable suspensions promote adsorption of organic and inorganic substances onto the colloids. Trace metals, organics from pesticides, phosphate anions, and other substances, get strongly bound to the soil colloids, making them highly mobile.

A stable colloidal suspension can become unstable in several ways, such as by

- Coagulation
- Change in pH
- Cation exchange

Coagulation

Gravity ultimately will cause the colloidal solids to settle out of the suspension. Coagulation is the process that results from the Brownian motions of particles, causing them to collide and form larger mass to facilitate settling under the gravitational force. Coagulation rate depends on the radius and number of colloidal particles, viscosity of water (or other liquid), and temperature. Coagulation that results in formation of bulky masses of particles having a high water content are called *floccules*; the process is known as *flocculation*. Large-scale agglomeration of colloidal particles giving rise to a dense, organized mass of settled solids is called *aggregation*.

Coagulation and flocculation are commonly used chemical methods for treatment of hazardous wastes, and are used to separate heavy metals, clays, and organic compounds from wastewater. Various water-soluble chemicals, such as alum [$Al_2(SO_4)_3$], ferric chloride ($FeCl_3$) and ferric sulfate [$Fe_2 (SO_4)_3$], and polymers are used to promote coagulation and flocculation of aqueous waste streams. Details of various treatment methods are discussed in Chapter 13.

Change in pH

Dispersion of colloids occurs at higher pH values. Therefore, lowering the pH in negatively charged clays and humus will cause a reduction in the negative charge and will initiate aggregation of the colloidal particles.

Cation Exchange

Cation exchange is the process of replacement of a positively charged ion (cation) by another cation in the soil mass. Figure 7–11 illustrates the concept of cation exchange in soils. Upon addition of lime to an acidic soil, the Ca ion in lime replaces the two H ions that were adsorbed on the surface of the soil solid. Thus, by loosing two H cations, the soil solid becomes less acidic. The reaction is

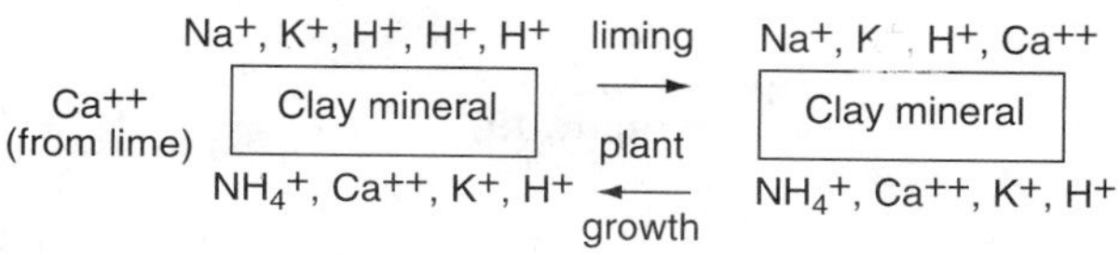

FIGURE 7–11 Mechanism of cation exchange in clayey soil with addition of lime (top reaction) and reversal of the exchange following plant growth (lower reaction)

Na^+, K^+, H^+, H^+, H^+ liming Na^+, K^+, H^+, Ca^{++}

Ca^{++} (from lime) | Clay mineral | → | Clay mineral |

plant

$NH_4^+, Ca^{++}, K^+, H^+$ ← $NH_4^+, Ca^{++}, K^+, H^+$

growth

reversible, because with the continued growth of plants on Ca-enriched soil there is an increase in the supply of carbonic acid, which causes the H ion to replace the Ca cation. This results in the soil solid becoming acidic once again.

All ions in a soil solid are not equally exchangeable; this is due to the position occupied by these ions in the soil solid. If the ions lie within the hexagonal crystal lattice, they are harder to replace than those occurring at the edge of soil solid. How tightly an ion is held in the soil solid is a function of ionic charge, surface charge on soil solid, ionic radius, and the exchange site. Thus, divalent ions (Ca^{++}, Mg^{++}) are held more tightly than the monovalent ions (K^+, NH^{4+}); similarly, smaller ions are held more tightly than larger ones. In general, the relative ease of cation exchange is

$$Al^{+++} > Ca^{++} > K^+ > Na^+$$

The ability of a soil to exchange the cations, called the cation exchange capacity (CEC), varies with soil type, pH, and temperature. In general, CEC is highest for organic matter, intermediate for the expandable clays, and lowest for the non-expandable clays and the hydrous oxides. Table 7–13 shows the CEC for various minerals in soils at pH 7.0.

Soil pH and CEC

The CEC in most soils increases with an increase in pH. Figure 7–12 shows the influence of pH on the CEC of soils. The CEC for montmorillonite is constant below pH 6.0, which is due to the fact that in this pH range the ions are held tight to resist exchange. Above pH 6.0, the charge on mineral colloid increases, allowing for greater CEC. In contrast to clays, CEC of organic colloid increases with increase in pH throughout.

TABLE 7–13   Cation Exchange Capacities of Soil Solids at pH 7.0 (adopted from Hausenbuiller, 1985)

Exchangeable Material	Cation Exchange Capacity (me/100 g)
Organic matter	100–300
Vermiculite	100–150
Allophane	100–150
Montmorillonite	60–100
Chlorite	20–40
Illite	20–40
Kaolinite	2–16
Hydrous oxides	0

FIGURE 7–12 Influence of pH on CEC of soils (adopted from Brady, 1974)

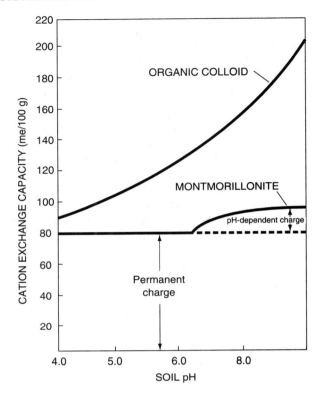

Thermophysical Properties of Earth Materials

In routine geotechnical practice, one does not usually encounter situations or projects that call for evaluation of thermophysical properties of earth materials. But in projects involving rocks as a construction medium for structures located deep in the subsurface, or in the vicinity of regions of high geothermal gradients such as active plate margins, proper knowledge and accurate characterization of the rocks in terms of their thermophysical properties are critical. Deep mines, tunnels, shafts, geothermal energy projects, and disposal facilities for high-level nuclear wastes are some examples of projects that require careful evaluation of the thermophysical properties of rocks.

After the passage of laws to control the management of hazardous wastes, and with the emergence of incineration as a viable alternative for destruction of some of the toxic compounds (such as dioxin and PCBs), thermophysical properties of soils have been receiving close attention. Because incineration plants are required to achieve the desired Destruction and Removal Efficiency (DRE) optimally, and because of the high energy requirements of an incineration system, knowledge of the thermophysical properties of soils will help a great deal in efficient utilization of the heat energy to destroy hazardous constituents in a contaminated soil.

A brief discussion of thermophysical properties is included here because of their relevance in thermal methods of treatment and disposal of hazardous wastes.

The four important thermophysical properties of geologic materials are

1. Thermal conductivity
2. Thermal diffusivity
3. Heat capacity
4. Thermal expansion

Thermal Conductivity

The relative ease (or difficulty) of flow of heat energy through an earth material is a measure of its thermal conductivity (cf: permeability relates to flow of fluid). If Q is the total amount of heat transferred during time t across the soil surface of area A, under a temperature gradient ΔT, then the thermal conductivity, K, is given by

$$K = QL/At\Delta T \qquad (7\text{–}11)$$
$$= QL/At\ (t_1\text{–}t_2)$$

where $t_1 > t_2$, and t_1 and t_2 = temperature at two measurement points in the soil, °C; L = distance between the temperature measurement points in the soil, cms; and K = coefficient of thermal conductivity, Watts/m/°C (SI System) or Cal/m/h/°C (conventional unit).

Thermal conductivity of earth materials varies with

1. *Mineralogy.* The higher the silica content, the higher the K
2. *Degree of saturation.* Saturated rocks/soils have higher K, because $K_{water} > K_{air}$

Thermal Diffusivity

The rate of change of temperature in a soil/rock mass is called thermal diffusivity; it is also known as temperature conductivity. It is expressed as

$$\alpha = K/C_p\,\rho \qquad (7\text{–}12)$$

where C_p = specific heat of the soil/rock; ρ = density of soil/rock; and K = coefficient of thermal conductivity of soil/rock. Thermal diffusivity is higher for soils with low ω, and vice versa.

Thermal Expansion

This property relates to both expansion (upon heating) and contraction (upon cooling) of materials. A special point is that while solids (minerals) contract upon cooling, water expands (about nine percent volume increase during cooling and conversion into ice).

Depending upon whether the expansion (or contraction) is being measured along the length (one dimension), area (two dimensions), or volume (three dimensions) of the material, we have three different measures: the coefficient of linear thermal expansion, the coefficient of surficial thermal expansion, and the coefficient of cubical thermal expansion, respectively.

If a soil/rock sample having an initial length l_0 is first heated to temperature t_1 such that its length becomes l_1 and then to the final temperature, t_2, at which the length is l_2, then the coefficient of thermal linear expansion, α_t, can be determined from

$$\alpha_t = (\Delta l / l_o)\ (1/\Delta_t) \times 100 \tag{7–13}$$

where $\Delta l = l_2 - l_1$; l_o = original length of the soil/rock sample; and Δt = temperature corresponding to l_1 and l_2, $(t_2 > t_1)$.

Similarly, the coefficient of thermal volume expansion, ß, is given by:

$$\text{ß} = (\Delta v / v_o)\ (1/\Delta t) \tag{7–14}$$

where v_o is the original volume of soil/rock sample, and Δv is the change in volume corresponding to temperature t_1 and t_2 $(t_2 > t_1)$.

Heat Capacity

The amount of heat energy required to change the temperature of a unit mass of soil/rock by one degree is its heat capacity. It can be calculated using the equation

$$C_p = Q / \Delta t \tag{7–15}$$

where Q = calories of heat required to raise the temperature by 1°C, and Δt = temperature change $(t_2 - t_1)$. Unit of heat capacity is cal/g/°C.

Specific heat, C_v, is the ratio of the heat capacity of the soil or rock to that of water, so that

$$C_v = C_p \text{ soil} / C_p \text{ water} \tag{7–16}$$

Specific heat of soil/rock varies with ω, ρ, and mineralogy. Values of C_p for common materials are shown below:

Material	C_p at 0°C, cal/g/°C
Water	1.0
Ice	0.5
Dry soil	0.2

The significance of the thermophysical properties of geologic materials in hazardous waste management has not yet been fully explored. However, with the intense research focused on developing an efficient and economical thermal remediation system for contaminated soils, more insight will be gained in this area.

STUDY QUESTIONS

1. What natural substances constitute the earth materials? What is a rock? List the three major groups of rocks.

2. What is the difference between an extrusive igneous and an intrusive igneous rock? How are sedimentary rocks different from igneous rocks?

3. Define discontinuity; give five examples of discontinuities found in rocks. How do discontinuities influence the strength and permeability of rocks?

4. What is rock weathering? Why is it important to assess the degree of weathering at a construction site?

5. What criteria could be used to distinguish rocks from soils for engineering construction?

6. Why is it not advisable to use the value of rock properties (strength, density, elasticity, etc.) from the standard handbooks for design of engineering structures? What, then, is the utility of such information?

7. What is the difference between a rock mass and a rock material? Why is it important to measure the permeability of rocks in the field (in situ) rather than the laboratory?

8. Distinguish between mass density and specific gravity of rocks. Which of the two is greater? Determine the mass density of a limestone that weighs 250 g in air and 259 in water. Report the value in (a) kg/m^3, and (b) lb/ft^3.

9. What is the RQD? What is its significance in hazardous waste management? Determine the RQD for a 20-ft core run in which the sum of all core pieces 4 in. or longer was 80 in. How would you rank the rock quality?

10. Why do the results of chemical analyses of a large number of igneous and sedimentary rocks from worldwide locations not show marked difference in their chemical composition? What are the two most abundant elements that occur in the lithosphere?

11. What factors control the process of soil formation? All other factors remaining the same, would you expect a soil profile to develop faster (shorter time) on quartzite or alluvial sediments? Explain.

12. What is the basis of classifying soils into the 15 engineering types? Give an example of a project where a silty clay may be the preferred material over a sandy soil.

13. A sample of clayey soil taken in the vicinity of a hazardous waste site was found to weigh 138 g in its natural state. The same soil after oven-drying weighed 96 g. What is its moisture content?

14. The liquid and plastic limit of the soil in question 13 were determined to be 33 and 18. What is its plasticity index?

15. Name the important group of minerals where Si and O are the two main constituents. How does the Si/O molar ratio relate to weathering?

16. What are the heavy minerals? What is their relative susceptibility to weathering? Would you expect the percentage of such minerals to be greater in a soil or in a partly weathered rock? Why?

17. How are the secondary minerals formed? List three of them. Discuss the significance of clay minerals in hazardous waste management.

18. What is the difference between humic and non-humic substances? Explain the importance of humic substances in bioremediation of hazardous waste.

19. What are soil colloids? How may a stable colloidal suspension become unstable?

20. Define the cation exchange capacity of a soil. How does pH control this property?

21. List the four thermophysical properties of rocks. In which method of hazardous waste remediation are they more relevant?

REFERENCES CITED

American Society for Testing and Materials, 1993, Annual Book of Standards, Vol. 04.08, Section 4: Philadelphia, PA, American Society for Testing and Materials, 1482 p.

Brady, N.C., 1974, The Nature and Property of Soils (8th ed.): New York, Macmillan Publishing Co., Inc., 639 p.

Deere, D.U., A.J. Hendron, F.D. Patton, and E.J. Cording, 1967, Design of Surface and Nearsurface Construction in Rock: Minneapolis, MN, Proceedings, 8th Symposium on Rock Mechanics, American Institute of Mining, Metallurgical and Petroleum Engineers, p. 237–302.

Deere, D.U., D.W. Deere, 1989, Rock Quality Designation (RQD) After Twenty Years: Vicksburg, MS, U.S. Army Engineer Waterways Experiment Station, Report GL-89-1, 67 p. plus appendix.

Geological Society, 1977, The description of rock masses for engineering purposes, Geological Society (London) Engineering Group Working

Party: Quarterly Journal of Engineering Geology, vol. 10, p. 355–388.

Hatcher, P.G., I.A. Breger, and N.M. Szeverenyi, 1985, Geochemistry of humin, *in* G.R. Aiken, D.M. McKnight, and R.L. Wershaw, (eds.), Humic substances in soil, sediment, and water, New York, Wiley Interscience, p. 278.

Hatheway, A.W., 1990, Rock Quality Designation (RQD: A Wonderful Shortcut): AEG News, vol. 33, no. 4, p. 28–30.

Hausenbuiller, R.L., 1985, Soil Science (3rd ed.): Dubuque, IA, Wm. C. Brown Publishers, 610 p.

Hunt, C.B., 1972, Geology of Soils: Their Evolution, Classification, and Uses: San Francisco, W.H. Freeman Co., 344 p.

Hunt, R.E., 1984, Geotechnical Engineering Investigation Manual: New York, McGraw-Hill Book Company, 983 p.

Keller, E.A., 1992, Environmental Geology (6th ed.): New York, Macmillan Publishing Company, 521 p.

Kiersch, G.A., and R.C. Treasher, 1955, Investigations, areal and engineering geology—Folsom Dam Project, Central California: Economic Geology, v. 50, p. 271.

Mason, B., and C.B. Moore, 1982, Principles of Geochemistry (4th ed.): New York, John Wiley & Sons, Inc., 344 p.

Mitchell, R.L., 1964, Trace elements in soils, *in* F.E. Bear, (ed.), Chemistry of the Soil: New York, Reinhold Publishing Corp., p. 320–368.

Sposito, G., 1989, The Chemistry of Soils: New York, Oxford University Press, 277 p.

Steelink, C., 1985, Implications of elemental characteristics of humic substances, *in* G.R. Aiken, D.M. McKnight, and R.L. Wershaw, (eds.), Humic substances in soil, sediment, and water, New York, Wiley Interscience, p. 457–476.

Williamson, D.A., 1984, Unified rock classification system: Bulletin, Association of Engineering Geologists, vol. 21, no. 3, p. 345–354.

SUPPLEMENTAL READING

Beiniawski, Z.T., 1984, Rock Mechanics Design in Mining and Tunneling: Boston, A.A. Balkema, 272 p.

Brady, N.C., 1990, The Nature and Property of Soils (10th ed.): New York, Macmillan Publishing Co; Inc., 621 p.

Das, B.M., 1990, Principles of Geotechnical Engineering (2nd ed.): Boston, PWS-KENT Publishing Co., 665 p.

Johnson, R.B., and J.V. DeGraff, 1988, Principles of Engineering Geology: New York, John Wiley & Sons, Inc., 497 p.

Hydrogeology

INTRODUCTION

Groundwater is the branch of hydrology—the science of study of water—that deals with the geologic aspects of subsurface water. Groundwater geology, up until about 30 to 40 years ago, was mainly limited to study of groundwater's geologic occurrence and exploitation. Water quality was treated in a very cursory manner in earlier texts and there was practically no discussion of groundwater pollution and its remediation. The science was essentially qualitative in the early years, but witnessed a gradual shift toward quantification. With increased environmental concern, recognition of the fact that our activities have been causing problems of groundwater contamination, and the government-mandated requirements to limit pollution and restore the quality of the nation's groundwater to safe levels, groundwater geology, now termed *hydrogeology* (Bates and Jackson, 1987), has become much more quantitative. *Geohydrology* is another term that has been used interchangeably with hydrogeology, but, as pointed out by Fetter (1994), it describes a specialty in the engineering field with emphasis on subsurface fluid hydrology.

Many textbooks on this subject are available, and new ones are frequently being added to the list. Some texts treat the subject in a descriptive way; others are highly quantitative and use mathematical approaches.

The availability of fast and inexpensive computing power led to mathematical modeling of groundwater systems, and currently there is a proliferation of new models to assist in both evaluation of groundwater contamination and design of groundwater remediation. Models, while useful in conceptualization of groundwater flow systems, contaminant movement, and the like, should be used with caution because of the inherent limitation in studies involving modeling of natural systems. Models tend to idealize the natural conditions and set limits or boundaries for mathematical analyses and syntheses. Invariably models require certain assumptions to be made; and that is where the problems occur. Natural aquifer systems and materials, such as soils, rocks, and sediments, vary greatly in their spatial and temporal characteristics. Geologic nature of sediments and rocks, with their lateral and vertical facies changes, structural peculiarities, and overall lack of homogeneity and isotropism, are difficult to incorporate precisely into the modeling studies. The result, therefore, does not completely represent the natural conditions. For this reason, it is prudent to use models only as an aid in understanding aquifer characteristics, contaminant transport, and groundwater flow regime, and to base models on observed field and laboratory data. Total reliance on models for designing the remediation plan, without taking into account the geologic aspects, may not be desirable.

FUNDAMENTAL CONCEPTS

Groundwater is the main source of drinking water for one-half of the entire population of the United States; and 95 percent of the country's rural population relies on groundwater to meet its drinking water needs. At the same time, thousands of wells are closed each year because of contamination from hazardous waste (NRC, 1991). It is therefore very important to have a clear understanding of the occurrence and movement of groundwater, along with the mode of contaminant transport, for evaluation of contaminated groundwater systems and their remediation. This section presents a discussion of the fundamental concepts and definitions in hydrogeology that are essential for acquiring a basic knowledge.

Types of Water

The waters that occur at, above, or below the Earth's surface can be grouped into five types:

1. *Meteoric water.* The water that is present in the atmosphere, usually in the form of water vapor; it is the source of all precipitation.

2. *Connate water.* The water that was trapped in sedimentary rocks during the time of their formation; it represents a very small percentage of the total water. Connate water is usually saline and is out of circulation within the hydrologic cycle.

3. *Juvenile water.* The water that forms from condensation of the steam released from the Earth's interior (along with other gases and lava) during volcanic eruptions; it accounts for a very small percentage of total water in the hydrosphere. Juvenile water, unlike connate water, becomes a part of the hydrologic cycle.

4. *Surface water.* All water that occurs at or above the ground surface; includes the water in rivers, lakes, streams, oceans, glaciers, and ice caps.

5. *Subsurface water.* The water that occurs below the land surface or within the lithosphere; however, not all subsurface water is groundwater.

Figure 8–1 shows various types of subsurface water. Two major zones—unsaturated and saturated—can be readily identified in the diagram. The zone where the voids in soil/rock are only partly full of water and the remainder may be occupied by air or soil gas (the latter is very significant in assessment of extent of contamination at a site) is the zone of *aeration,* or *vadose* zone. The lowermost region in the unsaturated zone is the *capillary fringe,* which represents a narrow zone above the groundwater table where water from below is drawn up (against gravity) by surface tension. The height of rise of capillary water is inversely proportional to the grain size—the smaller the grain size, the higher the water will rise. Below the capillary fringe all openings in the earth material are full of water (no air); this is the saturated, or *phreatic,* zone.

Groundwater Table

Water that occurs below the zone of saturation is groundwater; the uppermost surface of the zone of saturation is the *groundwater table* (GWT). The GWT can be defined as the elevation of the water surface with reference to a standard datum, mean sea level (MSL), below which all voids in the earth material are full of water. For example, the statement "the GWT at location X is at 340 m" means that all pores in the soil/rock below the elevation of 340 m are full of water. Sometimes the GWT is reported in relation to the depth below the surface where it occurs; for example, "groundwater occurs 12 m below the ground surface" or "the GWT is 12 m deep" means that the voids in earth materials 12 m below the ground surface *at that particular location* are saturated with water. Hydraulic pressure at the GWT equals the atmospheric pressure, and increases with increasing depth in the zone of saturation.

FIGURE 8–1   Various types
of subsurface waters

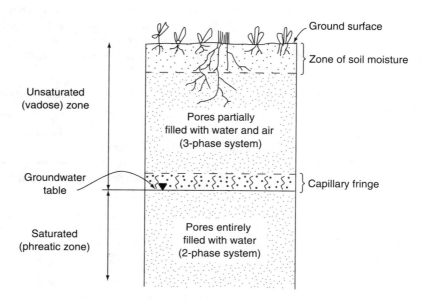

TABLE 8–1   Rate of Movement of Water (adapted from L'vovich, 1979)

Type of Water	Rate of Movement
Atmospheric	100s of km/day
Surface water	10s of km/day
Subsurface water	meters/day
Ice caps and glaciers	meter/day

The Hydrologic Cycle

Groundwater is a part of the hydrologic (water) cycle. Runoff begins when infiltration ceases. Infiltration rates vary with land use, soil type, and its moisture content, as well as the intensity, frequency, and duration of precipitation (a generalized hydrologic cycle is shown in Figure 2–3).

The rate of water movement is the key concept in the hydrologic cycle. Table 8–1 gives the typical values of rate of movement for the four general types of water.

The infiltration component of precipitation, after first moving through the overlying layers of earth materials, ultimately reaches the zone of saturation. Groundwater travels from areas of high hydraulic pressures to areas of lower pressures. Unlike water in rivers or other open channels, which flows in an unobstructed path, groundwater flows through the openings in earth materials; it has to overcome the obstruction from solid minerals along its flow path. This explains why the velocity of groundwater flow is extremely slow—generally a few cm/day to m/day—compared to that of surface waters. Because of this slow rate of movement, it takes an extremely long time for existing groundwater at a particular location to be replaced by fresh groundwater. This exchange time, also referred to as *residence time*, is up to 1/9000 the rate of exchange of river water (Table 8–2).

Groundwater accounts for 96 percent of all available *freshwater* on the Earth (Table 8–2); it provides the drinking water supply for many people all over the

TABLE 8–2   Residence Time and Water Volume (adapted from Nace, 1971)

Type of Water	Volume (km³)	Volume (%)	Residence Time (yr)
Oceans and seas	137,000,000	94	4000
Ice sheets and glaciers	30,000,000	2	10–1000
Groundwater	60,000,000	4	0.04–10,000
Lakes and reservoirs	130,000	<0.01	10
Soil moisture	70,000	<0.01	0.04–1
River	12,000	<0.01	0.04
Vapors in atmosphere	14,000	<0.01	0.027

world, and is also used for agricultural and industrial purposes. Consequently, it is imperative that groundwater contamination be avoided and every possible measure be taken to maintain its quality. The current emphasis in the United States and other countries on cleaning up contaminated groundwater illustrates the critical role groundwater plays in our daily lives.

Rate of Groundwater Movement

The rate of groundwater movement is highly variable. It is a function of the physical nature of the aquifer material and the hydraulic head. While the average rate of groundwater movement is generally under 1 m/day, extreme departures from this value have been recorded. For example, in the Mississippian carbonate aquifers in the Great Plains, the rate of groundwater movement was found to vary between 0.0055 m/day and 0.055 m/day (2–20 m/yr); the mean was 0.03 m/day (10.3 m/yr) (Back et al., 1983). In karst terrain in the Ozarks region in the Midwestern United States, the rate was 266.7 m/day (8 km/month). In a gravel aquifer in the Vienna basin in Austria, dye tracers gave groundwater flow rates of up to 20 m/day (see "Case History: Quaternary Aquifer Remediation in Austria" in Chapter 14), and the rate can be over 70 m/day. Not counting these extremes, the typical rate of groundwater movement is on the order of 0.082 m/day (30 m/yr) (Fetter, 1994).

Water Budget

The universal hydrologic equation is also used to estimate the *water budget* of a geographical area. Water budget is the quantification of the elements of the hydrologic cycle, i.e., the input, the output, and the storage of water in a system. For example, the water budget for the 48 conterminous states in the United States is: Precipitation [4200 billion gallon/day (bgd)] = Runoff (1352 bgd) + Infiltration (61 bgd) + Evapotranspiration (2787 bgd), from the equation

$$P = R + I + E \tag{8–1}$$

Knowledge of the water budget is very important in planning water resource developments in an area. Water budgets for the continents are given in Table 8–3. A water budget is also of great significance in hazardous waste management as it

TABLE 8–3   Water Budget of Selected Continents (adapted from Keller, 1992)

Continent	Precipitation (cm/yr)	Evaporation (cm/yr)	Runoff	
			(cm/yr)	(km^3/yr)
Africa	69	43	26	7700
Asia	60	31	29	13,000
Australia	47	42	5	380
Europe	64	39	25	2200
North America	66	32	34	8100
South America	163	70	93	16,600

quantifies the amount of groundwater that may become contaminated, the chances of dilution of the contaminants, and the quantity of contaminated groundwater that will have to be withdrawn during the remediation stage at a given location.

EARTH MATERIALS AND SUBSURFACE WATER

The geologic aspects of earth materials, including thickness, porosity, permeability, and spatial location, determine whether the material will serve as a groundwater reservoir or as a barrier to hold the groundwater in the overlying or underlying formations. Various terms used to designate these layers of earth materials are discussed in the following sections.

Aquifer, Aquiclude, and Aquitard

An *aquifer* is an underground mass of saturated rock or soil material that is capable of yielding a water supply. Aquifers may comprise sand and/or gravel or other soil materials, as well as rock formations. Sand and gravel, however, constitute about 90 percent of all aquifers developed for water supply. Porous sandstone, limestone, and highly fractured crystalline and volcanic rocks (e.g., granite and basalt) are examples of aquifers in rocks. In order for a geologic formation to be an aquifer, it must possess adequate porosity and permeability. This enables the aquifer to retain (a function of porosity) and transmit (controlled by permeability) groundwater.

An *aquiclude* is a geologic formation that, because of its extremely low or negligible porosity and permeability, has little or no capacity to hold or transmit groundwater. Cemented sandstone and unfractured granites are examples of aquicludes.

Aquitards are geologic materials characterized by high porosity and low permeability. Due to high porosity, such materials can hold significant quantities of water but, because of their low permeability, lack the capacity to transmit it. Shale and clay are good examples of earth materials that constitute aquitards. They have high porosity (45–55 percent) but very low permeability (1×10^{-7} to 1×10^{-9} cm/s).

Confined and Unconfined Aquifers

A *confined aquifer* is one that occurs sandwiched between aquitards. The hydraulic pressure in such aquifers is greater than the atmospheric pressure. If a well drilled into a confined aquifer causes the groundwater to rise above the top of the aquifer, then it is called an *artesian aquifer*. In some cases, when the confining aquifer is under high hydraulic pressure, the groundwater level may rise above the ground surface, causing the water to flow; such water wells are called *flowing artesian wells*. Confined aquifers occur at some depth below the ground and are not directly exposed at the surface. They receive recharge by slow percolation through the confining layers or, occasionally, from a remote location where the geologic conditions cause the aquifer to be exposed at the surface as an unconfined aquifer (Figure 8–2).

The groundwater level in wells completed in confined aquifers usually rises above the top of the (confined) aquifer; this surface is called the *potentiometric surface*.

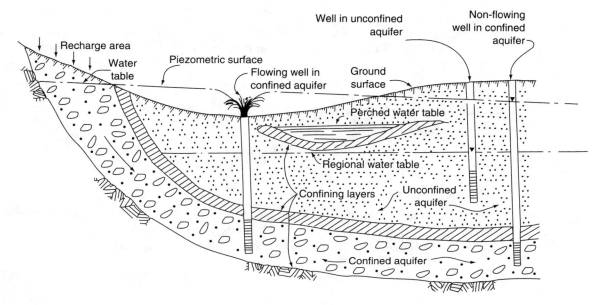

FIGURE 8–2   Types of aquifers and their recharge areas (adopted from U.S. Bureau of Reclamation, 1985)

Unconfined aquifers consist of permeable earth materials that extend upward to the earth's surface. In such aquifers, known as *water table aquifers,* the water table marks the upper boundary of the zone of saturation. Water does not fill the entire thickness of the porous and permeable material. The upper part is under partial saturation (three-phase system); the lower layer, below the groundwater table, is fully saturated with water (two-phase system). Because of this condition, the GWT is free to rise and fall in response to seasonal changes. Unconfined aquifers occur near the surface. Recharge into such aquifers is by way of direct percolation of precipitation.

Unconfined aquifers are most commonly affected by waste disposal practices, because their upper part is directly connected to the surface.

HYDRAULIC HEAD AND HYDRAULIC GRADIENT

Hydraulic head (h) is the height of the water column, or elevation of the water surface above a reference plane, usually mean sea level (MSL). In a well the hydraulic head, h, can be determined by subtracting the depth-to-the-GWT from that of the ground surface, i.e.,

$$h = \text{elevation of ground surface} - \text{depth-to-the-GWT}$$

For example, in Figure 8–3 the elevation of well X is 578 m, and the GWT was found to occur at a depth of 25 m below the surface; therefore

Elevation of the GWT = 578 – 25 = 553 m = h, the hydraulic head in well X

Similarly, for well at Y,

$$h = 400 - 10 = 390 \text{ m}$$

Hydraulic gradient is a measure of rate of change of hydraulic head over unit distance; or the head loss over a horizontal distance. For example, in Figure 8–3 the hydraulic head in well X is 553 m, and in well Y it is 390 m; the two wells are separated by a horizontal distance of 1990 m. The hydraulic gradient, i, then is

$$i = \Delta h / L \qquad (8\text{--}2)$$

where Δh = the head difference between X and Y, and L = the horizontal distance between them. Consequently,

$$i = 553 - 390 \text{ m}/1990 \text{ m}$$
$$= 163 \text{ m}/1990 = 0.082 \text{ (dimensionless)}$$

Darcy's Law

Darcy's law relates to the flow of water through porous media. The law (named after Henri Darcy, 1856) states that the velocity of flow of water through a porous medium is proportional to the hydraulic gradient. Mathematically it is expressed as

$$V \propto i$$
$$\text{or}$$
$$V = Ki \qquad (8\text{--}3)$$

where V = velocity of flow of water, i = hydraulic gradient, and K = constant, called the coefficient of permeability.

Permeability relates to the ability of an aquifer (or any earth material) to transmit fluids, usually water and gas, through it. From fluid mechanics we know that discharge (Q) is related to flow velocity (V) and the cross sectional area (A) of the medium, expressed as

$$Q = VA$$
$$\text{or}$$
$$V = Q/A \qquad (8\text{--}4)$$

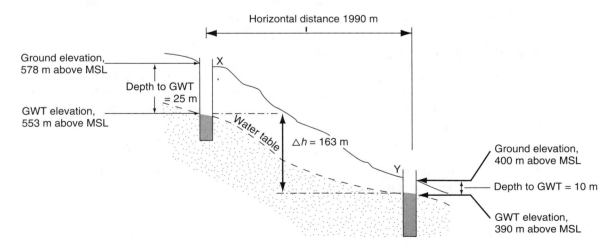

FIGURE 8–3   Hydraulic head and hydraulic gradient

Then, from equations 8–3 and 8–4, we have

$$Q = KiA \qquad (8\text{–}5)$$

Equation 8–4 gives a superficial value of groundwater flow velocity which is less than the actual flow velocity. This is because groundwater flow occurs only through the pores and interstices in the rock or soil, and not through the entire porous medium. Because the porous soil or rock is made up of solids and pores, and only the pores allow flow of water, equation 8–3 should be modified. In Figure 8–4, if the total x-sectional area of the soil is A_t, the area of pores A_p, and the area of the solids A_s, then

$$A_t = A_p + A_s$$

Also, assuming a unit volume of the soil, porosity *(n)* can be expressed as

$$n = A_p/A_t \qquad (8\text{–}6)$$

In reality, however, the actual flow velocity (V_a) is a function of *n*; therefore

$$V_a = Q/A_p \qquad (8\text{–}7)$$

Example. For a sandy soil having x-sectional area $A_t = 100 \text{ m}^2$, pore area $= 25 \text{ m}^2$, and $Q = 1000 \text{ m}^3/\text{day}$, determine V_d and V_a (V_d is the Darcy velocity and V_a is the actual flow velocity).

$$V_d = Q/A_t$$
$$= 1000 \text{ m}^3/\text{day} \div 100 \text{ m}^2$$
$$= 10 \text{ m/day}$$

and

$$V_a = 1000 \div 25$$
$$= 40 \text{ m/day}$$

Therefore $V_a > V_d$.

The actual groundwater flow velocity (V_a) is related to V_d and *n* by the equation

$$V_a = V_d/n \qquad (8\text{–}8)$$

Therefore, if V_d and *n* are known, V_a can easily be determined from equation 8–7.

Example. The Darcy velocity in a sandy soil having a porosity of 25% was found to be 10 m/day; determine the actual flow velocity.

FIGURE 8–4 Effective porosity of earth materials

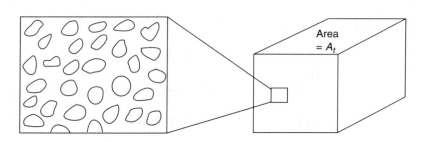

TABLE 8–4 Hydraulic Conductivity of Earth Materials (from Domenico and Schwartz: *Physical and Chemical Hydrogeology* © 1990, John Wiley & Sons, Inc., reprinted with permission)

Material	Hydraulic Conductivity (m/sec)
Sediments	
Gravel	$3 \times 10^{-4} - 3 \times 10^{-2}$
Coarse sand	$9 \times 10^{-7} - 6 \times 10^{-3}$
Medium sand	$9 \times 10^{-7} - 5 \times 10^{-4}$
Fine sand	$2 \times 10^{-7} - 2 \times 10^{-4}$
Silt, loess	$1 \times 10^{-9} - 2 \times 10^{-5}$
Till	$1 \times 10^{-12} - 2 \times 10^{-6}$
Clay	$1 \times 10^{-11} - 4.7 \times 10^{-9}$
Unweathered marine clay	$8 \times 10^{-13} - 2 \times 10^{-9}$
Sedimentary Rocks	
Karst and reef limestone	$1 \times 10^{-6} - 2 \times 10^{-2}$
Limestone, dolomite	$1 \times 10^{-9} - 6 \times 10^{-6}$
Sandstone	$3 \times 10^{-10} - 6 \times 10^{-6}$
Siltstone	$1 \times 10^{-11} - 1.4 \times 10^{-8}$
Salt	$1 \times 10^{-12} - 1 \times 10^{-10}$
Anhydrite	$4 \times 10^{-13} - 2 \times 10^{-8}$
Shale	$1 \times 10^{-13} - 2 \times 10^{-9}$
Crystalline Rocks	
Permeable basalt	$4 \times 10^{-7} - 2 \times 10^{-2}$
Fractured igneous and metamorphic rock	$8 \times 10^{-9} - 3 \times 10^{-4}$
Weathered granite	$3.3 \times 10^{-6} - 5.2 \times 10^{-5}$
Weathered gabbro	$5.5 \times 10^{7} - 3.8 \times 10^{-6}$
Basalt	$2 \times 10^{-11} - 4.2 \times 10^{-7}$
Unfractured igneous and metamorphic rocks	$3 \times 10^{-14} - 2 \times 10^{-10}$

Using equation 8–8, we have

$$V_a = 10 \text{ m/day} \div 0.25$$
$$= 40 \text{ m/day}$$

Hydraulic Conductivity and Permeability

The constant K in equation 8–3 is called the hydraulic conductivity, and is valid for most hydrogeologic situations where the fluid involved is water. But for other fluids, such as petroleum or other chemicals, whose densities and viscosities vary over a wider range in response to temperature changes than those for water, permeability *(k)* is used. Equation 8–9 gives the relationship between the fluid density, viscosity, and hydraulic conductivity.

$$k = \mu K / \rho g \qquad (8–9)$$

where k = intrinsic (or specific) permeability, K = hydraulic conductivity, μ = dynamic viscosity of the fluid, ρ = density of the fluid, and g = acceleration due to gravity. Permeability has the dimensions L^2, whereas hydraulic conductivity has the dimensions LT^{-1}.

For natural earth materials, K ranges over 12 orders of magnitude. Table 8–4 gives typical K values for various earth materials.

AQUIFER PARAMETERS

Any remediation program involving aquifers contaminated from hazardous waste requires an accurate characterization of the aquifer. If reliable aquifer test results are not available, aquifer pumping tests have to be done in order to obtain the values for the various parameters. These parameters form the basis for developing a remediation program and constitute the critical input data in modeling studies. These parameters are also very significant when an aquifer is evaluated for its water supply potential.

Specific Yield and Specific Retention

For a porous and permeable material, theoretically, porosity is a measure of the maximum yield. But in reality all water in the pores of an aquifer will not drain under the influence of gravity. This is due to the force of surface tension and other hygroscopic forces that hold the water next to the solid mineral grains. These forces are related to the size of mineral solids and the void spaces in between, and are stronger in fine-grained materials, such as clay and silt, than in coarse-grained soils. Referring to Figure 8–5, specific yield (S_y) can be expressed as

$$S_y = n - S_r \qquad (8–10)$$
$$= 0.30 - 0.1 = 0.2 \text{ m}^3$$

FIGURE 8–5   Specific yield

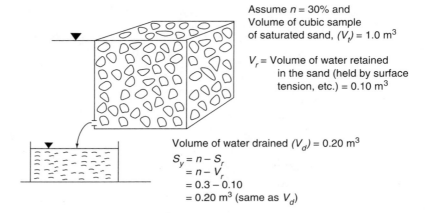

Assume n = 30% and
Volume of cubic sample
of saturated sand, (V_t) = 1.0 m³

V_r = Volume of water retained
in the sand (held by surface
tension, etc.) = 0.10 m³

Volume of water drained (V_d) = 0.20 m³
$S_y = n - S_r$
$\quad = n - V_r$
$\quad = 0.3 - 0.10$
$\quad = 0.20$ m³ (same as V_d)

where S_r = specific retention, i.e., the water retained in the aquifer by capillary and hygroscopic forces.

Rearranging terms in equation 8–10, we have

$$n = S_y + S_r \qquad (8–11)$$

Knowing the volume of water drained under gravity (V_d), the total volume of the aquifer (V_t), and the volume of water retained (V_r), specific retention and specific yield can also be expressed as

$$S_y = V_d / V_t$$

and

$$S_r = V_r / V_t \qquad (8–12)$$

Values of n, S_y, and S_r for common earth materials are given in Table 8–5. From the data given, the following generalizations can be made:

- Earth materials having higher porosity usually have lower specific yield and greater specific retention.
- Low-porosity materials generally have greater specific yield but smaller specific retention.

These relationships are illustrated in Figure 8–6.

Transmissivity

The capacity of an aquifer to transmit water is known as its transmissivity (T). It is a function of hydraulic conductivity and aquifer thickness, and is given by

$$T = Kb \qquad (8–13)$$

where b = aquifer thickness and K = its hydraulic conductivity.

TABLE 8–5 Porosity, Specific Yield, and Specific Retention of Common Earth Materials (adapted from Heath, 1983)

Earth Material	Porosity (%)	Specific Yield (%)	Specific Retention (%)
Clay	50	2	48
Gravel	20	19	1
Sand	25	22	3
Soil (general)	55	40	15
Basalt	11	8	3
Granite	0.1	0.09	0.01
Limestone	20	18	2
Sandstone	11	6	5

Also, from equation 8–5, we have

$$Q = KiA$$

and, since $A = wb$ (w is the width of the aquifer), by substituting for A in equation 8–5, we have

$$Q = Kiwb$$

But,

$$T = Kb$$

(from equation 8–13), therefore

$$Q = Tiw$$

and

$$T = Q/wi \qquad (8\text{–}14)$$

Dimensions of T are area over time, i.e., L^2/T or L^2T^{-1}.

Example. Determine the transmissivity (T) of a confined sandy aquifer 100 m thick, having hydraulic conductivity of 50 m/day. The difference in hydraulic head in two wells 1200 m apart was found to be 2 m; if the width of the aquifer is 1000 m, calculate the discharge (Q).

$$T = Kb$$
$$= (50 \text{ m/day})(100 \text{ m})$$
$$= 5000 \text{ m}^2/\text{day}$$

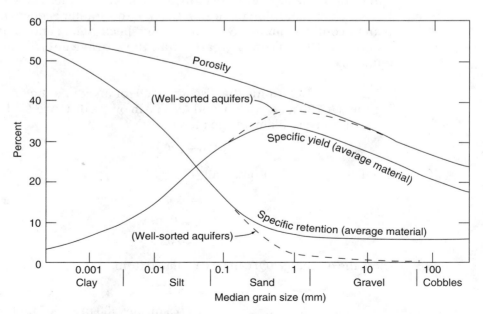

FIGURE 8–6 Relationship between grain size, porosity, specific yield, and specific retention (adopted from Rahn, 1986)

And

$$Q = Twi$$
$$= (5000 \text{ m}^2/\text{day})(1000 \text{ m})(2 \text{ m}/1200 \text{ m})$$
$$= 4166.7 \text{ m}^3/\text{day}$$

Storage Coefficient (Storativity)

Storage coefficient or storativity *(S)* is the quantity (volume) of water that an aquifer releases from, or takes into, storage per unit surface area of the aquifer per unit change in head. It can be expressed as

$$S = (\text{volume of water}) \div (\text{unit area} \times \text{unit head change})$$
$$= \text{m}^3 \div (\text{m}^2 \times \text{m}) = \text{m}^3/\text{m}^3 \text{ (cancels out)}$$

Storage coefficient is, therefore, a dimensionless parameter.

The value of *S* depends upon whether the aquifer is confined or unconfined. If the aquifer is confined, the water released from storage in response to decline in head comes from expansion of water and compression of the aquifer. However, the expansion of a given volume of water in response to a decline in head is very small. For example, in a confined aquifer having $n = 20\%$ (0.2), containing water at a temperature of 15°C, expansion of water in response to a 1-m-decline of head will release about $3 \times 10^{-7} \text{ m}^3$ (0.3 cm^3) of water for each cubic meter of aquifer volume. *S*, then, for most confined aquifers ranges from about 10^{-5} to 10^{-3} (0.00001 to 0.001).

For an unconfined aquifer, gravity is primarily responsible for the flow of water out of the aquifer in response to a head decline. The volume of water derived from expansion or compression of the aquifer is negligible. Therefore, in an unconfined aquifer, *S* is same as the specific yield and ranges from 0.1 to 0.3 (100 to 10,000 times greater than the storage coefficient of confined aquifers).

Example. Determine the discharge *(Q)* for an aquifer having $K = 25$ m/day, thickness $b = 30$ m, $w = 5$ km, and $\Delta h = 20$ m over a distance $L = 1$ km.

First, convert kilometers to meters

$$1 \text{ km} = 1000 \text{ m}$$

Calculate the hydraulic gradient *(i)*

$$i = \Delta h/L$$
$$= 20 \text{ m}/1000 \text{ m}$$
$$= 0.02$$

Then, the discharge *(Q)* can be calculated from

$$Q = KAi$$
$$= (25 \text{ m}/\text{day}) (30 \text{ m} \times 5000 \text{ m}) (0.02)$$
$$= 75,000 \text{ m}^3/\text{day}$$

WATER MOVEMENT THROUGH THE UNSATURATED ZONE

Water falling from the atmosphere, as it begins to move below the ground, first travels though the unsaturated zone and, aided by the pull of gravity, ultimately makes its way down to the GWT and enters the zone of saturation. Two physical forces control movement of water through the unsaturated zone: capillarity and gravity.

Capillarity

Capillarity is related to surface tension, which is a physical force that results from the intramolecular attractions among air, water, and soil solids. The surface tension force acts opposite to gravity and causes water to rise above the GWT. Strong surface tensions are generated when a fine-grained wet soil, such as clay, dries. The characteristic polygonal cracks seen in clays are produced by surface tension. Capillary stress in fine-grained soils can be very high, up to 245 kPa (~2.5 atmosphere). Capillary forces are only set up in moist soils, and become zero when the soil becomes fully saturated or completely dry, i.e., when the soil is a two-phase system—either solids and water or solids and air. Capillary forces in moist soils vary inversely with the diameter of the pores. This means water will rise higher in fine-grained soils having smaller pores than in coarse-grained soils having larger pore diameters. Table 8–6 shows the height of rise of capillary water, *h*, in different soils.

There is some confusion regarding the position of the capillary fringe in respect to the groundwater table—some authors consider it part of the zone of saturation, which is not exactly the case. As stated in the preceding paragraph, capillary forces become zero when the soil becomes completely saturated or completely dry, meaning that the capillary fringe can occur only in partially saturated soil. This, in turn, implies that the capillary fringe must always be above the GWT and not below it. Soil physicists have recognized this fact and use terms like tension head or suction head to indicate the partially saturated zone and the negative capillary pressure. Freeze and Cherry (1979) have recommended using the term *tension-saturated zone* in place of capillary fringe. The GWT marks the uppermost surface of the zone of saturation and this definition should remain as such. In the conventional cross sectional diagrams, such as Figure 8–1, the

TABLE 8–6   Capillary Rise in Different Soils

Soil Type	Average Pore Diameter (mm)	Height of Capillary Rise (m)	Capillary Pressure (kPa)
Coarse sand	0.95–0.084	0.12–0.18	1.0–1.5
Fine sand	0.084–0.015	0.3–1.2	3.0–10.0
Silt	0.015–0.0004	0.76–7.6	10.0–100.0
Clay	<0.0004	7.6–23	>100.0

TABLE 8–7   Features of the Saturated, Unsaturated, and Tension-Saturated Zones

Zone	Occurrence	Degree of Saturation	Fluid Pressure	Hydraulic Conductivity	Remarks
Saturated	Below the GWT	100%	>atmospheric	constant	h measured with a piezometer
Unsaturated	Above the tension-saturated zone	<100%	<atmospheric	varies with pressure head	h cannot be measured with piezometer; tensiometer needed
Tension-Saturated	Above the GWT	<100%	<atmospheric	varies with pressure head	h measured with tensiometer

inverted triangle that marks the GWT should be placed at the top of the zone of saturation, and not above the capillary fringe. The tension-saturated or capillary zone lies above the GWT; water cannot flow out of this zone under the influence of gravity, such as springs on banks of rivers and streams, or into wells drilled in the capillary zone. Such flows occur only from the zone of saturation. Table 8–7 summarizes important features of the zone of saturation, the tension-saturated zone, and the unsaturated zone.

Neither the unsaturated nor the tension-saturated zones can produce natural outflow of subsurface water into the atmosphere. Water movement can occur only through evaporation or transpiration. This tells us that contamination of water in the vadose zone will not produce outflow of contaminated water, thereby eliminating the possibility of contamination of other water sources. The only movement will be in the vertical direction, as infiltration.

GEOLOGIC FACTORS AFFECTING GROUNDWATER FLOW

A number of geologic factors may affect movement of groundwater. These factors can be grouped under the categories (a) topographic, (b) lithologic, (c) stratigraphic, and (d) structural.

Topography and Erosion

In general, the groundwater table mimics the surface topography, and groundwater flow is from the point of higher hydraulic head to the point of lower head, which, in a general sense, corresponds with the slope of the land surface.

It is well known that the geologic process of erosion that leads to formation of stream valleys and intermountain divides also results in development of stress-relief joints (fractures) in rocks. A study by LeGrand (1954) in crystalline rock terrain in North Carolina showed that well yield was highest in valleys and broad ravines, and lowest at or near the hill crests; yield from wells in flat upland areas and on slopes was intermediate. This can be explained by the fact that the stress-relief joints form close to the valley walls and disappear some distance away from the walls. Figure 8–7(A) shows the formation of stress-relief joints in a river valley. Figure 8–7(A) represents original condition of the land before erosion set in; the horizontal component of the ambient tectonic stress is shown by the arrows. Figure 8–7(B) shows the configuration of the land after the valley was entrenched. The rocks were removed from the valley area. Erosion and removal of rocks from the valley area caused the rock mass close to the valley walls to become unrestrained, thereby creating an imbalance of stress—moderate to high inside the slope, and zero at the outcrop points of the rock in the valley walls. In order to adjust to the changed stress conditions, the rocks close to the valley wall began to fracture (open out), forming the stress-relief joints. Both the number and aperture of such joints are greater close to the valley walls, progressively lessening away from it. At some depth inside the slope, they become nonexistent; Figure 8–7(C) is a representation of this condition.

Lithology

Lithology relates to the mineral composition, grain size, shape, and packing in sediments or rocks that comprise geologic formations. Sandstone, shale, and carbonate rocks are common sedimentary rocks. Some prolific aquifers are located in sandstone (the Dakota aquifer) and limestone (the Floridan aquifer). The shape and size of mineral grains making up the rocks and the degree of cementation and recrystallization control the porosity and permeability of sandstone and limestone. Removal of minerals by solution action plays a dominant role in the formation of large voids and the development of secondary porosity and permeability in dolomites and limestones.

Shales, despite their high porosity (up to 50 percent) and ability to hold plenty of water, are incapable of readily transmitting water. However, the presence of fractures and other discontinuities may allow for significant movement of groundwater through shales. Such rocks, therefore, should not be considered impermeable because of their extremely low permeability, and must be properly evaluated when assessing the extent of contamination of aquifers underlain by shale.

Among the unconsolidated materials, sand and gravel are the best sources of groundwater supply. Some highly productive, regional aquifers in the United States, such as the High Plains aquifer (about 65 m thick), are in sand and gravel. The High Plains aquifer encompasses the states of Wyoming, South Dakota, Nebraska, Colorado, Kansas, Texas, Oklahoma, and New Mexico, and is tapped by nearly 17,000 wells and accounts for 30 percent of the groundwater used for irrigation in the United States.

Stratigraphy

Stratigraphy deals with the order of deposition, mode of occurrence, geometry, and age relationships of geological units of sedimentary origin. The presence of lensoidal bodies of different hydrologic characteristics in an aquifer, lateral change in character of sediments making up an aquifer (lateral facies change),

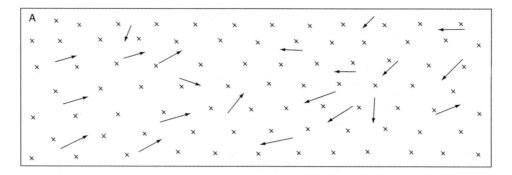

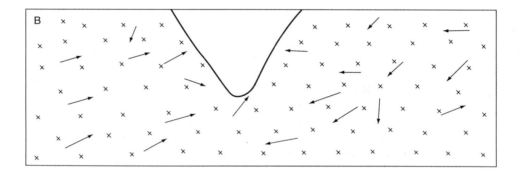

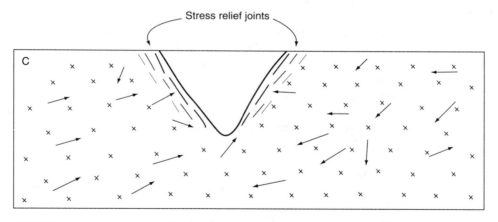

FIGURE 8–7   Conceptual diagram illustrating development of stress-relief joints in valleys (arrows show ambient stresses): (A) Distribution of stresses in rock formation before erosion; (B) erosion leading to entrenchment of the valley in the rock formation; (C) removal of rock mass from area occupied by valley causes rocks on valley sides to become laterally unrestrained because there is no rock mass to balance the horizontal component of the stress. Stress imbalance causes rock mass in the vicinity of the valley to fracture, resulting in development of stress-relief joints. Joints become less frequent away from the valley wall.

FIGURE 8–8 Stratigraphic control on groundwater occurrence (adopted from Freeze and Cherry, 1979)

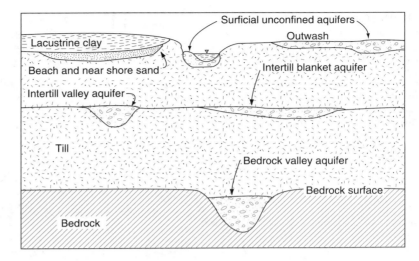

pinching and swelling of units, presence of paleochannels, and the like, all are stratigraphic details that control groundwater movement. A *perched water table* occurs when a lens of an impermeable material occurs in a permeable layer; for example, a lens of clay in a bed of sand and gravel. Some of these features and how they control the occurrence of groundwater are shown in Figure 8–8. Any remediation plan for a contaminated groundwater system must be based on a thorough understanding of these stratigraphic features.

Geologic Structures

Flow of groundwater is controlled to some extent by the geologic structure. Many times the structure of the aquifer itself dictates its nature and flow pattern. Figure 8–9 shows various geologic structures and how they control groundwater occurrence.

DISCONTINUITIES AND GROUNDWATER FLOW

Discontinuities are planes of weakness in rocks that generally tend to be open. Discontinuities in rocks result in the development of appreciable porosity and permeability in an otherwise impermeable rock. Some discontinuities are exclusive to one rock class; others are common to all three major classes of rocks. For example, bedding plane discontinuities are common in sedimentary rocks; foliation, schistosity plane, and rock cleavage discontinuities are common in metamorphic rocks; and sheeting is common in intrusive igneous rocks. Joints, fault planes, fold axes, fracture planes, shear zones, and cracks are examples of discontinuities that are found in all three rock classes.

In terms of dimension, discontinuity planes have a wide range: from very fine and microscopic, such as microfaults and mineral cleavage planes, to megascale, such as large faults and fractures that may extend over kilometers.

It is well recognized that had it not been for the development of secondary openings (joints, fractures) in crystalline rocks (granite, gabbro, quartzite, gneiss, and schist), their permeability would be negligible and they would not

FIGURE 8–9   Control of geologic structures on groundwater occurrence: (A) Fracture zone in crystalline rocks; (B) anticlinal fold resulting in artesian aquifer; and (C) folds and faults resulting in occurrence of groundwater in arid regions. ((A) and (B) adopted from Rahn, 1986; (C) adopted from Hamblin, 1975)

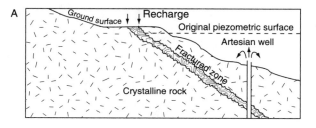

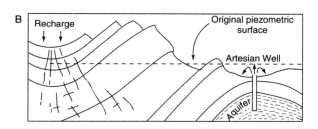

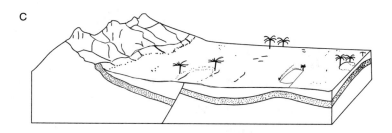

serve as aquifers. Studies have shown that there is a marked contrast in the permeabilities of unfractured crystalline rocks and their fractured varieties—10^{-13} to 10^{-8} cm/s for the former and 10^{-6} to 10^{-2} cm/s for the latter. The aperture (width) of fractures also plays a very important role in secondary permeability of fractured crystalline rocks. Snow (1968) gave the following relationship between fractures and hydraulic conductivity

$$K = (\rho g / \mu)\,(Nb^3/12) \qquad\qquad (8\text{–}15)$$

where ρ = the density of water, μ = the dynamic viscosity of water, g = the gravitational constant, N = the number of joints (fractures) per unit face of the rock, and b = the joint aperture.

The above equation shows that for a given hydraulic gradient, flow through a fracture is proportional to the cube of the fracture aperture. This means that there is an enormous difference in groundwater flow in a rock mass having a fracture aperture of less than 1 mm compared to flow in the same rock mass with an aperture of a few millimeters.

Measurement of groundwater flow in wells drilled in crystalline rocks has shown that the yield progressively decreases with increasing depth. This is due to the fact that fractures in crystalline rocks become tighter at depth, and eventually close at depths of 300–500 m.

Near-horizontal fractures in granite, parallel to the ground surface, called *sheeting*, may cause such rocks to hold significant quantities of groundwater. Sheeted granite in Georgia is an important source of groundwater supply down to depths of 100 m (LeGrand, 1949).

Special attention must be paid when evaluating groundwater movement through fine-grained geologic materials such as shales and clays. Both have high porosity but are nearly impermeable. However, because of the development of extensive joints, often closely spaced and open, shales are capable of transmitting significant quantities of water—a fact that must be kept in mind when evaluating aquifers overlain and underlain by shale.

Clays do not have joints, but contain numerous openings, such as worm holes, root holes, fissures, and desiccation cracks. These secondary openings dramatically alter the hydraulic conductivity of clays. Whereas a sample of clay tested in the laboratory may give a hydraulic conductivity value of 10^{-7} to 10^{-9} cm/s, the same material, when tested in situ, may give hydraulic conductivity values 5 to 10,000 times greater than the laboratory values (Daniel, 1984). This is a result of the presence of desiccation cracks and other openings. Clay is commonly used as a liner material in waste-containment structures, but the possibility of compacted clay developing extensive desiccation cracks in climatic environments characterized by alternating wet and dry cycles must not be overlooked in evaluating the long-term performance of clay liners. A laboratory study by Hasan and Hoyt (1992) found that desiccation cracks penetrated as deep as 19.1 cm in a clayey soil confined on the sides and subjected to alternating cycles of wetting and drying.

STUDY QUESTIONS

1. Define hydrogeology. Why is it important to protect aquifers from contamination?
2. Draw a sketch to show the unsaturated zone, the saturated zone, the capillary fringe, and the groundwater table (GWT). How would you define the GWT?
3. What is the universal hydrologic equation? How is it used in determining the water budget of a geographical area?
4. Define aquifer, aquitard, and aquiclude. Give one example of each.
5. What is Darcy's law? What is the difference between the actual flow velocity and the Darcy velocity? How are the two related to each other?
6. The difference in the GWT elevation in two wells 1.5 km apart is 35.5 m. The wells have been drilled in a sandstone aquifer that has a hydraulic conductivity of 9.4×10^{-4} cm/s and a porosity of 20%. What is the actual groundwater flow velocity?
7. Calculate the transmissivity of a 15m-thick aquifer comprised of sandy gravel that has K

of 5.1×10^{-4} cm/s. What is the transmissivity in m^2/day?
8. If the width of the aquifer in question 7 was found to be 500 m and Δh were 20 m over a distance of 1 km, what is the discharge in m^3/day?
9. Discuss the significance of topographic factors in groundwater occurrence. Why do the frequency and aperture of joints increase toward the valley walls?
10. Despite its high porosity (40–45%), why is shale not a good source of groundwater?
11. Define secondary porosity and secondary permeability. How do they affect the occurrence and movement of groundwater through crystalline rocks, such as granites and quartzites, that have a negligible primary porosity?
12. Calculate the hydraulic conductivity for a granitic rock that has two sets of joints per m^2 of surface area. The joints have an aperture of 3 mm. What will be the value of K if the aperture is 1.5 mm? Assume the value of ρ and μ for water to be 1.0 g/cm^3 and 0.016 g/cm.s; g is 981 cm/s^2.

REFERENCES CITED

Back, W., B.B. Hanshaw, L.N. Plummer, P.H. Rahn, C.T. Rightmire, and M. Rubin, 1983, Process and rate of dedolomitization: mass transfer and ^{14}C dating in a regional carbonate aquifer: Geological Society of America Bulletin, vol. 94, p. 1415–29.

Bates, C.C., and J.A. Jackson, 1987, Glossary of Geology, 3d ed.: Alexandria, VA, American Geological Institute, 788 p.

Daniel, D.E., 1984, Predicting hydraulic conductivity of clay liners: Journal of Geotechnical Engineering, American Society of Civil Engineers, vol. 110, no. 2, p. 285–300.

Domenico, P.A., and F.W. Schwartz, 1990, Physical and Chemical Hydrogeology: New York, John Wiley & Sons, 824 p.

Fetter, C.W., 1994, Applied Hydrogeology, 3d ed.: New York, Macmillan Publishing Co., 691 p.

Freeze, R.A., and J.A. Cherry, 1979, Groundwater: Englewood Cliffs, NJ, Prentice-Hall, Inc., 604 p.

Hamblin, W.K., 1975, The Earth's Dynamic Systems: A Textbook in Physical Geology; Minneapolis, MN, Burgess Publishers, 578 p.

Hasan, S.E., and A.L. Hoyt, 1992, Model experiment on leachate migration through a clayey soil: Bulletin, Association of Engineering Geologists, vol. 29, no. 3, p. 311–327.

Keller, E.A., 1992, Environmental Geology, 6th ed.: New York, Macmillan Publishing Company, p. 244.

LeGrand, H.E., 1949, Sheet structure, a major factor in the occurrence of groundwater in the granites of Georgia: Economic Geology, v. 44, p. 110–118.

LeGrand, H.E., 1954, Geology and groundwater in the Statesville area: North Carolina Department of Conservation and Development, Division of Mineral Resources, Bulletin 68.

L'vovich, M.I., 1979, World Water Resources and Their Future (English translation, R.L. Nace, ed.): Washington, D.C., American Geophysical Union, 415 p.

Nace, R.L., (ed.), Scientific Framework of World Water Balance: UNESCO Technical Papers, Hydrology, vol. 7, 27 p.

National Research Council, 1991, Environmental Epidemiology: Washington, D.C., National Academy Press, 282 p.

Rahn, P. H., 1986, Engineering Geology: An Environmental Approach: Amsterdam, Elsevier Science Publishing Company, 589 p.

Snow, D.T., 1968, Rock fracture spacings, openings, and porosities: Journal of Soil Mechanics and Foundation Division, Proceedings, American Society of Civil Engineers, v. 94, p. 73–91.

Todd, D.K., 1964, Groundwater *in* V.E., Chow (ed.), Handbook of Applied Hydrology: New York, McGraw-Hill Book Company, p. 4–14 – 4–20.

SUPPLEMENTAL READING

Domenico, P.A., and F.W. Schwartz, 1990, Physical and Chemical Hydrogeology: New York, John Wiley & Sons, 824 p.

Fetter, C.W., 1994, Applied Hydrogeology, 3d ed.: New York, Macmillan Publishing Co., 691 p.

Heath, R.C., 1983, Basic Ground-water Hydrology (7th printing, 1993): U.S. Government Printing Office, U.S. Geological Survey Water Supply Paper 2220, 84 p.

C H A P T E R

9

Contaminant Transport in the Subsurface

INTRODUCTION

Mismanagement of hazardous materials and unsound waste disposal practices of the past have impacted groundwater more severely than any other media. A study done by Fred C. Hart Associates (1984) on behalf of the EPA examined 927 sites and found the following trend in media contamination (Goldman et al., 1986):

Groundwater	32 percent
Soil	31 percent
Surface water	29 percent
Air	8 percent

Another report by the EPA (U.S. EPA, 1992) indicates that groundwater contamination accounted for 31 percent of the 193 Superfund sites targeted for remediation at the end of 1991. It is clear that groundwater contamination is very prevalent; it is also the most complex and expensive to remediate.

Groundwater contamination, in all cases, is the result of human activities. Various anthropogenic sources that cause groundwater contamination have been identified by the U.S. Geological Survey (1988). These include waste disposal, materials storage and handling, petroleum-related activities; and urban activities. Table 9–1 gives a list of such activities, the number of contaminated sites in various states, and the nature of the contaminants. During the 1980s, interest in remediation of contaminated groundwater systems and preservation of aquifer quality led to intensive research and investigations in the areas of transport and

TABLE 9–1 Anthropogenic Sources of Groundwater Contamination in the United States (adapted from USGS, 1988)

Activity	States Reporting	Number of Sites (estimated)	Common Contaminants
Waste disposal			
Septic systems	41	22 million	Bacteria, viruses, nitrate, phosphate, chloride, and organic compounds such as trichloroethylene
Landfills (active)	50 plus territories*	16,400	Dissolved solids, iron, manganese, trace metals, acids, organic compounds, and pesticides
Surface impoundments	32	191,800	Brines, acidic mine wastes, feedlot wastes, trace metals, and organic compounds
Injection wells	10	280,800	Dissolved solids, bacteria, sodium chloride, nitrate, phosphate, organic compounds, pesticides, and acids
Land application of wastes	12	19,000 land application units	Bacteria, nitrate, phosphate, trace metals, and organic compounds
Storage and handling of materials			
Underground storage tanks	39	2.4–4.8 million	Benzene, toluene, xylene, and petroleum products
Aboveground storage tanks	16	Unknown	Organic compounds, acids, metals, and petroleum products
Material handling and transfers	29	10,000–16,000 spills per year	Petroleum products, aluminum, iron, sulfate, and trace metals
Mining activities			
Mining and soil disposal— coal mines	23	15,000 active; 67,000 inactive	Acids, iron, manganese, sulfate, uranium, thorium, radium, molybdenum, selenium, and trace metals
Oil and gas activities			
Wells	20	550,000 in production; 1.2 million abandoned	Brines
Agricultural activities			
Fertilizer and pesticide applications	44	363 million acres	Nitrate, phosphate, and pesticides
Irrigation practices	22	376,000 wells; 49 million acres irrigated	Dissolved solids, nitrate, phosphate, and pesticides
Animal feedlots	17	1900	Nitrate, phosphate, and bacteria

*Puerto Rico, Virgin Islands, Guam, and American Samoa.

TABLE 9–1 *(continued)*

Activity	States Reporting	Number of Sites (estimated)	Common Contaminants
Urban activities			
Runoff	15	47.3 million acres	Bacteria, hydrocarbons, dissolved solids, lead, cadmium, and trace metals
De-icing chemical storage and use	14	Not reported	Sodium chloride, ferric ferrocyanide, sodium ferrocyanide, phosphate, and chromate
Other			
Saline intrusion or upconing	29	Not reported	Dissolved solids and brines

fate of contaminants in the subsurface. The studies focused on prediction of the arrival times and concentration levels of contaminants from their source to their receptors, such as water supply wells, monitoring wells, or surface water bodies. The processes involved in the transport of contaminants in both porous and fractured (rock) media under conditions of saturation or undersaturation must be understood. While aqueous-phase transport is more common, it is also important to understand the processes controlling contaminant transport in the nonaqueous phase for effective remediation of sites contaminated with hydrocarbons and related chemicals.

In this chapter we discuss the basic concepts of contaminant transport and the various processes and factors that control the movement of chemicals in porous media. We begin with the definition of important terms, then discuss contaminant transport in the aqueous and nonaqueous phases. Because of the increasing use of computer models in assessing aquifer conditions and contaminant transport, a brief review of model studies is also included.

DEFINITIONS

Aqueous Flow

Aqueous flow occurs when the medium of transport is water and the contaminants are carried by ground (or surface) water in a dissolved state. The contaminants, whether solid, liquid or gas, dissolve in water and thus remain in the aqueous phase. Aqueous flow is the most common mode of contaminant transport in the subsurface. Transport of such contaminants is closely related to the physico-chemical processes that control flow of groundwater through porous media. Many of the concepts discussed in Chapter 8 are applicable to transport of contaminants in the aqueous phase.

Nonaqueous Flow

Another situation arises when the contaminants are insoluble in water. In such cases, solids remain in suspension and the contaminating fluids do not readily dissolve in groundwater, but exist as a separate phase. Hydrocarbons usually are the common immiscible liquid. Flow of nonaqueous-phase fluids is controlled by factors different from those that control aqueous flow. A full understanding of the processes and mechanics of transport of nonaqueous-phase liquids (NAPLs) is lacking. The quantity of NAPLs that enter the groundwater through spills and leakage from underground storage tanks far exceeds the quantity that is being remediated. NAPLs have been identified at four out of five hazardous waste sites in the United States (Plumb and Pitchford, 1985). In addition, NAPLs are known to persist in the subsurface for a long time and are capable of contaminating a relatively large volume of groundwater in comparison to their own volume. For example, seven liters of TCE can contaminate 10^8 liters (10,000 m^3) of groundwater at a concentration level of 100 ppb (Feenstra and Cherry, 1987).

Light Nonaqueous-Phase Liquids

If the density of NAPL is less than that of water (<1 g/mL; 1000 kg/m^3), the liquid may be classified as a light nonaqueous-phase liquid (LNAPL). Common LNAPLs include acetone, gasoline, heating oil, kerosene, jet fuel, and benzene. Table 9–2 lists common LNAPLs with their densities and water solubilities.

Dense Nonaqueous-Phase Liquids

If the density of the nonaqueous-phase liquid is greater than that of water, it is classified as dense nonaqueous-phase liquid (DNAPL). Chlorinated hydrocarbons, PCBs, anthracene, pyrene, 1,1,1-TCE, phenol, and coal tar are chemical compounds that have densities greater than 1 g/mL. Table 9–3 lists DNAPLs commonly found in contaminated groundwater with their densities and water solubilities. Most DNAPLs belong to the general chemical class of halogenated/non-halogenated semivolatiles and halogenated volatiles. DNAPLs are used in the wood preservation and pesticides industries, as well as those industries involving the use of solvents and coal tar. Chlorinated solvents are the main contaminants that contain DNAPLs. Current technology to achieve permanent cleanup of DNAPLs in groundwater to the acceptable drinking water standard is

TABLE 9–2 Density and Water Solubility of LNAPLs (adapted from Fetter, 1994)

Compound	Density (mg/L)	Solubility (mg/L water at 20°C)
Acetone	0.79	Infinite
Benzene	0.88	1780
Ethyl benzene	0.87	152
Toluene	0.87	515
Methyl ethyl ketone	0.81	353 (at 10°C)
Vinyl chloride	0.91	1.1 (at 25°C)

TABLE 9–3 Common DNAPLs Found at U.S. Superfund Sites (adapted from U.S. EPA, 1991)

Compound	Density (g/mL)	Dynamic Viscosity (cp*)	Water Solubility (mg/L)
Halogenated Semivolatiles			
1,4-Dichlorobenzene	1.2475	1.2580	8.0
1,2-Dichlorobenzene	1.3060	1.3020	1.0×10^2
Aroclor 1242	1.3850		4.5×10^{-1}
Aroclor 1260	1.4400		2.7×10^{-3}
Aroclor 1254	1.5380		1.2×10^{-2}
Chlordane	1.6	1.1040	5.6×10^{-2}
Dieldrin	1.7500		1.86×10^{-1}
2,3,4,6-Tetrachlorophenol	1.8390		1.0×10^3
Pentachlorophenol	1.9780		1.4
Halogenated Volatiles			
Chlorobenzene	1.1060	0.7560	4.9×10^2
1,2-Dichloropropane	1.1580	0.8400	2.7×10^3
1,1-Dichloroethane	1.1750	0.3770	5.5×10^3
1,1-Dichloroethylene	1.2140	0.3300	4.0×10^2
1,2-Dichloroethane	1.2530	0.8400	8.69×10^3
Trans-1,2-Dichloroethylene	1.2570	0.4040	6.3×10^3
Cis-1,2-Dichloroethylene	1.2480	0.4670	3.5×10^3
1,1,1-Trichloroethane	1.3250	0.8580	9.5×10^2
Methylene Chloride	1.3250	0.4300	1.32×10^4
1,1,2-Trichloroethane	1.4436	0.1190	4.5×10^3
Trichloroethylene	1.4620	0.5700	1.0×10^3
Chloroform	1.4850	0.5630	8.22×10^3
Carbon Tetrachloride	1.5947	0.9650	8.0×10^2
1,1,2,2-Tetrachloroethane	1.6	1.7700	2.9×10^3
Tetrachloroethylene	1.6250	0.8900	1.5×10^2
Ethylene Dibromide	2.1720	1.6760	3.4×10^3
Non-Halogenated Semivolatiles			
2-Methyl Napthalene	1.0058		2.54
o-Cresol	1.0273		3.1×10^4
p-Cresol	1.0347		2.4×10^4
2,4-Dimethylphenol	1.0360		6.2×10^3
m-Cresol	1.0380	21.0	2.35×10^4
Phenol	1.0576		8.4×10^4
Naphthalene	1.1620		3.1
Benzo(a)Anthracene	1.2250		1.4×10^{-2}
Flourene	1.2030		1.9
Acenaphthene	1.2250		3.88
Anthracene	1.2500		7.5×10^{-2}
Dibenz(a,h)Anthracene	1.2520		2.5×10^{-3}
Fluoranthene	1.2520		2.65×10^{-1}
Pyrene	1.2710		1.48×10^{-1}
Chrysene	1.2740		6.0×10^{-3}
2,4-Dinitrophenol	1.6800		6.0×10^3
Miscellaneous			
Coal Tar	1.028	18.98	
Creosote	1.05	1.08**	

*Centipoise (cp); water has a dynamic viscosity of 1 cp at 20°C.

**Varies with creosote mix.

rarely feasible. The thrust of current research is to develop new technologies that will meet the desired standard of cleanup at a reasonable cost.

Advection

Advection is the process by which dissolved substances (solutes) are transported along with the moving groundwater; the direction and rate of transport are the same as that of the groundwater. Advective velocity *(Va)* is related to the hydraulic conductivity *(K)*, porosity *(n)*, and hydraulic gradient *(i)*, by the equation

$$Va = (Ki) \div n \tag{9–1}$$

Dispersion

The process that causes mixing of contaminated groundwater with uncontaminated groundwater as it moves through the porous medium is called dispersion. (The terms *mechanical dispersion* and *hydraulic dispersion* are synonyms for this process.) Dispersion is caused entirely by the motion of the fluid; it causes dilution and a reduction in the concentration level of contaminated water. Dispersion is a physical process that results in spreading of the contaminated water. Dispersion does not lower the total quantity of the contaminant in the aquifer system; it causes the contaminant to spread over a large volume of the aquifer.

Diffusion

The process that causes solutes (ions and molecules) to move from zones of higher concentration to zones of lower concentration is called diffusion. Diffusion is a chemical process; it may cause solute movement in groundwater even if the hydraulic gradient is zero and the groundwater remains stationary. In such situations, solute transport occurs due to ionic transfer in the aqueous medium. Diffusion is an important process for contaminant transport in earth materials of very low permeability that allow groundwater to move very slowly; the process of diffusion allows solutes to move at a rate much greater than that of groundwater flow.

Hydrodynamic Dispersion

Groundwater flow involves both dispersion and diffusion; hydrodynamic dispersion is the term used for the combined processes. Hydrodynamic dispersion is responsible for the spreading out of solute from the advective front—dilution is the net effect of hydrodynamic dispersion. According to Fetter (1994), the process of molecular diffusion cannot be separated from mechanical dispersion in flowing groundwater.

Retardation

Retardation is a general term for any of the several processes that retard or delay the movement of solute in groundwater. Retardation causes a slowdown in the rate of advance of a reactive contaminant. Filtration, dilution, chemical reaction, and biotransformation are some of the common processes that may cause retardation. Dilution resulting from dispersive mechanisms often causes marked retardation in the movement of groundwater constituents; for this reason it is included as one of the processes that cause retardation.

MECHANICS OF TRANSPORT IN THE AQUEOUS PHASE

Aqueous-phase transport of contaminants is relatively common. The contaminants dissolve in groundwater and move in the subsurface medium along with the flow of groundwater.

Movement of solute in the aqueous phase occurs through (a) mass transport, and (b) mass transfer. Mass transport involves movement of solute through a porous medium along with the flow of groundwater (i.e., advective flow) or through dispersion. Mass transfer, on the other hand, involves some kind of reaction—chemical, biological, or nuclear—that causes a gain or loss in the quantity of matter. Mass transport includes

- Advection
- Dispersion
- Diffusion
- Particle transport

Mass transfer includes

- Acid–base reactions
- Solution, volatilization, and precipitation
- Complexation
- Sorption reactions
- Oxidation-reduction reactions
- Hydrolysis reactions
- Ion exchange
- Isotopic reactions
- Biochemical reactions

In this chapter we will focus on mass transport. (For details of mass transfer, see Domenico and Schwartz, 1990.)

Mass Transport

Contaminants in groundwater may be in the form of solids (both dissolved and undissolved), liquid, or gas. Soluble chemical compounds such as NaCl, nonsoluble mineral particles such as clays and bacteria, miscible organic liquids such as benzene and alcohol, immiscible liquids such as petroleum, and gases that include petroleum and halogenated organic vapors may be transported by moving groundwater from one location to another. With the exception of the transport of suspended solids, all materials are transported by advection, diffusion, dispersion, or a combination of the three. Advection is the most important process of contaminant transport in groundwater.

Advective velocity can be determined from the equation

$$Va = Vd \div n \qquad (9\text{--}2)$$

where Va = the actual flow velocity of the groundwater (the advective velocity), Vd = the Darcy velocity, and n = the porosity of the medium.

Advective velocity increases with decreasing porosity; it is assumed that the solutes are nonreactive with the groundwater and that no precipitation, dissolution, adsorption, or partitioning occurs during advection.

In contaminated water, one-dimensional advective flux of one of the several pollutants through a representative volume of a porous medium is given by the equation

$$J_n = V_x C_n n \tag{9-3}$$

where J_n = the mass flux of the nth constituent/unit area/unit time, V_x = the actual groundwater velocity (V_a) in the x direction, C_n = the concentration of pollutant (n) per unit volume of the solution, and n = the porosity.

Units of J are $ML^{-2}T^{-1}$ (a vector quantity).

While in most cases the advective velocity is the same as the liner groundwater velocity, there are situations where they may be different. When the density of the contaminated groundwater is significantly greater than that of the ambient groundwater, for example, the result will be a divergence in the flow of groundwater and the solute. Other conditions involve the presence of clay minerals and resulting electrical charges, the presence of semipermeable membrane in the geologic formation, and retardation resulting from sorption (Domenico and Schwartz, 1990).

Dispersion

Dispersion is the process that results in the spreading of a solute by causing it to mix with the ambient groundwater. Dispersion is a time-dependent phenomenon. Figure 9–1 shows introduction of a dye at point A at time t_0. After time t_1, the dye will spread around A such that its concentration outside the large oval will be zero, progressively increasing toward the center. The concentration at the center itself will be less than what it was at A at time t_0. After time t_2, the concentration at the center will be less than it was at time t_1, and the zone of spreading will have grown larger but weaker in concentration. If the process of hydrodynamic dispersion continues for some time, all dye will be dispersed so that no significant concentration will be detected.

If mixing occurs along the streamline of fluid flow it is termed *longitudinal dispersion*; if it occurs at right angles to the fluid-flow direction, it is called *lateral* or *transverse dispersion*.

The shape, size, and packing of mineral grains—*fabric*—in the porous geologic medium control dispersion. Figure 9–2 illustrates how the pore size, length of flow path, and pore friction affect longitudinal dispersion.

Lateral dispersion occurs when the flow path of the contaminated fluid, instead of moving in streamlines, splits, branches off, and joins again during longitudinal flow (similar to the flow of water in a braided stream). Figure 9–3 shows an example of lateral dispersion.

FIGURE 9–1 Dispersion resulting in attenuation of chemicals in groundwater

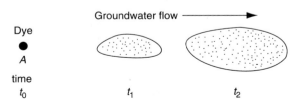

FIGURE 9–2 Control of mineral fabric on longitudinal dispersion (adopted from Fetter, 1994)

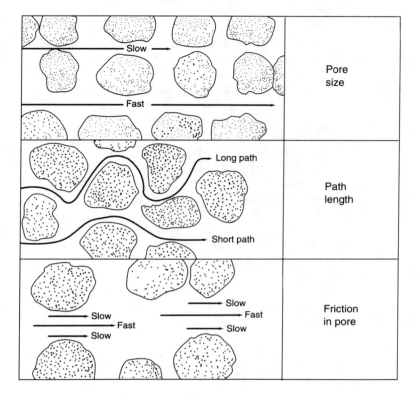

Dispersivity is a measure of the solute tendency to mix; it controls the shape and size of the contamination plume. Dispersivity does not include velocity. Dispersivity can be converted into dispersion by multiplying it by flow velocity.

FIGURE 9–3 Lateral dispersion in aquifer material (adopted from Fetter, 1994)

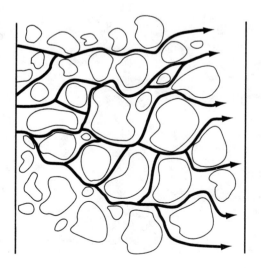

Principle of Dispersion

The classical experiment using a cylindrical tube open at both ends (for entry and exit of liquid) and filled with sand (Figure 9–4) can be used to explain the mechanism of dispersion. Initially, distilled water is allowed to flow at a constant rate through the tube, from time t_0 (beginning of flow) to time t_1, when the flow of distilled water is stopped and a weak chemical solution of NaCl (brine) is allowed to flow through the tube. As we monitor the effluent for chloride, we find that it has a zero chloride concentration at the beginning because distilled water was coming out at the exit end. Eventually, though, we find chloride present in the effluent. Its concentration will be low initially, but with time will gradually increase to the same level as in the influent. If we plot the values of relative NaCl concentration (C/C_0) against time, we get the curve shown in Figure 9–4C. C_0 is the chloride concentration at time t_1, and C is its concentration at any time (t). If no brine is added to the water, the condition after time t_1 can be represented by the straight-line plot shown in Figure 9–4B.

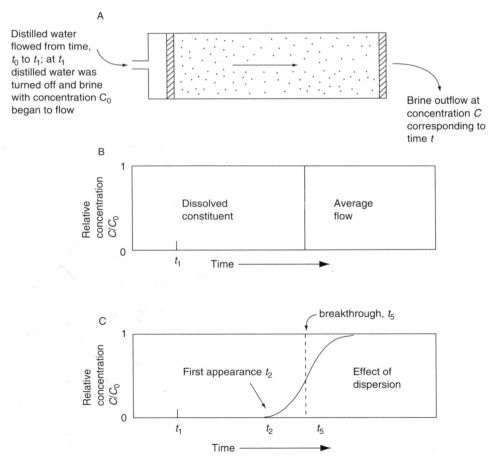

FIGURE 9–4   (A) Longitudinal dispersion of brine in a porous column; (B) advancing front without brine; (C) breakthrough curve at t_5 corresponding to C/C_0 ratio of 0.5

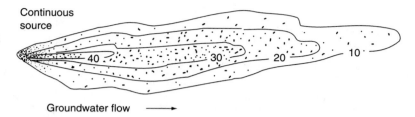

FIGURE 9–5 Continuous source of contamination resulting in development of a plume

In a one-dimensional, homogeneous system, advection is the sole process responsible for arrival of brine at the exit end of the tube. However, natural porous media, because of the varied size, shape, and packing of the mineral solids, are nonhomogeneous. This aspect is responsible for both horizontal and transverse dispersion of the solute relative to the groundwater flow direction. As a result, the advancing front becomes spread out or "smeared" (Figure 9–4C). The plot of relative concentration against time resulting from the combined advection-dispersion (Figure 9–4C) is called the *breakthrough* curve. The tube experiment has led to formulation of some important concepts in developing the advection-dispersion theory for contaminant transport. It helps in predicting the time when an action limit—the concentration level set by the EPA for drinking water or other standards—will be reached following a spill or leak of hazardous material in a groundwater recharge area. It is also useful in selection of the appropriate and cost-effective measures for remediation of contaminated aquifers.

As mentioned previously, dispersion is caused by (a) diffusion, and (b) hydromechanic mixing. The former is important only at very low flow velocities; hydromechanic mixing is important at higher velocities. Because of hydromechanic dispersion, concentration of a contaminant will decrease with distance from the source. There will be a greater spreading of the contaminant along the groundwater flow direction (longitudinal dispersion) than perpendicular to it. This is because the longitudinal dispersivity is greater than the lateral dispersivity. Consequently, a continuous source of contamination will produce a plume (for example, leachate from a leaking landfill) (Figure 9–5), whereas a transient or one-time source (a spill) will produce a slug (Figure 9–6). Dispersion occurring along both the longitudinal and transverse directions with respect to the groundwater flow will result in the formation of a cone-shaped waste plume downstream from a continuous pollution source. The figures also show that the contaminant concentration is less at the margin of the plume and increases toward the source. A plume will increase in size and be elongated when the

FIGURE 9–6 Development and movement of a contaminant slug from a point source

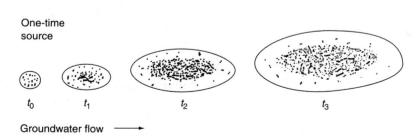

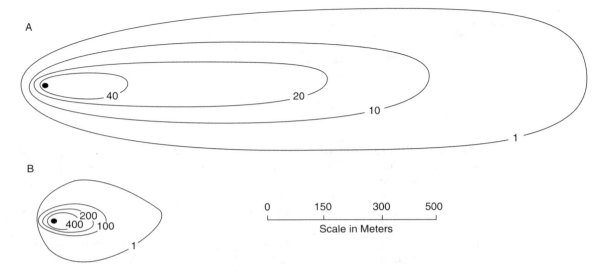

FIGURE 9–7   (A) Effect of groundwater flow velocity on the size and shape of contaminant plumes at a velocity of 1 m/day, and (B) at a velocity of 0.5 m/day. Numbers are concentration of contaminant in the groundwater.

groundwater flows rapidly, because dispersion is directly related to flow velocity. Lower flow velocity will result in a localized plume (Figure 9–7). A simple equation can be used to determine the approximate length of the plume for nonreactive contaminants, if the flow velocity and the time during which dispersion has been taking place are known (assuming no retardation has taken place).

$$PL = vt \qquad\qquad (9\text{--}4)$$

where PL = the plume length, v = the groundwater flow velocity, and t = the time of spreading.

   Example. At an abandoned landfill, leachate was first detected to be entering groundwater in 1971. The groundwater flow velocity was found to be 350 ft/yr. What will the plume length be in 1993?
   First find the time *(t)*

$$1993\text{--}1971 = 22 \text{ years}$$

Using equation 9–4, we have

$$PL = (350 \text{ ft/yr}) \times (22 \text{ yr})$$
$$= 350 \times 22 = 7700 \text{ ft}$$

Retardation of Contaminants

A reactive contaminant will undergo chemical reaction with water. Chemical reactions between contaminants and the aqueous medium during transport slow down, or retard, their movement. This causes the contaminants to move at a rate substantially less than that of the average groundwater flow (Palmer and

FIGURE 9–8 Effect of retardation on contaminant (solute) transport

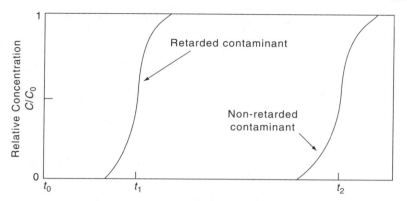

Johnson, 1989). Precipitation, adsorption, ion exchange, organic-inorganic partitioning, and biologic reactions are examples of chemical reactions that may cause contaminant retardation. Figure 9–8 shows the influence of retardation on movement of a reactive solute as compared to that of a nonretarded solute. There is an inverse relationship between retardation and flow velocity. For this reason, a contaminant with a lower retardation *(R)* factor will travel longer distances over a given period of time than a contaminant with a higher R factor. Figure 9–9 shows that of the four contaminants with R values of 1, 2, 3, and 4 leaching out of an abandoned waste pile, the one with the lowest R factor ($R = 1$) will be transported farthest; the plume boundary of this contaminant will be at $R = 1$. This contaminant will also occupy the largest volume of the aquifer. The contaminant with the highest R value ($R = 4$) will be transported over the shortest

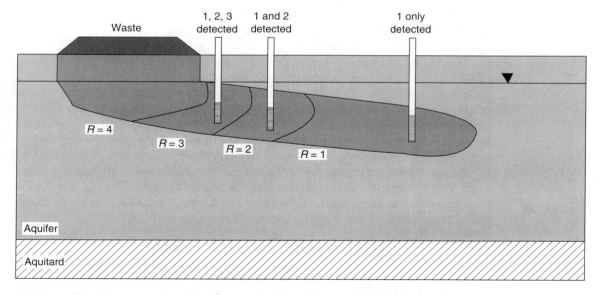

FIGURE 9–9 Influence of retardation factor on transport and detection of contaminants (adapted from U.S. EPA, 1989)

distance during the given time; its plume will be at $R = 4$. Based on the nature of the contaminants (reactive vs. nonreactive) and the R factor for reactive contaminants, certain generalizations can be made in relation to their detection in monitoring wells and the time required for remediation.

- In a network of groundwater monitoring wells, the contaminants with low R values are more likely to be encountered because they will spread farther in a given time period; consequently they will also occupy the largest volume of the aquifer.
- Estimate of removal time for a nonreactive contaminant will be far more accurate than that of a reactive contaminant because of retardation.
- In general, remediation of aquifers contaminated with reactive substances will require longer times for removal and will entail greater costs.

BIOLOGIC AND GEOLOGIC FACTORS

We have already discussed how the physical (advection and dispersion) and the chemical (diffusion) processes control contaminant transport. We will now look into the biologic and geologic factors and their influence in contaminant transport in a subsurface aqueous medium.

Biologic Factors

Until about ten years ago, the long-held belief was that microorganisms are confined to the top few feet of the soil and do not occur in aquifers. The common notion was that the soil cover provides an environment conducive to survival and growth of microorganisms and, because groundwater generally occurs at some depth below the surface and beyond the soil profile, that these microorganisms are nonexistent in groundwater. Studies done mainly during the 1980s showed that the subsurface environment was not abiotic and that a surprisingly rich assemblage of *procaryotic* (simple organisms, e.g., bacteria) and some *eukaryotic* (higher order organisms) life forms occur here (Suflita, 1989). These conclusions are based on direct microscopic, cultivation, metabolic, and biochemical studies conducted in carefully sampled aquifer materials. These studies confirmed the presence of microorganisms in the groundwater and sediments comprising the aquifers. Table 9–4 gives a partial listing of microbe populations in aquifer materials drawn from various locations in the United States. It should be noted that microbes may occur in both contaminated and unpolluted aquifers at varying depths; there is good evidence to show that even deep geologic formations may be populated with microbes (Updegraff, 1982).

In general, microorganisms that occur in the subsurface are small, primarily attached to solid surfaces, and capable of reacting with an influx of nutrients. Currently the consensus is that aquifers less than 100 m deep (i.e., those most susceptible to contamination) possess diverse microbial communities (Suflita, 1989). Under the right conditions, these microorganisms cause biodegradation of hydrocarbons and various other contaminants. Extremes of temperature, pH, salinity, hydrostatic or osmotic pressure, radiation, free-water limitations, contaminant concentration, and the presence of toxic heavy metals or other toxic

TABLE 9–4   Microbe Count in Different Aquifer Materials (adapted from Suflita, 1989)

Aquifer Material and Location	Sample Type*	Depth (m)	Aquifer Contaminant	Total Count ($\times 10^6$ cells/mL)
Sand/Gravel Lula, OK	S	5.0	None	3.9–9.3
Sand Pickett, OK	S	5.5	None	5.2
Loamy Clay Fort Polk, LA	S S	5.0 5.5	None None	9.8 1–10
Sand/Clay Conroe, TX	S	7.5	Creosote	5–49
Sand Ontario, Canada	W	10.0	Septage	0.14
Gravel Dayton, OH	W	10–12	None	0.036–0.06
Glacial Till St. Louis Park, MN	W	25–35	Creosote	0.07–10
Sand/Gravel Marmot Basin Alberta, Canada	W	1.5	None	0.05–2.5
Bacatunna Clay Pensacola, FL	S	410	Acid waste	10
Sand/Gravel Cape Cod, MA	S	12–32	Sewage	11–34
Sand Norman, OK	S	1.8	Landfill leachate	11–17

*S = Aquifer Solids; W = Groundwater.

substances can adversely influence microbial growth and biodegradation processes. In addition, the very nature of some contaminating chemicals would preclude biodegradation. Such chemicals are called *recalcitrant*—their chemical structure is unique in the sense that it is not analogous to any naturally occurring substance. On the other hand, there are chemicals (called *labile*) that possess structures similar to natural materials. Biodegradation is easier in labile chemicals than in recalcitrant chemicals. The continuum of biodegradational ease is shown in Figure 9–10.

The fate of contaminants in an aquifer should be evaluated in relation to the nature of microorganisms occurring in the subsurface, the characteristics of the

Labile Chemicals   Recalcitrant Chemicals
◄ -
High   Susceptibility to biodegradation   Low

FIGURE 9–10   Biodegradability of chemicals

contaminants, and the ambient physico-chemical conditions in the aquifers and along the transport pathways. It can be said that biologic factors play a significant role in controlling the transport and fate of contaminants in aquifers. New insights gained in recent years indicate that microbiologists, especially those with expertise in subsurface microbiology, along with the geologists, chemists, and engineers, will make significant contributions to the formulation of remediation systems for contaminated groundwater.

Geologic Factors

The nature of sediments and bedrock comprising the aquifer—their environment of deposition, geologic history, and structural features—has significant influence on contaminant transport in the subsurface. The shape, size, packing, porosity, and permeability of mineral grains are the important physical characteristics that control contaminant transport. The mineralogy of the solids comprising the aquifer, the extent of chemical alteration of the minerals, and the hydrogen ion concentration (pH) are significant chemical factors that affect contaminant transport. Finally, the nature of the source of contamination itself greatly influences transport of contaminants in the aquifer. The critical role of the various geologic factors in transport of contaminants is presented after the discussion of DNAPLs.

NONAQUEOUS-PHASE LIQUIDS

Nonaqueous-phase liquids, as they move through the subsurface, can displace air and water present in the pores of the solids. Before we discuss the mode of transport of the NAPLs, we need to point out that, unlike the aqueous flow that occurs in the zone of saturation, NAPL transport involves both the saturated and the unsaturated zones.

Physical Factors Affecting NAPL Transport

Figure 9–11 shows the common three-phase system that occurs in the unsaturated (vadose) zone; in the saturated zone only two phases are present, because all mobile air in the pores is replaced by water. When NAPLs enter the subsurface, an additional phase is developed. For example, in the saturated zone contaminated with NAPLs, three possible phases may occur—solids, NAPL, and water. In the unsaturated zone, entry of NAPL will result in the occurrence of four phases (Figure 9–12). The four phases that may be present in an aquifer contaminated with NAPLs are

- *Air phase.* Contaminants present in vapor (gas) phase
- *Aqueous phase.* A portion of the contaminants may dissolve in water, depending upon their solubility
- *Immiscible phase.* Contaminants are present as immiscible phase
- *Solid phase.* Contaminants may adsorb onto the soil surface

All four phases may or may not be present in the subsurface. When a spill of NAPL occurs at the surface, it will eventually move downwards—first through the unsaturated zone, and later through the saturated zone. Above the

FIGURE 9–11 Three phases in the unsaturated (vadose) zone (adapted from U.S. EPA, 1989)

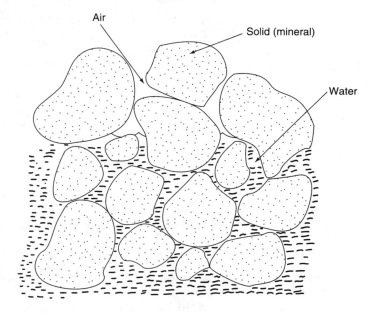

groundwater table, the NAPL will tend to coat the mineral grains and form a film around the pores. The films will eventually coalesce to occupy the entire pore space if a sufficient quantity of NAPL is available. This is because, of the two fluids (air and NAPL) in the unsaturated zone, the NAPL has a greater attraction for soil solids (inorganic mineral grains) than air. This affinity, called *wettability*, refers to the relative attraction of various fluids—air, water, or NAPLs—for soil solids (inorganic mineral grains). The fluid that coats the mineral grain is called the *wetting fluid*; the other fluid is the *non-wetting fluid*. Table 9–5 lists the wetting and non-wetting fluids in a four-phase system. The

FIGURE 9–12 Four phases (solid, air, water, and NAPL) occurring in the unsaturated zone (adapted from U.S. EPA, 1991)

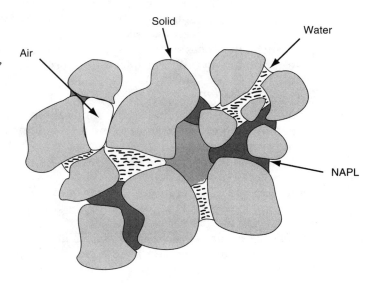

TABLE 9–5   Wetting Fluid Relationship in a Four-Phase System (adapted from U.S. EPA, 1991)

System	Wetting Fluid	Non-Wetting Fluid
Air-water	Water	Air
Air-NAPL	NAPL	Air
Water-NAPL	Water	NAPL
Air-NAPL-water	Water>organics>air	

following generalizations on the relative wettability of common fluids encountered in contaminated groundwater can be made:

- In a three-phase system—comprising inorganic solids (minerals), water, and oil—water is the wetting fluid and will coat the solid. If air is present in place of oil, water will still be the wetting fluid (Albertsen et al., 1986).
- In a three-phase system comprising mineral solids, oil, and air—oil is the wetting fluid. In a saturated medium, such as below the groundwater table, where water and oil represent the two fluid phases, water is the wetting fluid (Albertsen et al., 1986).
- In a three-phase system comprising oil, water, and organic solids (humus or peat)—oil is the wetting fluid. (Albertsen et al., 1986).
- In a four-phase system comprising air, oil, water, and organic solids—oil is the wetting fluid (Albertsen et al., 1986).
- Most organic contaminants, such as TCE and benzene, show wetting characteristics similar to those of oil (Domenico and Schwartz, 1990).

Saturation-Dependent Factors

Depending upon the relative percentages of pore volume occupied by different liquids, the degree of saturation and permeability of the liquids will be different than it would be if the pores were occupied by one liquid. The physical processes responsible for such differences significantly influence the transport and fate of the NAPLs. Some important physical principles are discussed in the following sections.

Residual Saturation

Residual saturation is defined as the volume of hydrocarbon (or any other NAPL) trapped in the pores in relation to the total volume of pores; mathematically, it can be expressed as

$$S_r = V_h/V_t \times 100 \tag{9-5}$$

where S_r = residual saturation, V_h = the volume of hydrocarbon in the pores, and V_t = the total volume of the pores.

This definition was developed as a result of studies related to occurrence and exploitation of petroleum. In the context of groundwater contamination, residual saturation has been described as the degree of saturation at which the NAPL

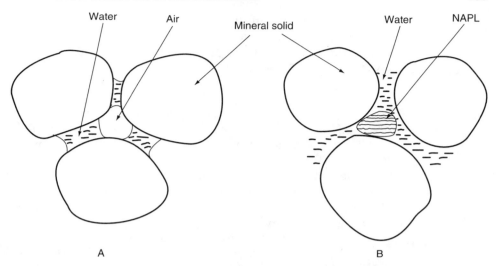

FIGURE 9–13 Residual saturation resulting in (A) pendular saturation (solid-air-water phase), and (B) insular saturation (solid-water-NAPL phase) (adapted from U.S. EPA, 1991)

becomes discontinuous and is immobilized by capillary forces (Mercer and Cohen, 1990).

Varying values of residual saturation for hydrocarbons (<1% to 50%) have been reported. However, it appears that the relative percentage of silt and clay in the porous and permeable medium plays a more dominant role than fluid properties in controlling residual saturation.

Because NAPL is the non-wetting fluid in most saturated media, it is trapped in the larger pores. In the unsaturated zone, however, NAPL, as the wetting fluid, spreads into adjacent pores. This spreading results in a reduced level of residual concentration at its original location, implying that a relatively larger amount of NAPL will be retained in the unsaturated zone than in the saturated zone.

Wetting fluids tend to occupy smaller spaces in the soil mass, while non-wetting fluids occupy the largest openings. Albertsen et al. (1986) used the term *pendular saturation* for the wetting fluid that is held in the narrowest part of the pore space (pore throats) by capillary forces (Figure 9–13A). *Insular saturation* relates to residual saturation of non-wetting fluids in the pore spaces, which tend to occur as isolated blobs in the center of the pores. Figure 9–13B shows occurrence of NAPL at its residual saturation in an aquifer; in this case, water, the wetting fluid, surrounds mineral grains.

Relative Permeability

As discussed previously, because of the relative wettability of water and NAPL, water forms a coating around the mineral solid while NAPL occupies the central portions of pores. In such situations, unlike a typical clean aquifer, neither the NAPL nor the water occupies the entire pore volume. This means that the permeability of the geologic medium in relation to the two liquids will be different, less than what it would be if the medium were occupied by one liquid alone. This reduction in permeability is called *relative permeability* and is defined as the ratio of permeability of a fluid at a given saturation to its permeability at 100% saturation. Relative permeability may range from 1.0 at complete (100%) saturation to 0 at no (0%) saturation. Mathematically it can be expressed as

$$k_{ri} = k_i(S_i) \div k_{si} \tag{9–6}$$

where k_{ri} = the relative permeability for the fluid i, S_i = the volume of pore space occupied by fluid i, $k_i(S_i)$ = the permeability of the medium for fluid i at saturation S_i, and k_{si} = the permeability of the medium when it is completely saturated with fluid i.

Studies have shown that when the two phases comprise NAPL and water, zero permeability does not necessarily occur at 0 percent saturation. Figure 9–14 shows permeability of a two-phase liquid, comprising NAPL and water, in a porous geologic medium. The relative permeabilities of NAPL and water at 100 percent NAPL saturation are 1.0 and 0.0, respectively. As the percentage of pore volume occupied by water increases, a corresponding decrease in the volume of NAPL in the pores takes place. This results in a decrease in the permeability of NAPL, because of its reduced volume in the pores. Continued increase of volume of water in the pores at the expense of NAPL will ultimately result in zero relative permeability of NAPL. This value of zero permeability does not occur at specific retention of zero for NAPL, i.e., when the volume of NAPL in the pores becomes zero. Indeed, zero permeability for NAPL occurs at a level of residual saturation above zero, corresponding to the point S_{rn}. The degree of NAPL saturation at S_{rn} is called *residual NAPL saturation*. The relative permeability of NAPL at S_{rn} is effectively zero, which means that NAPL is immobile. The immobile NAPL poses special problems in remediation in that it cannot be easily removed from the pores, except through dissolution by flowing groundwater.

FIGURE 9–14 Relative permeability curves for water and LNAPL (adapted from Schwille, 1988)

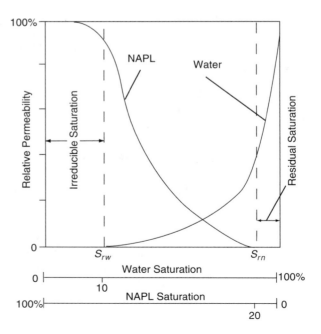

Relative permeability of water is affected in the same way as NAPL: With increasing percentage of NAPL in the pore volume, relative permeability of water decreases, until the point S_{rw} is reached. The relative permeability of water becomes zero when the water saturation equals S_{rw}. Water saturation corresponding to S_{rw} is called *irreducible water saturation*. At irreducible water saturation water becomes effectively immobile and there is no significant flow of water. Schwille (1988) determined the values of zero residual saturation for water and DNAPL, based on model studies. These values are 10 percent and 20 percent respectively. This means that in a two-fluid system, at a degree of saturation of 10 percent for water, only DNAPL flow will occur; at DNAPL saturation of 20 percent the mobile fluid will be water. Mixed flow (of water and DNAPL) may occur within the two limiting saturation levels (Figure 9–15).

Transport of LNAPL

Transport of LNAPL through the unsaturated and saturated zones depends on the quantity of LNAPL released: If the quantity of LNAPL is small, it will flow through the unsaturated zone until residual saturation is reached. This is because a four-phase system is developed as a result of LNAPL entry into the unsaturated zone, with the following order of fluid wettability: Water>LNAPL>air. The infiltrating water will dissolve soluble components present in LNAPL, such as benzene, toluene, and xylene, and transport them down to the GWT. These dissolved contaminants will form a plume that will radiate from the residual products. Since many of the products found in LNAPL tend to be volatile, they will partition into soil gas and air, and will be transported to other parts of the aquifer by molecular diffusion. Volatiles will first move through the unsaturated zone and will ultimately enter the surface soil layer where, if the physico-chemical conditions are right, they may partition back into the liquid phase. This process of volatilization, transport, and devolatilization will result in transport of contaminants over much larger areas. The upward transport of volatiles can have one of the two fates: If a

FIGURE 9–15 Relative permeability graph showing the region of DNAPL flow (I), water flow (III), and mixed flow (II) (adapted from Williams and Wilder, 1971)

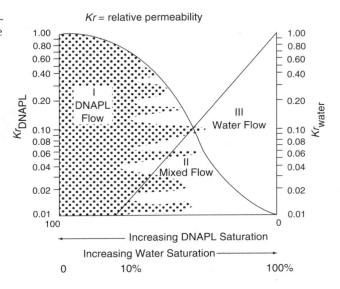

permeable layer exists in the soil mass, the vapors will diffuse through it and ultimately escape into the air; on the other hand, if a relatively impermeable layer is present in the soil mass, volatiles will not escape into the atmosphere, resulting in the development of a significantly high concentration in the soil mass.

When a large volume of LNAPL is released, the LNAPL will travel down through the unsaturated zone to the top of the capillary fringe above the GWT. Soluble components of the LNAPL will travel ahead of the less soluble components. Upon entering the capillary fringe, where water saturates a large percentage of the pores, the advected mass will cause a reduction in relative permeability of LNAPL. Unable to move downward rapidly, the LNAPL will tend to spread out across the top of the capillary fringe. Once a sufficient quantity of LNAPL has accumulated within the capillary fringe, it will begin to flow in the same direction as the groundwater (Figure 9–16). LNAPLs, because of their lower density, will float above the GWT in the capillary zone. This explains why some products, such as gasoline, kerosene, and some oils, occur as free products on top of the capillary fringe. With large spills, however, a continuous supply of LNAPL will be available in the unsaturated zone; this will lead to a progressive buildup of hydraulic head to the point that it will result in depression of the GWT–NAPL interface. The NAPLs will begin to accumulate in the depression. In the event of the removal of the NAPL source, or the exhaustion of the source of supply, NAPLs still present in the unsaturated zone will continue their downward migration, which will stop when the NAPLs' concentration reaches the residual saturation level. At this point the NAPL will become immobile, and may not travel further. Otherwise it will continue to move downward and will recharge the LNAPL pool occurring in the depression over the GWT (Figure 9–17A), causing it to spread laterally over the capillary fringe. If the LNAPL supply is exhausted, the accumulated LNAPL will keep moving downward until residual saturation is reached. Drainage of LNAPL from the upper parts of the unsaturated zone will result in reduction of total head at the LNAPL–GWT inter-

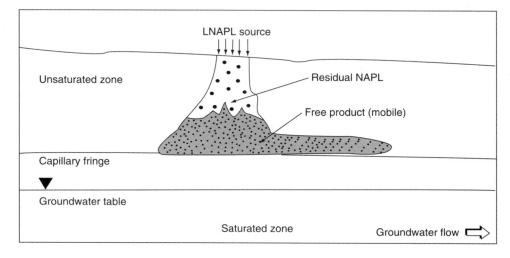

FIGURE 9–16   Movement of LNAPL in the unsaturated zone and the capillary fringe (adapted from U.S. EPA, 1991)

face. This reduction of head will cause the GWT to rebound (Figure 9–17B). Since the LNAPL above the GWT is at residual saturation, the rebounding groundwater cannot displace all LNAPL (it is relatively immobile because of residual saturation). Part of the rebounding groundwater that will pass through the LNAPLs will, nonetheless, be able to dissolve materials within the residual LNAPLs, creating a contaminant plume. Water infiltrating from the surface will also be able to

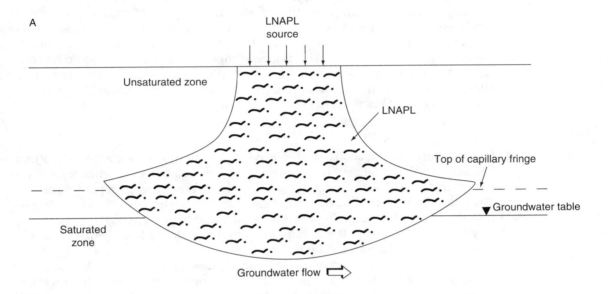

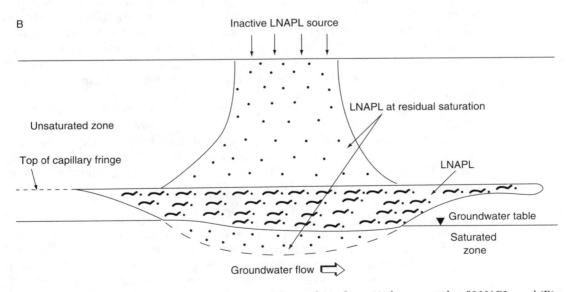

FIGURE 9–17   Depression of the groundwater table resulting from (A) large supply of LNAPL, and (B) rebound of the groundwater table due to depletion of LNAPL source (adapted from U.S. EPA, 1991)

dissolve material from the residual LNAPL, thus resulting in a larger level of contamination of the subsurface materials.

Pumping of water from the aquifer and seasonal changes also result in lowering of the GWT. If LNAPLs are present at the top of the GWT, depression of the GWT related to the above causes will also result in further lowering of the LNAPL–groundwater front. When the water table rises up, due to recharge or establishment of the steady state conditions in the aquifer, part of the LNAPL will be pushed upward, but not all of it—because of residual saturation some of it will remain in the pores below the new GWT. This will result in LNAPLs occurring over greater thickness of an aquifer, causing contamination of a much larger volume of the aquifer. This is a very important aspect that should be considered when designing an LNAPL remediation plan: LNAPLs should not be allowed to move to uncontaminated parts of the aquifer, because doing so will allow for more NAPLs to be held back as a result of residual saturation.

Transport of DNAPL

High density, low viscosity, and relatively low solubility cause the DNAPL to become greatly mobilized in the subsurface. Low solubility prevents ready mixing of DNAPL with water, which results in the two existing as separate phases. The combination of high density and low viscosity causes DNAPL to sink deeper into the aquifer by displacing the low-density and higher viscosity fluid—water. Physically this results in the development of destabilizing forces and the occurrence of viscous fingering (Kueper and Frind, 1988). Like the LNAPL, the transport of DNAPL is a function of the quantity of the spill or leak—small or large.

If the quantity of DNAPL is small, it will migrate downward through the unsaturated zone, under gravity. Downward movement will cease when DNAPL reaches residual saturation. Viscous fingering will occur in the lower portions of the unsaturated zone, where relatively larger amounts of water may be available (Figure 9–18). In the dry parts of the unsaturated zone, no such fingering should occur. During its downward transport, some of the DNAPL may partition into the vapor phase, and the denser vapor will sink down to the capillary zone. Residual DNAPL or its vapors can be dissolved by infiltrating water, which will transport it to the GWT, from where it may spread to form a plume of dissolved chemicals in the aquifer (Figure 9–19).

If the amount of DNAPL is large, it will keep flowing downward across the capillary fringe and the GWT into the aquifer. While moving through the capillary zone, it will experience resistance from the capillary forces. If a large quantity of DNAPL is present, its head will ultimately overcome capillary forces. The *critical height* of DNAPL needed to overcome the capillary forces can be calculated from the equation

$$z_c = 2Y \cos ß \, (1/r_t - 1/r_p) \div (\Delta \rho g) \tag{9–7}$$

where z_c = the capillary forces, Y = the interfacial tension between water and DNAPL, $ß$ = the contact angle between the fluid and the solid surface (mineral grain), r_t = the radius of the pore throat, r_p = the pore radius, $\Delta \rho$ = the density difference between water and DNAPL, and g = the gravitational constant (Villaume et al., 1983).

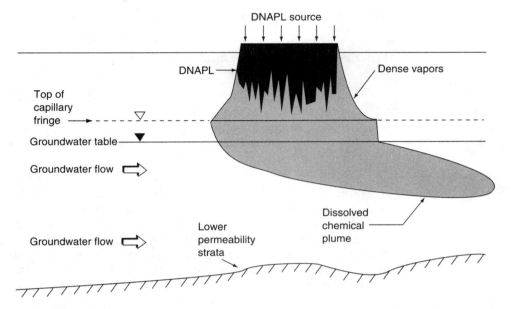

FIGURE 9–18 Viscous fingering of DNAPL in the unsaturated zone (adapted from U.S. EPA, 1991)

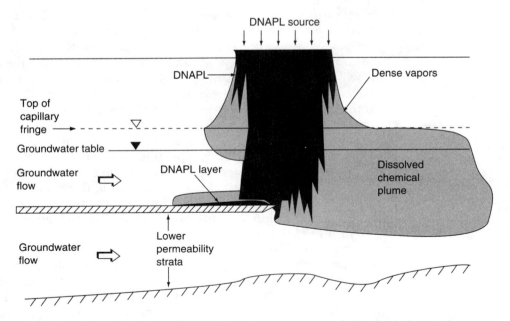

FIGURE 9–19 Partitioning of DNAPL into dense vapors and dissolved chemicals (adapted from U.S. EPA, 1991)

TABLE 9–6   Critical Height for Perchloroethylene as a Function of Particle Size (adopted from Anderson, 1988)

Material	Diameter (mm)	Critical Height (cm)
Coarse sand	1.0	13
Fine sand	0.1	130
Silt	0.01	1300
Clay	0.001	13,000

Anderson (1988) determined critical heights for perchloroethylene in various textural classes (grain size) of soil. Table 9–6 shows the relationship between critical height and grain size. The data indicate that the smaller the grain size, the larger the critical height—a generalization similar to the height of rise of capillary water discussed in Chapter 7. A consideration of data in Table 9–6 also shows that soil materials of low permeability—silt and unfractured clays—can be more effective barriers to DNAPL transport than sands.

Once within the zone of saturation, DNAPLs will continue to move downward until they reach residual saturation. The condition of residual saturation represents a three-phase system comprising DNAPL, water, and the mineral solids. Water-soluble components of DNAPL constitute the mobile phase. Residual saturation and the adsorbed components of DNAPL on the aquifer solids represent the immobile phases. The main mobilization mechanism during the residual-saturation stage is the removal of soluble components of DNAPL by the groundwater. Moving groundwater, as it passes through the contaminated zone, will transport the dissolved DNAPL, creating a contaminant plume that can cover a large volume of the aquifer.

If the geologic conditions are such that beds of lower permeability or discontinuous lens-shaped beds of fine-grained material occur in the aquifer, the infiltrating DNAPL will accumulate on top of such beds, forming a perched DNAPL pool (Figure 9–20). It has been found by observation that a perched DNAPL pool exists whenever a discontinuous layer of low-permeability material, such as clay or silt, occurs in the aquifer. Lateral migration of DNAPL along the upper boundary of these discontinuous impermeable layers will continue to occur until (a) residual saturation is reached, or (b) it encounters a geologic trough or other low spots in the impermeable layer, which will result in accumulation of DNAPL to form a large reservoir. Under both conditions, the soluble components in the DNAPL will partition into the groundwater from both the residual saturation and the DNAPL pools. Molecular diffusion may cause the dissolved DNAPL to move vertically across the aquifer. The advecting groundwater will become further contaminated as it flows through or around the DNAPL zone.

When a much larger quantity of DNAPL is spilled, DNAPL will migrate to the bottom of the aquifer and, depending on the geologic structure of the boundary, can either form pools in bedrock troughs or flow in a direction opposed to that of

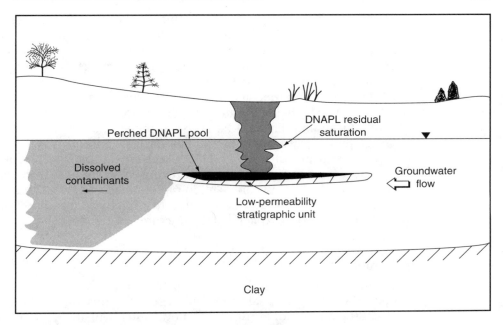

FIGURE 9–20 Perched DNAPL pool (adapted from U.S. EPA, 1991)

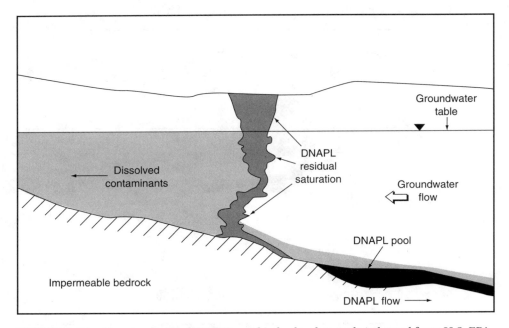

FIGURE 9–21 Development of DNAPL pool in bedrock trough (adapted from U.S. EPA, 1991)

groundwater flow (Figure 9–21). Dipping beds with different permeabilities can also result in DNAPL moving in a direction opposite to that of groundwater flow (Figure 9–22). Under such conditions, standard hydraulic principles cannot be used to predict the direction of plume movement; geologic information is critical in determining this direction.

Fractures in impermeable materials, such as shales, clays, and crystalline rocks, that occur below a water table aquifer or as the impermeable layer above a confining aquifer (Figure 9–23), serve as preferential pathways for DNAPL transport. Other pathways that may act as vertical conduits for DNAPL include root holes, disposal wells, unsealed geotechnical boreholes, improperly sealed hydrogeologic sampling and monitoring wells, and old uncased or unsealed water supply wells. All can result in rapid transport of DNAPL, and even a small amount of DNAPL may travel deep into a fractured formation due to the relatively low retention capacity of the fractured system.

Geologic Factors

Geologic variations in the porous formation (pinching and swelling of strata, beds of varying permeabilities, and discontinuous lenses of low permeability) affect the pattern of groundwater flow, which will always follow the most permeable pathways. This implies that contaminant transport will occur along the layers of higher permeabilities. The geologic factors related to stratigraphy, lithology, and structure of the porous media are discussed in the following sections.

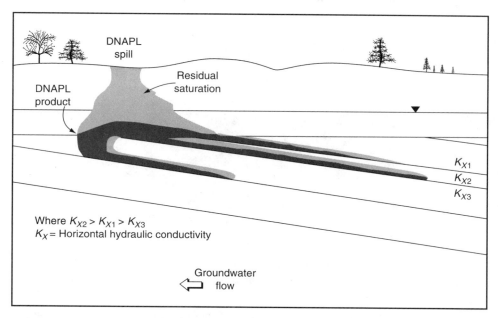

FIGURE 9–22   Dipping beds with varying hydraulic conductivities resulting in DNAPL transport in a direction opposite to groundwater flow direction (adapted from U.S. EPA, 1991)

Stratigraphic Control

The history and sequence of deposition of sediments, and subsequent modification during structural disturbances and/or during cementation and hardening (diagenesis and lithification), will influence the geometry and hydraulic characteristics of the layers. A deposit of sand and gravel is usually associated with a shallow depositional environment of high energy; on the other hand, finer sediments—silt and clay—represent deeper and low-energy environments. In addition, irregularity in preexisting bedrock topography (the result of previous cycles of uplift-erosion-submergence) may result in the formation of low spots—fossil valleys, paleochannels, unconformities, and solution features in carbonate rocks. Lateral variation in the nature of sediments in a sequence of beds is not uncommon. Such lateral changes, called *facies change*, result in occurrence of discontinuous bodies of material of different composition and texture within a layer characterized by sediments of a certain composition. For example, a deposit of sand and gravel of a certain thickness may be found to contain lens-shaped layers of clays and/or silt. Such layers have a different texture, and their hydraulic conductivity may be substantially different from that of the enclosing bed. Attention to stratigraphic details is extremely important, and such details must be evaluated as fully as possible before a remediation plan is finalized.

Desiccation of soft clays and shales is known to be caused by erosion and unloading, in addition to other factors. During the Pleistocene, when most of the land in the northern hemisphere was under a cover of ice, the underlying beds were subjected to tremendous vertical stresses. This resulted in consolidation of the soft layers. When the climate changed and the glacial ice melted, the

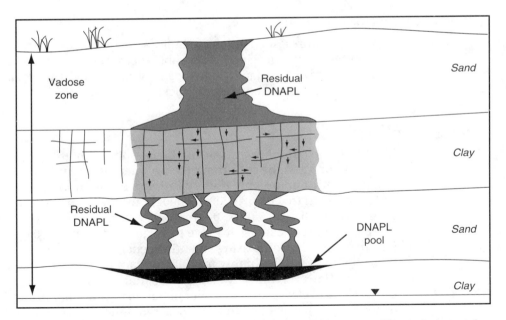

FIGURE 9–23 Transport of DNAPL through fractures in impermeable earth material (adapted from U.S. EPA, 1991)

high overburden stresses were relieved and the underlying beds developed desiccation cracks while readjusting to the changed loading conditions. Such cracks are found in beds of clays and shales, and need careful evaluation when studying the extent of aquifer contamination. If a bed of clay occurs below an aquifer, or above it in the case of a confined aquifer, it will be erroneous to assume that these beds will act as an aquiclude and provide effective barriers against contaminant migration. Apertures of desiccation cracks vary from a fraction of a mm to several cm. DNAPLs, because of their high density and low viscosity, may enter very fine cracks that otherwise may be nearly impermeable for water and other fluids.

Lithologic Control

Carbonate rocks—limestone and dolomite—present special problems because of their high water solubility. Under the right climatic conditions, minerals comprising such rocks are dissolved away by the water percolating through joints and other discontinuities in the rock mass. Over a period of time, well-defined solutioned openings may develop in carbonate rocks. Such openings can range in size and shape from a planar feature, such as a solution-enlarged joint having an aperture of few mm, to large voids, such as caves and passages, measuring several 100 to 1000 m^3 in volume. Occurrence of such karst features in carbonate rocks needs careful study and evaluation in relation to contaminant transport. As a general rule, whenever limestone or dolomite is found at a hazardous waste site, one should always suspect the presence of karst features; investigations based on this assumption are more likely to result in identification of such features than investigations where this lithologic aspect is ignored.

Structural Control

Structural features, such as joints, faults, and shear zones, serve as pathways for contaminant migration in impermeable rocks. Depressions resulting from erosion of the bedrock surface occurring in crystalline rocks affect DNAPL transport in the same way as in sedimentary rocks.

PARTICLE TRANSPORT

In addition to contaminants occurring as solutes in groundwater, suspended solids also occur in the subsurface environment. These include bacteria, viruses, and inorganic and organic materials that are usually extremely small—clay to colloid size (2μ to 0.01 μ, 0.002–0.00001 mm). Bacteria and viruses, because of their very small size, 5 μ to 200 nm (0.005–0.0002 mm) and <250 nm (0.00025 mm), respectively, can easily move through sand grains (4–0.074 mm in diameter). In an aquifer comprising essentially coarse sediments—sand and gravel—these bacteria will be highly mobile and can be freely transported over long distances. However, the processes can be hindered by the presence of fine particles—silts and clays—in the aquifer material. Contaminants may be transported as suspended particles by trace metals, radionuclides, and organic compounds. The important point is that, because of their very small size, the suspended solids possess a large surface area that facilitates adsorption, making them excellent contaminant collectors. Transport of such suspended particles is influenced

FIGURE 9–24 Filtration of suspended solids through a porous medium resulting in (A) mat formation, and (B) closure of pores (adapted from U.S. EPA, 1989)

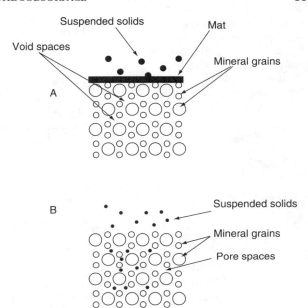

to a greater degree by the physico-chemical processes operating in the subsurface than by the nature of the contaminants.

Physical filtration and straining are two processes that can cause suspended materials to stop moving. Figure 9–24A shows how suspended particles (either larger than or the same size as pore openings) that cannot penetrate into the porous medium are filtered out, forming a cake or mat and inhibiting transport of contaminants associated with these solids. Another mechanism, called straining, occurs when small suspended particles pass through larger pore spaces in the medium but ultimately get trapped in smaller pores (Figure 9–24B). This process also affects contaminant transport. The net effect of both filtration and straining is the alteration of pore openings, with a resulting decrease in hydraulic conductivity of the medium following clogging of the pores by the suspended solids. Changes in physico-chemical conditions may result in aggregation of small particles into a larger mass, thus blocking their movement by filtration or straining.

CONTAMINANT TRANSPORT THROUGH FRACTURED MEDIA

In special cases, fractures in rock formations may constitute the only pathway for fluid transport. Crystalline rocks, such as granite, gneiss, basalt, and quartzite, lack primary porosity and permeability, but fractures that developed subsequent to the formation of the rocks may act as effective pathways for contaminant transport. Contaminant transport in fractured rock may occur by advection and dispersion. Advection is common along fractures; dispersion occurs when there is a change in the aperture of the fracture and at fracture intersections. However, diffusion across the wall rock enclosing the fracture may become an important transport mechanism when high concentration gradients

FIGURE 9–25 Contaminant transport by advection and diffusion in fractured rocks (enlarged view) (adapted from U.S. EPA, 1989)

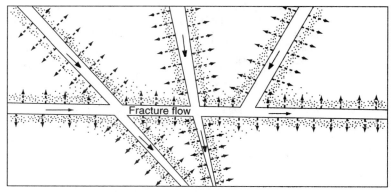

→ Flow in fractures - advective
↓ ↑ Flow by diffusion through rock pores

exist in the fractured medium (Figure 9–25). Four general models of contaminant transport through fractured media have been proposed by Palmer and Johnson (1989): (a) continuum, (b) discrete fracture, (c) hybrid, and (d) channel.

The *continuum model* has two categories, depending on whether the media permeability is the result of fractures only—the *single porosity continuum model*—or is a function of both fractures and primary porosity and permeability—the *double porosity continuum model*. Granite, basalt, and quartzite are some of the rocks that are good examples of the single porosity model. Sandstones represent a rock type that possesses double porosity. The pores formed during lithification and diagenesis of sands result in the primary porosity. Subsequent formation of joint planes and other discontinuities result in development of the secondary (double) porosity in the rock.

Complete information on fracture dimension, density, persistence, aperture, and nature of infilling material must be obtained in order to develop a realistic model. The importance of fractures in controlling rock porosity and permeability is discussed in Chapter 7. RQD is a very important parameter related to fracturing in rocks, and when used in conjunction with guidelines for recording fracture data in the field (Figure 11–3) will provide as complete a characterization of the fractures as possible. This will, in turn, help to predict their influence on contaminant transport.

Discrete fracture models describe contaminant transport in individual fractures. Use of statistical data to develop stochastic models may obviate the practical limitation of obtaining information about each fracture in the rock mass.

Hybrid models represent a combination of discrete fracture and continuum models. *Channel models* relate to solute transport as small fingers or channels rather than as a uniform flow along the width of a fracture.

Case History: *Contaminant Transport Through Fractured Dolomite*

This case history illustrates the significance of structural and karst features in carbonate rock and their control on contaminant migration.

An operating sanitary landfill is located about 0.5 km (0.3 mi) west of a former chemical recycling facility in the U.S. Midwest. A creek running approximately

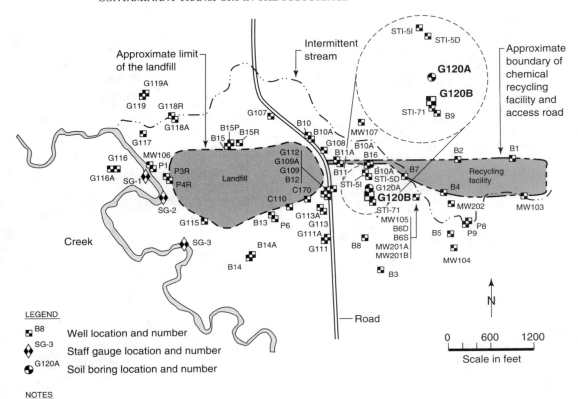

FIGURE 9–26 Site plan showing the locations of operating landfill, chemical recycling facility, water wells, and bore holes (illustration provided by Warzyn, Inc., Addison, IL)

NW–SE flows west of the landfill (Figure 9–26). Land use in the vicinity of the site comprises a mix of agricultural, industrial, commercial, and rural residential types. Wastes accepted at the landfill consist of municipal solid waste and sewage treatment sludge. The presence of chlorinated ethenes and ethanes was detected in private water supply wells, and it was presumed that the landfill was the source of chemical compounds. However, detailed geologic and hydrogeologic investigations revealed that this was not the case; the source of the chlorinated compounds was the chemical recycling facility. The presence of fractures and solutioned features in the dolomite was determined to be the key factor in contaminant transport.

The type, origin, and quantities of wastes disposed of at the chemical recycling facility were generally undocumented, but were known to include solvent still-bottom sludge, non-recoverable solvents, paints, and oils that were disposed of in four lagoons. It was believed that some 10,000 to 15,000 drums may have been present at the 8.1 ha (20 acre) site when the facility was closed in 1973. Cleanup and removal of buried drums and contaminated soils at the chemical recycling facility was due for completion in 1995.

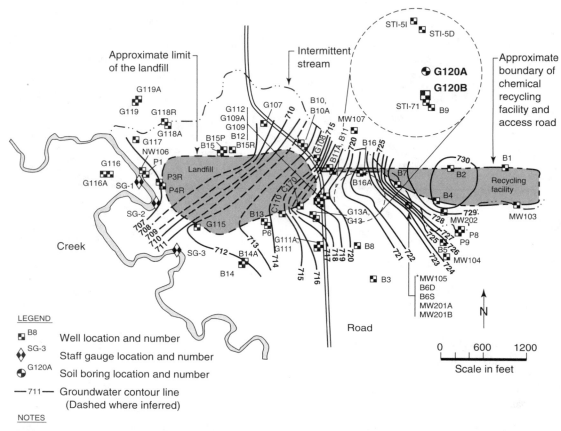

FIGURE 9–27   Potentiometric map showing groundwater mound at the chemical recycling facility (illustration provided by Warzyn, Inc., Addison, IL)

Medium- to coarse-grained dolomite of the Ordovician Galena Group occurs at a shallow depth of 3 to 6.1 m (10 to 20 ft) below the ground surface. Thin [up to 3.8 cm (1.5 in.)] layers of nodular chert occur interbedded with the dolomite. The rock formation is horizontally bedded and is traversed by steep to vertical fractures. Solutioning is not widespread, but vugs [pinhole size to 3.8 cm (1.5 in.) across] are common in the dolomite. RQD values ranging between 0 and 63 for fractured and solutioned dolomite and 0 and 98 for unfractured dolomite were computed for drill core samples.

A series of monitoring wells were installed for hydrogeologic observations and to sample groundwater. Results showed that the groundwater flow direction was generally from east to west. A zone of high permeability was noted between the chemical recycling facility and the landfill. Hydrogeologic data also indicated the presence of a groundwater mound near the center of the chemical recycling facility (Figure 9–27). Because the general flow of the groundwater is from east to west, the presence of the groundwater mound causes a local reversal in the flow direction. High localized recharge from the intermittent stream flowing by the

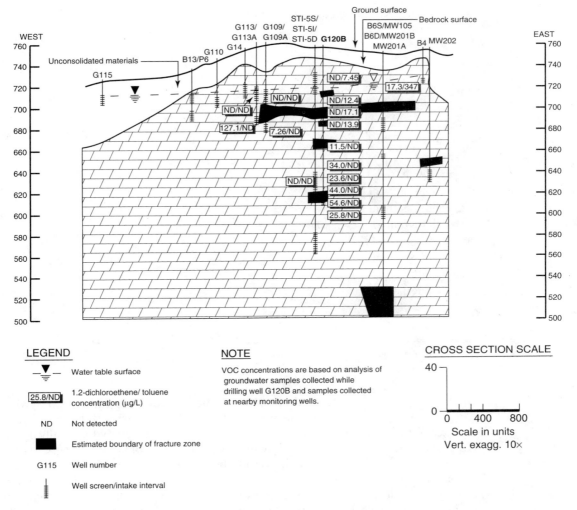

FIGURE 9–28 Geologic cross section showing the locations of bore holes and fracture zones in dolomite (illustration provided by Warzyn, Inc., Addison, IL)

northern boundary of the chemical recycling facility may be the reason for the occurrence of the groundwater mound.

In order to determine the source of contaminants and their mode of transport, water wells were drilled around the existing landfill and the chemical recycling facility (Figure 9–26). As the borings into the bedrock were advanced, successive intervals below the GWT were isolated by using packers and samples collected for chemical analysis using an onsite gas chromatograph. Bore Hole G120B was drilled to a total depth of 49.4 m (162 ft) below the ground surface. Several fracture zones within the dolomite were noted in the borings (Figure 9–28). Various intervals were isolated in this bore hole, and successive samples were taken and analyzed. Results indicated the presence of a VOC plume occurring at two different levels, each with distinct contaminant characteristics: The upper plume in the

TABLE 9–7   VOCs Present in Groundwater (adapted from Ihm and Schmidt, 1993)

| VOCs | Well Number | | | |
	B-4	G-109	G-113A	G-120B
Benzene	9.7*	—	—	—
Toulene	347	—	—	11.58
Xylenes	590	—	—	—
Ethylbenzene	104	—	—	—
1,1 Dicholoroethane	52.5	—	17.1	—
1,2 Dicholoroethane	—	—	7.2	—
Tetrachloroethane	10.8	—	12.4	—
Trichloroethene	—	—	36.2	—
1,2 Dichloroethane**	17.3	7.62	54.2	255.3
1,1 Dichloroethane	—	—	—	45.6
1,1,1 Trichloroethane	5.5	—	—	20.3
PCE	—	—	—	16.0
TCE	—	—	—	34.0

*Concentrations in mg/L.
**Includes cis and trans isomers.

shallow fracture zone was dominated by toluene, ethylbenzene, and xylenes, while the plume in the deeper zones was dominated by chlorinated ethenes and ethanes. Toluene was also found in another boring, B4, located hydraulically upgradient of G120B and immediately downgradient of an area of known solvent disposal at the chemical recycling facility. Similarly, in the deep well MW202 (122 ft; 37.2 m), located in close proximity to well B4, the presence of chlorinated ethenes and ethanes in the lower fracture zones was detected (Table 9–7). Based on the hydrogeologic, structural, and chemical data, it was concluded that the fracture zones in the dolomite acted as pathways for transport of the contaminants that originated at the chemical recycling facility. It was also concluded that VOC-contaminated groundwater was migrating from the chemical recycling facility toward the southeastern corner of the landfill (Ihm and Schmidt, 1993). Figure 9–29 shows a conceptual model of contaminant transport through the fractured dolomite.

(Materials for this case history were provided by Sandra Ihm, Warzyn, Inc., Addison, IL.)

MODEL STUDIES

A model is a representation of the real physical system. Because of the practical limitations of obtaining all needed information within reasonable limits of time and cost, models are used to simulate conditions occurring at the site. This can be done by developing one of the several approaches, such as the (a) conceptual model, (b) scale model (sand tank), or (c) mathematical (analytical) model. Use of models in

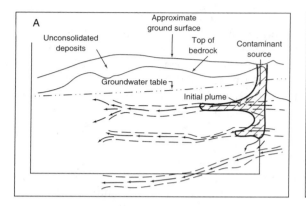

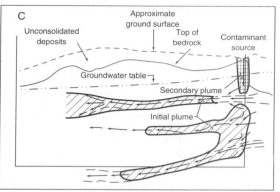

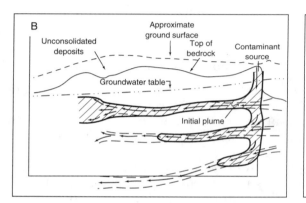

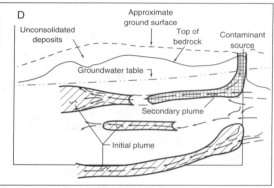

FIGURE 9–29 Conceptual model showing various stages of contaminant migration through fractures in dolomite (illustration provided by Warzyn, Inc., Addison, IL)

hydrogeology has been on the increase in recent years. Not only their number but also their sophistication level is increasing. One important point about models is that the resulting picture that emerges is directly related to the accuracy of the input information: The model is not going to be accurate and reliable if the input data are inaccurate or the assumptions made are unrealistic. In such cases the GIGO principle applies—garbage in, garbage out. Oversimplification of the site conditions and generalization of subsurface hydrogeologic parameters result in inaccurate models with little or no applicability. A workable and realistic model must avoid the following pitfalls:

- Misinterpretation of hydrogeologic data
- Ignoring uncertainty in the subsurface conditions
- Oversimplification of biologic, chemical, geologic, and physical conditions prevailing in the subsurface

The basic purpose of all modeling studies is (a) to gain understanding of the nature and behavior of the aquifer, and (b) to predict how the system will function in the future. In the context of hazardous waste management, the second aspect of modeling studies—prediction—is more relevant. Numerous models have been developed for use in this field; the majority of them are designed for use on personal computers. These models use a set of codes to simulate physical and chemical processes operating in the subsurface. These codes, when properly manipulated in relation to field conditions, can provide estimates of how far, how fast, and in what directions contaminants will travel in the subsurface medium. This information, in most cases, forms the basis for developing remediation alternatives and selecting the most suitable remediation plan.

Input Data

The accuracy and thoroughness of the input information determines how successfully a model will serve its intended purpose. Models used to predict aquifer flow characteristics and contaminant transport require information related to (a) parameters affecting groundwater flow, and (b) parameters affecting contaminant transport. Each includes a set of six parameters.

Parameters Affecting Groundwater Flow

Several important parameters have to be input into the model. These are

- *Aquifer Type* (whether confined or unconfined). In general, modeling of confined aquifers is easier than modeling of unconfined aquifers.
- *Media Characteristics.* Groundwater flow in porous media is easier to model than flow occurring in fractured and/or solutioned-rock aquifers.
- *Aquifer Homogeneity and Isotropism.* A homogeneous and isotropic aquifer is the easiest to model. In reality, though, such ideal aquifers do not exist; geologic materials are nonhomogeneous and anisotropic. However, from a practical standpoint we generalize the *overall* features of an aquifer to characterize it as homogeneous and isotropic or as nonhomogeneous and anisotropic. The latter requires three-dimensional simulation and requires more extensive site-characterization data.
- *Number of Phases Present.* Contaminant plumes that do not greatly differ from the unpolluted aquifers in their physical and chemical characteristics are easy to model. Multiple phases, such as NAPLs together with their physical features—whether viscous fluids denser or lighter than water, partitioned vapor, and other phases—complicate modeling studies.
- *Number of Aquifers.* Flow and contaminant transport involving single aquifers are easier to simulate than multiple-aquifer systems.
- *Flow Conditions.* Groundwater flow under steady-state conditions is easier to model than flow under transient conditions.

Parameters Affecting Contaminant Transport

In addition to the parameters controlling groundwater flow, the factors that affect contaminant transport constitute critical input data. These include

- *Nature of the Source of Contamination.* Contaminants may come from a point, line, area, or volume source. Depending upon the source of release, modeling can be relatively complicated. Simple two- or three-dimensional models work well for point, line, or area sources, but volume source requires complex three-dimensional modeling. Examples of point, line, and area sources include pipe outflow or injection well; contaminants leaching from the bottom of a trench, and leachate from a waste lagoon or an agricultural field, respectively. Volume source occupies three dimensions in an aquifer and may include DNAPL that has penetrated the entire thickness of the aquifer.

- *Mode of Contaminant Release.* An instantaneous release of contaminant, known as pulse or slug, is easier to model than a continuous release.
- *Dispersion.* The portion of contaminant transport occurring by dispersion is important input data for model study. This information is generally obtained by using the advective-dispersion equations. However, it has the limitation of not accurately predicting the in situ dispersion.
- *Adsorption.* Adsorption that involves a single partition coefficient or single distributions is easier to model than adsorption that is nonlinear and subject to spatial and temporal variations.
- *Degradation.* Contaminants with a single, first-order degradation coefficient or single distribution are easier to model than those involving a second-order degradation coefficient. pH changes, substrate concentration, and microbial population are some of the factors that result in second-order degradation.
- *Density and Viscosity.* Temperature differences between the contaminant plume and the unpolluted aquifer complicate modeling. Because density and viscosity are temperature dependent, effects of temperature changes must be incorporated into the model.

Modeling is a complicated exercise that is multidisciplinary in nature. Therefore, the model developer must seek input from other experts involved in hazardous waste site characterization and remediation design.

This discussion presents only a bare outline of models related to hazardous waste management. Fetter (1994) includes a detailed discussion of various models and data requirements, and a critique of available models and case histories. Gavanasen and Hussain (1993) present an excellent summary of groundwater modeling.

STUDY QUESTIONS

1. Differentiate between aqueous and nonaqueous flow. How are LNAPLs different from DNAPLs? Give three examples of each.
2. How is mass transfer different from mass transport? Explain with suitable examples.
3. Define advection. How does porosity affect the advective velocity? A sandy aquifer was found to have an average porosity of 28% and hydraulic conductivity of 2.5×10^{-4} cm/s. Two wells, 750 m apart, encountered the GWT at depths of 22 and 13 m, respectively. Determine the advective velocity.

4. What geologic factors control dispersion? What is the most important conclusion to be drawn from the tube experiment? What is its practical significance?
5. In an aquifer located in the vicinity of an abandoned landfill, leachate was first detected in a well in 1984. Measurements in two monitoring wells, spaced at 1200 ft, gave Δh of 35 ft. The hydraulic conductivity and porosity of the aquifer material was found to be 20×10^{-5} ft/s and 30%, respectively. What will the plume length be in 1995?

6. What is contaminant retardation? What factors cause retardation? Of the two contaminants having R values of 1 and 2, which will be detected first in a monitoring well? Why?

7. "Biodegradation of pollutants in groundwater does not occur because microorganisms do not thrive in groundwater." Do you agree or disagree with this statement? Explain your answer.

8. A leaking underground storage tank released benzene that traveled down to the groundwater. What phases would you expect to occur in the saturated and the unsaturated zones?

9. Define residual saturation in the context of groundwater contamination. What is the difference between pendular and insular saturation?

10. Explain the concept of relative permeability involving water and NAPL. Why doesn't the zero permeability for NAPL correspond to zero saturation? In a DNAPL–water system, what is the value of zero residual saturation for the two liquids? If pumping is done to remove the DNAPL, how much of the DNAPL will remain behind in the aquifer? Why?

11. What will the fate and transport of LNAPL through the saturated and unsaturated zones be when a large quantity of the LNAPL was spilled but further spill was controlled?

12. What will happen if a large quantity of DNAPL leaks into the ground where the GWT is 10 ft below the surface?

13. Under what geologic conditions will DNAPL move in a direction opposite to the groundwater flow direction? Use suitable sketches to illustrate your answer.

14. Discuss the factors that control contaminant transport in limestone. What may be the best way to assess the nature and extent of contamination in carbonate rocks?

15. Describe the processes of physical filtration and straining. How may they impede or inhibit movement of suspended solids?

16. Discuss the significance of discontinuities in contaminant transport through fractured quartzite. Besides advection-dispersion, which other process may cause migration of contaminants? Explain.

17. What factors must be taken into account in developing models for contaminant transport? Why is it important to have input from various experts in the design of a remediation program?

REFERENCES CITED

Albertsen, M., and an ad hoc task force, 1986, Beurteilung und behandlung von Mineralölschadensfallen im Hinblick auf den Grundwasserschutz, Jeil 1, Die wissenschaftlichen Grundlagen zum Verständnis des Verhaltens von Mineralöl im Untergrund: Berlin, Federal Office of the Environment, LTws no. 20, 178 p.

Anderson, M.R., 1988, The dissolution and transport of dense nonaqueous phase liquids in saturated porous media, Ph.D. Dissertation, Department of Environmental Engineering, Oregon Graduate Center, Beaverton, OR, 278 p.

Domenico, P.A., and W.F. Schwartz, 1990, Physical and Chemical Hydrogeology: New York, John Wiley & Sons, 824 p.

Feenstra, S., and J.A. Cherry, 1987, Dense chlorinated solvents in Groundwater: An Introduction, *in* Dense Chlorinated Solvents in Groundwater; Institute of Groundwater Research, University of Waterloo, Ontario, Progress Report No. 0863985.

Fetter, C. W., 1994, Applied Hydrogeology, 3d ed.: New York, Macmillan Publishing Co., 691 p.

Fred C. Hart Associates, 1984, Assessment of Hazardous Waste Mismanagement Damage Case Histories: Washington, D.C., U.S. Government Printing Office, U.S. EPA Report No. EPA/530/SW-84-002, 419 p.

Freeze, R.A., and J.A. Cherry, 1979, Groundwater: Englewood Cliffs, NJ, Prentice-Hall, Inc., 604 p.

Gavanasen, V., and S.T. Hussain, 1993, Modeling: The right tool for the job: Environmental Protection, vol. 4, no. 12, p. 24–29.

Goldman, B.A., J.A. Hulme, and C. Johnson, 1986, Hazardous Waste Management: Washington, D.C., Island Press, 314 p.

Ihm, S.C., and A.J. Schmidt, 1993, Tracking Chlorinated VOCs in a Fractured Dolomite Aquifer: Unpublished report prepared for the National Groundwater Association Meeting, Kansas City, MO, October 17–20, 5 p. plus tables and figures.

Kueper, B.H., and E.O. Frind, 1988, An overview of immiscible fingering in porous media; Journal of Contaminant Hydrology, vol. 2, p. 95–110.

Mercer, J.W., and R.M. Cohen, 1990, A review of immiscible fluids in the subsurface: Properties, models, characterization and remediation: Journal of Contaminant Hydrology, v. 6, p. 107–163.

Palmer, C.D., and R.L. Johnson, 1989, Physical processes controlling the transport of contaminants in the aqueous phase, *in* Transport and Fate of Contaminants in the Subsurface; U.S. Environmental Protection Agency, Seminar Publication EPA/625/4-89/019, p. 5–22.

Plumb, R.H., Jr., and A.M. Pitchford, 1985, Volatile organic scans: implications for groundwater monitoring: Proceedings, Petroleum Hydrocarbons and Organic Chemicals in Groundwater, National Water Well Association, November 13–15, Houston, TX, p. 207–222.

Schwille, F., 1988, Dense Chlorinated Solvents in Porous and Fractured Media: Model Experiments (in German: J.F. Pankow, trans.): Chelsea, MI, Lewis Publishers, 688 p.

Suflita, J.M., 1989, Microbial ecology and pollutant biodegradation in subsurface systems, *in* Transport and Fate of Contaminants in the Subsurface, Seminar Publication: U.S. EPA, Report No. EPA/625/4-89/019, p. 67–84.

Updegraff, D.M., 1983, Plugging and penetrating of petroleum reservoir rock by microorganisms, *in* E.C. Donaldson, and J.B. Clark, (eds.), Proceedings, 1982 International Conference on Microbial Enhancement of Oil Recovery, May 16–21, 1982, Afton, OK: Springfield, VA, NTIS; p. 80–85.

U.S. Environmental Protection Agency, 1989, Seminar Publication, Transport and Fate of Contaminants in the Subsurface: Report No. EPA/625/4-89/019, 148 p.

U.S. Environmental Protection Agency, 1991, Groundwater Issue: Report No. EPA/540/4-91/002, 21 p.

U.S. Environmental Protection Agency, 1991, Handbook, Groundwater, Vol. II: Methodology: Report No. EPA/625/6-90/016b, 141 p.

U.S. Environmental Protection Agency, 1992, ROD Annual Report, FY 1991, Vol. I: EPA Report No. OOWER 9355.6-05-02, vol. II, 406 p.

U.S. Geological Survey, 1988, National Water Summary 1986: Events and Groundwater Quality, USGS Washington, DC; U.S. Government Printing Office, Water Supply Paper 2325, 140 p.

Villaume, J.F., P.C. Lowe, and D.F. Unites, 1983, Recovery of coal gassification gas: an innovative approach, *in* Proceedings, 3rd National Symposium on Aquifer Restoration and Ground Water Monitoring: Worthington, OH, National Water Well Association, p. 434–445.

Waterloo Center for Groundwater Research (WCGWR), 1989, Short Course Notes: Dense immiscible phase liquid contaminants in porous and fractured media: University of Waterloo, Kitchener, Ontario, Canada, November 6–9, 1989, unpaginated.

Williams, D.E., and D.G. Wilder, 1971, Gasoline pollution of a groundwater receiver—a case history: Groundwater, vol. 9, no. 6, p. 50–54.

SUPPLEMENTAL READING

Domenico, P.A., and W.F. Schwartz, 1990, Physical and Chemical Hydrogeology: New York, John Wiley & Sons, 824 p.

Fetter, C.W., 1994, Contaminant Hydrogeology, 3d ed.: New York, Macmillan Publishing Co., 691 p.

Schwille, F., 1988, Dense Chlorinated Solvents in Porous and Fractured Media: Model Experiments (in German, J.F. Pankow, trans.): Chelsea, MI, Lewis Publishers, 688 p.

U.S. Environmental Protection Agency, 1989, Seminar Publication, Transport and Fate of Contaminants in the Subsurface: Report No. EPA/625/4-89/019, 148 p.

Chemical Analysis and Quality Control

INTRODUCTION

Chemical analysis is the backbone of any project involving hazardous waste management; it provides data used to determine the chemical makeup of both the waste stream and the hazardous waste. From the instant chemicals appear at the end of an industrial process, chemical characterization becomes important. First and foremost, chemical analyses determine if the product will be classified as hazardous or nonhazardous, thus deciding the treatment and/or disposal options. If analyses indicate that the product does not contain hazardous materials, it can be disposed of in a conventional manner: if it is found to be hazardous, then all applicable regulations have to be complied with before disposal can be accomplished—usually at a much higher cost. Chemical characterization also helps in risk assessment, in remediation design, and in setting regulatory standards for public health and safety.

The nature of the chemicals and their concentration also sheds light on the former use of the site. If, for example, high concentrations of PCBs are found at a site, it is very likely that the site was the location of a facility for manufacturing electric capacitors. On the other hand, if the nature of the chemicals is highly variable, it could mean that the site was a landfill that accepted both MSW and hazardous wastes.

The following discussion of analytical methods is intended to provide a general familiarity with the common techniques used in chemical analysis of hazardous wastes.

ANALYTICAL METHODS

The EPA has developed a comprehensive manual that contains details of the various physical and chemical tests used to evaluate waste materials. This manual is widely used in the hazardous waste industry and is commonly referred to as the SW-846.

Methods and equipment used for chemical analysis have become very accurate and highly sophisticated during the past 25 years. A great deal of improvement has occurred since the passage of the RCRA and TSCA legislation, and detailed guidelines have been developed by the EPA, in collaboration with the industry. Most analytical laboratories are now capable of detecting chemical concentrations in the parts per billion (ppb; 10^{-9}) range; for some chemicals the resolution has gone to parts per trillion (ppt; 10^{-12}) and even parts per quadrillion (ppq; 10^{-15}). The availability of cheaper and superior electronic components, faster computing power, and use of robotics, combined with the incentives derived from the regulations, have contributed to the improvement in analytical technology.

Of the various methods of chemical analysis, gas chromatography-mass spectrometry (GC-MS) and inductively coupled plasma mass spectrometry (ICP-MS) are the most common techniques used in the hazardous waste management industry. GC-MS is useful for determination of the types and quantities of organic compounds; ICP-MS, on the other hand, is suitable for detection and quantitation of inorganic compounds. In the GC-MS technique a very small amount (a few microliters) of the sample is injected into the instrument, where the sample instantly vaporizes. Individual molecules travel through a narrow column in the instrument at different speeds, exit out of the tube and enter the mass spectrometer, where an electronic beam breaks them into fragments. By studying the pattern formed by the broken molecules and their abundance, the nature and quantity of the chemicals that were present in the sample can be determined.

For inorganic substances, ICP-MS or ICP-AES (inductively coupled plasma-atomic emission spectometry) is used. The sample is ionized at high temperature in a plasma (a high-temperature ionized gas) and then sent to the spectrometer, which detects and quantifies individual atoms. The data are fed into a computer that matches the signatures of various chemicals with those of the known chemicals, and the analytical results are printed out. Common techniques are described in more detail in the following sections.

Gas Chromatography

Gas chromatography (GC) is a quantitative technique useful for analysis of organic compounds that can be volatilized without undergoing any chemical rearrangement or decomposition. GC, also referred to as vapor phase chromatography (VPC), represents two subcategories: (1) Gas-solid chromatography (GSC), and (2) Gas-liquid chromatography (GLC) or gas-liquid partition chromatography (GLPC).

GLC is the most widely used method for analyzing organic constituents in hazardous waste. It can determine a large number of halogenated VOCs in concentrations of 0.34 ppb to 0.006 ppb. GC instruments are not expensive, are easy to operate, and give excellent quantitative results (Johnson and James, 1989).

Gas Chromatography-Mass Spectrometry

The gas chromatography-mass spectrometry (GC/MS) method may be used to identify a large number of both halogenated and non-halogenated VOCs and semivolatile organic compounds in extracts. It is particularly suitable for quantitation of VOCs occurring in a variety of hazardous waste matrices such as groundwater, aqueous sludges, caustic liquors, acid liquors, wastes, solvents, oily wastes, tars, fibrous wastes, polymeric emulsions, filter cakes, spent carbons, spent catalysts, soils, and sediments. VOCs present in very small concentrations—5 ppb to 100 ppb—can be determined by the GC/MS method. GC/MS instruments are moderately expensive and require an experienced operator.

High-Resolution Gas Chromatography/High-Resolution Mass Spectrometry

The high-resolution gas chromatography/high-resolution mass spectrometry (HRGC/HRMS) method can be used to identify various types of dioxins, such as PCDDs and PCDFs, in the ppt to ppq range in all matrices, including bottom and fly ash and human adipose tissues.

Atomic Absorption

Atomic absorption (AA) is a simple and rapid method for determination of metals in solution. It is used to identify concentrations of metals in drinking water, surface water, groundwater, saline water, and in domestic and industrial wastes.

Inductively Coupled Plasma-Mass Spectrometry

This technique is suitable for determination of very small concentrations (ppm–ppb) of a large number of elements in water and wastes. It is routinely used to determine trace concentrations (detection limits in ppm are in parentheses) of the following elements: Al (0.1), Sb (0.02), As (0.4), Ba (0.02), Be (0.1), Cd (0.07), Cr (0.02), Co (0.01), Cu (0.03), Pb (0.02), Mn (0.04), Ni (0.03), Ag (0.04), Tl (0.05), and Zn (0.08).

Inductively Coupled Plasma-Atomic Emission Spectrometry

This technique is used for identification of trace elements, including metals in solution. It can detect metals occurring in concentrations approximately two orders of magnitude greater than what can be determined by the ICP-MS method. Samples require digestion prior to analysis.

High-Performance Liquid Chromatography

High-performance liquid chromatography (HPLC) is a popular method for chemical analysis of less volatile compounds that are not amenable to GC. It is used for identification of polynuclear aromatic hydrocarbons, thioureas and other residues from herbicides and pesticides, metal organic species, and various inorganic compounds. The test sample may be in solid or liquid form. Three different detectors—ultraviolet and visible (UV/VIS), fluorescence, and electrochemical—are used in the HPLC method. These detectors can identify trace quantities of selected compounds in the 10^{-9} to 10^{-12} range.

LIMITING FACTORS

The usefulness of the results of chemical analysis is directly related to the accuracy of sampling. The samples taken must be representative of the mass being sampled. Care should be taken to ensure that samples are not contaminated during sampling, storage, transportation to the laboratory, sample preparation, or handling prior to analysis. As far as possible, the constituents to be analyzed and their quantitation limits should be decided before beginning the sampling and analysis program; it is usually easier to confirm the presence of a *suspected* chemical than it is to identify a completely *unknown* chemical. For this reason, any available information about the site where the sample was taken, as well as the site history, should be provided to the personnel performing sample analyses. Analytical instruments have a lower limit of quantitation, below which they can only detect presence or absence of a chemical.

QUALITY ASSURANCE/QUALITY CONTROL

The main purpose of quality assurance (QA) is to ensure that all data obtained from analysis of air, soil, drinking water, wastewater, hazardous waste, and sludge are scientifically valid, defensible, and of known precision and accuracy. To achieve this, an analytical laboratory develops a QA procedure and a QA plan. The QA plan is management's tool for achieving QA goals. Quality assurance is the item of highest priority in analysis of environmental samples.

The QA plan is an orderly compilation of detailed procedures designed to produce data of the desired quality for a specific waste management project. The QA plan contains detailed information on quality control (QC) of the samples at all stages—field collection, homogenization, stabilization, containerization, transportation to the lab, storage, sample preparation for analysis, analytical procedures used, documentation, and reporting of the results. The plan also addresses the objectives, management structure, responsibilities of personnel, procedures for data generation, reduction, validation, storage, assessment of the quality of data produced, quality control checks and frequency, performance reviews, and corrective actions.

To ensure QC, it is very common to introduce controlled samples in a batch of samples to be sent for analysis to validate various procedures used in the QC program and to ensure accuracy of analysis and data. EPA recommends the following minimum number of samples for QC:

- Field duplicate—one per day per matrix type sampled. Matrix is the medium, i.e., soil, rock, air, surface water, groundwater, biota, or artifact.
- Equipment rinsate—one per day per matrix type. An equipment rinsate represents a sample of analyte-free medium which has been used for rinsing the sampling equipment. The sample is collected after completion of decontamination and prior to sampling. The blank is useful in documenting adequate decontamination of the sampling equipment.
- Trip blank—one per day (for volatile organic compounds only). The trip blank consists of a sample of analyte-free medium taken from the lab to the sampling site and returned to the lab unopened; it is used to document contamination attributable to shipping and field handling procedures.
- Matrix spike—one per batch (20 samples) per matrix type. Matrix spike is an aliquot of sample spiked with a known concentration of target

analyte(s). Such samples are used to document the bias of a method in a given sample matrix.

- Matrix duplicate or matrix spike duplicate—one per batch. These are intralaboratory samples spiked with identical concentrations of target analyte(s). Such samples are used to document the precision and bias of a method in a given sample matrix.

In addition to the samples above, method blanks and control samples are also used for QC. A method blank is a sample of an analyte-free matrix to which all reagents are added in the same volumes or proportions used in sample processing. Method blanks are used to document contamination resulting from analytical procedures. A control sample is a sample introduced into a process to monitor performance of the system.

STUDY QUESTIONS

1. What is the importance of chemical analysis in hazardous waste management?
2. What factors led to the development of sophisticated and accurate analytical equipment? What are the detection limits of modern equipment?
3. What are GC and GC/MS? Compare and contrast them.
4. Which method can be used for the detection of dioxins? What is the detection limit of this method?
5. What is quality assurance? Why is it important in waste management? How does a quality assurance plan help in achieving the desired goal?

REFERENCES CITED

Johnson, L.D., and R.H. James, 1989, Sampling and analysis of hazardous wastes *in* H.M. Freeman, ed., Standard Handbook of Hazardous Waste Treatment and Disposal: New York, McGraw-Hill, Inc., p. 13.3–13.44.

U.S. Environmental Protection Agency, 1986, Test Methods for Evaluating Solid Wastes (3d ed.), Vol 1A and 1B, Laboratory Manual, Physical/Chemical Methods: Washington, D.C., Report No. SW-846, Office of Solid Waste and Emergency Response, 4 chapters plus appendices, not sequentially paginated.

Hazardous Waste Site Selection and Assessment

INTRODUCTION

This chapter is a discussion of geologic and related factors that play a major role in selection of sites for disposal of hazardous waste and assessment of abandoned hazardous waste sites. Geologic processes are responsible for the formation, occurrence, and composition of earth materials, and also influence the important physical and engineering properties of earth materials, such as permeability, strength, workability, and overall engineering performance. Because of the intimate relationships between earth materials and their geologic origins, it is critical that the geologic aspects of a site be thoroughly evaluated in order to select the most suitable site for the intended use.

Basic geologic principles used in site selection, including geologic modeling and use of geophysical methods in hazardous waste management, are discussed. Sources of geologic and related earth sciences information useful in site selection are provided, as is a brief discussion of drilling and sampling.

GEOLOGIC MODELING

A common saying among geotechnical professionals is that "each site is different and no two sites are alike." The implication is that there cannot be a set of rules-of-thumb to follow in the site-selection process, especially the field investigation part, because of inherent variation in the nature and characteristics of geologic materials and processes. Much disappointment and waste of money can be avoided by developing a geologic model of the site based on the information that is gathered during the office phase of the site-selection process and revised as more data become available from field investigations and/or laboratory testing.

Geologic modeling is an exercise in developing a conceptual three-dimensional picture of the geologic conditions likely to occur at the proposed site. This seemingly simple task is highly complicated and requires superior geologic knowledge, skills, and experience. Unlike mathematical modeling, geologic models are semiquantitative to qualitative due to the extreme variability in geologic conditions and processes, which results in non-homogeneity and anisotropism of the earth materials. Despite these limitations, geologic modeling is very useful during both the site-selection and remediation stages, and should always precede the final design.

Elements of a Geologic Model

A geologic model should be based on the nature of the project and the site geology.

Nature of the Project

Site investigation must be designed in relation to the specific project; each site has inherent constraints and intrinsic capabilities that make it suitable for one project but unsuitable for another. Modeling studies and site investigations should aim at determining which inherent site features are favorable for the particular project and which are not.

Site Geology

The aspects of site geology that must be considered in the development of the geologic model include

- Geologic *materials* occurring both at the surface and in the subsurface: sediments (engineering soils) and rock formations (their depositional environment and history of formation), stratigraphy, and geologic structures.

- Past geologic *history* of the site at both the local and regional scales. Special consideration should be given to the geomorphic processes that operated during the Quaternary Period (last two million years) and the resulting landforms and deposits: buried valleys, channeled bedrock surfaces (paleochannels), influence of past climatic conditions on the geologic materials (weathering and erosion), and the landscape (karst features).

- *Hazardous processes*, such as landslides, flooding, active faults, earthquakes, subsidence, volcanic activity, and coastal erosion, at or near the proposed site. Not all hazards will occur at one place; some locations may be prone to several of these hazards, while others may be susceptible to just one or two. In the Midwestern United States, for example, one would consider flooding, landslides, and subsidence as major hazards, but in the western United States major emphasis would be on earthquakes, faults, landslides, and volcanic activity; coastal hazards are a major consideration in coastal states and in states surrounding the Great Lakes.

- *Existing physical environment* of the area and potential impact from the proposed project is a critical aspect of any large construction project; knowledge of this aspect helps in the preparation of an environmental impact statement (EIS), which is generally required for major projects.

The importance of geologic information is illustrated in the following case history.

Case History: *Conservation Chemical Company*

The Conservation Chemical Company (CCC) site is located on the west bank of the Missouri River in Kansas City, Missouri, in the floodplain and on the river side of a flood control levee (Figure 11–1). The site, about 2.4 ha (6 ac), was used between 1960 and 1980 for storage, treatment, and disposal of a variety of hazardous wastes, including caustic metal sludges, finishing solutions, liquid and solid cyanides, organic solvents, halogenated organic compounds, elemental phosphorus, and pesticides and herbicides. Six unlined sludge mixing and storage basins received an estimated 24,500 m³ (32,000 yd³) of unstabilized sludges during the 20-year period the facility was in operation (Hasan et al., 1988). It is interesting to note that at the time of active operations (1960–1980), CCC was granted a permit to operate the facility; in fact, it was openly coordinating its operations with the Missouri Department of Natural Resources and other regulatory agencies. The new environmental laws, RCRA and CERCLA, do not consider conformity to prior existing laws a defense, and

FIGURE 11–1   Area and vicinity map showing location of the Conservation Chemical Company (CCC) site

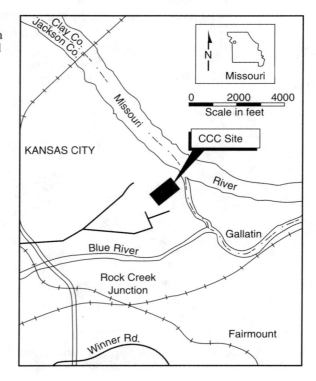

may be viewed as harsh and draconian. Nevertheless, CERCLA has consistently been upheld retroactively in the courts (O'Reilly, 1992).

Hazardous wastes were disposed of at depths of as much as 6.7 m (22 ft) below the ground surface, which had itself been graded to elevations of 3.4 to 4.6 m (11–15 ft) above the floodplain. Earthworks and fill were constructed of the uppermost silty floodplain soils around the site.

The Environmental Protection Agency began a remedial investigation in March 1979 to determine the nature and extent of contamination and the rate and direction of groundwater flow, and to develop the most suitable remediation plan. Extensive investigations were conducted to study the subsurface geology and hydrogeology at the site. These studies revealed the presence of a weathered bedrock (shale) surface (Figure 11–2) with scour channels, overlain by 27.4–29 m (90–95 ft) of alluvial materials comprising two distinct layers of sediments. The lower sediments are generally coarse—sand and gravel with boulders—while the upper layer consists of finer particles—clay and silt with fine to medium sands. This layering of the sediments reflects the depositional history of the valley. A wide valley was carved out by the Pleistocene glaciers over the bedrock surface, which had been weathered and channelized earlier. Part of the glaciated valley was filled with coarse sediments during the retreat of the ice sheet. The present-day Missouri River carved its valley in the glacial sediments.

The valley fills constitute highly productive aquifers. A well located within 610 m (2000 ft) of the site had a yield of about 1892 m³/day (500,000 gpd). The water table is shallow, from 1.5 m to 4.6 m (5–15 ft) below ground surface. Water-level data indicate that, for the area south of the river, groundwater discharges to the river, but during periods when the river is high, groundwater flows from the river into the aquifer.

Geologically, the CCC site is situated on the floodplain of a river flowing on a filled glaciated valley. The upper sediments represent the hydraulic characteristics

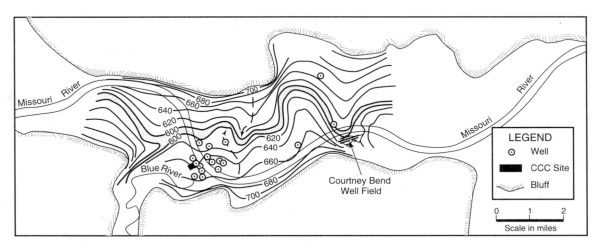

FIGURE 11–2 Structure contour map on top of the bedrock (Pleasanton Group) showing paleochannel (dashed line with arrows) (adapted from Simms, 1975)

of the present river, and the lower sediments reflect the hydraulic regimen of the Pleistocene drainage.

Various remediation alternatives were evaluated including (a) construction of a slurry wall cutoff keyed into the unweathered shale at depths of 24.4 m to 30.5 m (80–100 ft), and pumping the contaminated groundwater at rates ranging from 0.038 m³/min to 0.38 m³/min (10–100 gpm) to create a negative gradient to ensure that the groundwater level inside the containment wall remained lower than outside the wall, and (b) extraction of contaminated groundwater by pumping, and its subsequent treatment. The first option was eliminated because of the occurrence of high-permeability sediments in the lower parts of the valley and of scour channels in the shale. In 1987, the EPA selected the pump-and-treat system, without confinement structure, because it was determined to be technically feasible, capable of swift application, cost effective, permanent, and protective of human health (U.S. EPA, 1987). Estimated cost of remediation is $20 million and will take 20 years.

THE SITE-SELECTION PROCESS

Site selection involves evaluation of all aspects of the project's physical environment in relation to the proposed project. It is important that *all* relevant features at the proposed site be fully evaluated because each site is different in terms of its geology and environmental characteristics. These characteristics may be favorable for one project, but totally unfavorable for another. For example, one of the important considerations in selecting a site for a sanitary landfill is the permeability of the earth material: A geologic material of low to very low permeability, such as clay or shale, is most suitable. On the other hand, the same clay or shale would be undesirable for a project involving construction of a subdivision with a septic tank absorption field for each home. In this case, the important consideration is availability of an earth material with good permeability, such as sandy soil or sandstone.

The following sections discuss the basic geologic principles and approach used in site selection. The information presented here constitutes general guidelines for all geologic/geotechnical site selection, and should always be included in any site-selection process. However, depending on the nature of a particular project, additional factors may have to be evaluated. For example, the site-selection process for a hazardous waste disposal facility must take into account various regulatory details in addition to the required geologic and engineering factors.

General Considerations

Site selection is an organized process that follows a logical sequence. Two important steps are office study and data collection and field investigations.

Office Study and Data Collection

Office study is an important prerequisite of the site-selection process. Valuable information can be acquired by reviewing pertinent reports and documents. During this phase of the study, all available data for the potential site should be gathered and carefully analyzed. Such data may be obtained from geologic reports and maps, topographic maps, air photos, soil survey reports (county soil survey reports prepared by the Soil Conservation Service of the U.S. Department of Agriculture are very useful at this stage of the project), climatic data and reports, engineering surveys and reports, and water supply reports (U.S. Geological Survey water supply reports are a valuable source of information on both the surface water and the groundwater in an area). The type of data available from various sources is discussed later in this chapter. It should be noted that this stage of the site-selection process does not require a site visit; in fact, all studies are done in the office, which is far more economical than conducting studies in the field. After the above information has been analyzed in relation to the given project, two to three potential sites should be tentatively selected and rank-ordered for further studies.

At the completion of this stage, the site or sites that appear to be more promising and that possess features and characteristics that are favorable for the proposed project can be targeted, and sites that have negative attributes can be eliminated. This simple and relatively inexpensive process allows the investigator to initiate detailed field investigations at promising sites with generally favorable attributes, thus maximizing the effort. At the same time, elimination of sites with unfavorable characteristics means that time and effort will not be expended, only to find out later that the site is unfavorable. *Remember:* It is far less expensive to make such decisions in the office than in the field.

Field Investigations

Armed with the information gathered during the office study phase, investigators can then visit the site. The goal at this step of site selection is to study the features as they occur at the site, and to obtain maximum information at minimum cost. This presents a challenge to the geologist, because there can never be enough money or time to get all the information. Budgets for such studies are limited, which means that every aspect of field studies must be carefully planned and executed. The training and experience of the individual play a very important part in the process and help to optimize the time and effort spent on such studies.

During the field visit, information should be gathered from *all* potentially suitable sites, not just the site that appeared most suitable during the office phase of the study. The rationale for this approach is that by focusing on one site to the exclusion of other potential sites, one runs the risk of dropping these sites from further consideration—a move that may later prove wrong. In some cases, the best site may have to be dropped because of problems unrelated to the physical characteristics; such problems may include legal difficulties in acquisition of the land, unexpected discovery of archeological significance, environmental constraint, and the like. In addition, it takes very little extra effort to gather data on the other potential sites while out in the field. Mobilization of field crew and equipment is expensive, so it makes sense to plan site investigations to get as much information as possible in order to avoid subsequent trips.

One of the objectives during the field investigations is to identify potential problems associated with the site and to determine how they may be corrected during later stages of project development.

Collection of soil, rock, and water samples, both from surface exposures and from the subsurface, should be done during the field visit. Typically, mobile subsurface sampling equipment is used to obtain samples from below the ground surface. It is essential that all samples be properly labeled, handled, and stored from the time they are collected in the field to the time they are studied and analyzed in the laboratory. Soil samples (for geotechnical studies) and water samples (for geochemical analyses) are particularly susceptible to changes during storage and transportation; proper care must be taken to prevent or minimize such alterations.

Defining Purpose and Scope

Any geologic investigation must be preceded by a careful statement of the purpose and a clear definition of the scope of the work. This must be done before the first person goes out in the field to commence the investigations. Unfortunately, this simple step is often bypassed, resulting in either the collection of too much data or missing important data. It is best to write down the purpose and scope of the work and to develop a plan of investigation based on this. Such written documentation helps keep the investigation focused and ensures that all necessary data are gathered. The following example illustrates the importance of defining the purpose and scope of an investigation.

Example. Granitic bedrock occurs below a shallow cover of alluvial materials at an abandoned hazardous waste site. Though not prolific aquifers, crystalline rocks, like granite, are known to contain enough groundwater to meet local demands for water supply. The goal is to determine the potential migration paths of the contaminated water leaching out of the hazardous waste dump. Assuming that complete characterization of the nature and concentration of chemicals, and of their toxicity and risk to environment and lives, has been done, the purpose and scope of the investigation can be stated thus:

The investigation is designed to assess the flow paths of groundwater through granitic rocks. The information will be gathered from a series of samples drawn from the subsurface and obtained by drilling, and from evaluation of the earth materials as seen in the drill cores or cuttings and the exposed bedrock occurring in the vicinity of the dump.

Having defined the purpose and scope, the investigative team will focus on these tasks and aim at obtaining all relevant data at minimal cost. Since granite does not possess any primary porosity and permeability, the study will concentrate on joints and fractures as the main flow paths in the rock. It is also well known that three to four sets of joints and fractures are common in crystalline rocks such as granite. Also, the frequency, aperture, and the nature of infilling material will be very significant in the movement of contaminated liquids through the rock. Based on these simple considerations, a checklist and a data recording sheet (Figure 11–3) can be designed that will ensure that all pertinent data on joints and fractures have been studied and recorded, both from the drill cores and the exposed bedrock, and that none of the important information has been missed.

A. *Discontinuities*
1. **Number of discontinuities** . . . .
   Description
   Set 1 . . . .
   Set 2 . . . .
   Set 3 . . . .
   Set 4 . . . .
   Set 5 . . . .

2. **Structural attitudes of discontinuities**

Set	Strike	Dip
1	. . . . .	. . .
2	. . . . .	. . .
3	. . . . .	. . .
4	. . . . .	. . .
5	. . . . .	. . .

3. **Spacing of discontinuities**

	Set 1	Set 2	Set 3	Set 4	Set 5
Very wide (> 2 m)	. . . . .	. . . . .	. . . . .	. . . . .	. . . . .
Wide (0.6–2 m)	. . . . .	. . . . .	. . . . .	. . . . .	. . . . .
Moderate (0.2–0.6 m)	. . . . .	. . . . .	. . . . .	. . . . .	. . . . .
Close (0.06–0.2 m)	. . . . .	. . . . .	. . . . .	. . . . .	. . . . .
Very close (<0.06 m)	. . . . .	. . . . .	. . . . .	. . . . .	. . . . .

4. **Continuity (persistence) of discontinuities**

	Set 1	Set 2	Set 3	Set 4	Set 5
Very low (<1 m)	. . . . .	. . . . .	. . . . .	. . . . .	. . . . .
Low (1–3 m)	. . . . .	. . . . .	. . . . .	. . . . .	. . . . .
Medium (3–10 m)	. . . . .	. . . . .	. . . . .	. . . . .	. . . . .
High (10–20 m)	. . . . .	. . . . .	. . . . .	. . . . .	. . . . .
Very high (>20 m)	. . . . .	. . . . .	. . . . .	. . . . .	. . . . .

5. **Discontinuity aperture (opening)**

	Set 1	Set 2	Set 3	Set 4	Set 5
Very tight (<0.1 mm)	. . . . .	. . . . .	. . . . .	. . . . .	. . . . .
Tight (0.1–0.5 mm)	. . . . .	. . . . .	. . . . .	. . . . .	. . . . .
Moderately open (0.5–2.5 mm)	. . . . .	. . . . .	. . . . .	. . . . .	. . . . .
Open (2.5–10 mm)	. . . . .	. . . . .	. . . . .	. . . . .	. . . . .
Very wide (>10 mm)	. . . . .	. . . . .	. . . . .	. . . . .	. . . . .

6. **Nature of infilling material**

	Set 1	Set 2	Set 3	Set 4	Set 5
Type*	. . . . .	. . . . .	. . . . .	. . . . .	. . . . .
Moisture condition**	. . . . .	. . . . .	. . . . .		

* secondary mineral/surface soil/water/air/combination
** flowing/seepage/drip/wet/moist/dry

7. **Condition of wall rock**

	Set 1	Set 2	Set 3	Set 4	Set 5
Unweathered	. . . . .	. . . . .	. . . . .	. . . . .	. . . . .
Slightly weathered	. . . . .	. . . . .	. . . . .	. . . . .	. . . . .
Moderately weathered	. . . . .	. . . . .	. . . . .	. . . . .	. . . . .
Highly weathered	. . . . .	. . . . .	. . . . .	. . . . .	. . . . .
Completely weathered	. . . . .	. . . . .	. . . . .	. . . . .	. . . . .
Residual soil	. . . . .	. . . . .	. . . . .	. . . . .	. . . . .

B. *Drill Core Quality (RQD)*

Excellent	90–100%
Good	75–90%
Fair	50–75%
Poor	25–50%
Very poor	<25%

C. *Remarks*
   (Record unusual odor, coloration, or any other pertinent information)

FIGURE 11–3 Data sheet for rock discontinuities records

Project Planning

Site selection is an expensive and complex process. It is therefore essential to study *every* aspect of the process and to develop a sound project plan. Such a plan will not only include the geotechnical details, but will also consider personnel requirements, health and safety considerations, legal compliance (there are many legal requirements in the area of hazardous waste management), financial aspects, and the like. All these considerations require input from a number of experts and the availability of a skilled manager who can put all these considerations into a workable and cost-effective plan that will guarantee success. Such people are rare; experienced managers with a good understanding of the technical aspects of hazardous waste management are in great demand and command high salaries.

SITE SELECTION FOR A HAZARDOUS WASTE DISPOSAL FACILITY

General Aspects

The process of site selection for a hazardous waste land disposal facility is highly complex because of

- Strict governmental requirements at all levels—federal, state, and local
- Multidisciplinary nature of the process involves close coordination among various experts, such as geologists, geotechnical engineers, chemists, geophysicists, environmental engineers, toxicologists, and others. The different backgrounds of these experts make the task of open and clear communication very challenging.
- Unique nature of each site and of hazardous waste requires that every aspect be thoroughly investigated

Investigations for a hazardous waste site can be likened to a dam site investigation, because both require full attention to details from beginning to end. Also, most engineering construction projects require satisfactory performance for 30 to 100 years or so, but secure landfills are expected to perform satisfactorily for hundreds of years. Secure disposal sites have only been in existence for 35 years, so long-term performance records are lacking.

In addition to the technical considerations, related factors have also to be evaluated, including

- Proximity to waste generators
- Size of the facility
- Availability of large parcels of land
- Population-centers distribution
- Potential for obtaining regulatory approval
- Zoning

These items are usually evaluated by the owner of the facility, and may not be a concern of the technical team. Knowledge of these factors is desirable, however, as it will help avoid any unexpected surprises at later stages of site investigation.

Design Considerations

The EPA, under RCRA, has set guidelines for location of hazardous waste disposal facilities. Details are included in 40 CFR Part 264 (1990). Special consideration should be given to the aspects described in the following paragraphs.

Location

Hazardous waste facilities cannot be located within 61 m (200 ft) of an active fault, defined as a fault that has had movement in the Holocene (past 10,000 years).

In general, sites should not be located on a 100-year floodplain, except if it can be demonstrated that the waste can be removed safely before flood water reaches the facility, or that no adverse environmental effects will occur as a result of washout (defined as the movement of hazardous wastes from the facility due to flooding).

Noncontainerized hazardous wastes cannot be placed in mines or salt caverns.

Groundwater Protection Standards

Two or more barriers must be constructed at the disposal facility in addition to a leachate and landfill gas management system. Monitoring wells should be installed to check the quality of groundwater; these wells should be located both upgradient and downgradient from the facility. The rate of flow of groundwater in the uppermost aquifer must be measured annually by the owner of the facility.

Groundwater is considered contaminated if it exceeds the concentrations listed in Table 11–1. If a "statistically significant" increase in hazardous waste constituents is detected downgradient and is ruled to be affecting the groundwater, compliance monitoring will be required. This means that all monitoring

TABLE 11–1   Groundwater Quality Standards
(adapted from 40 CFR, 1990)

Constituent	Maximum Limit (mg/L)
Arsenic	0.05
Barium	1.0
Cadmium	0.01
Chromium	0.05
Lead	0.05
Mercury	0.002
Selenium	0.01
Silver	0.05
Endrin	0.0002
Lindane	0.004
Methoxychlor	0.1
Toxaphene	0.005
2,4-D	0.1
2,4,5-TP-Silvex	0.01

wells must be tested for more than 200 hazardous waste constituents (listed in Appendix I). The EPA has specified Permitted Quantitation Limits (PQL) for each of these constituents, as well as the method to be used for their detection and analysis (40 CFR, 1990).

Other Design Requirements

A secure landfill must meet the following design requirements:

- Two or more liners should be installed in a secure landfill. The lower liner must be at least 0.91 m (3 ft) thick, of an earth material having a permeability of $\leq 1 \times 10^{-7}$ cm/s
- Permeability of the leachate collection system should be $\geq 1 \times 10^{-2}$ cm/s
- Leachate depth over the liner should not exceed 0.3 m (1 ft)
- Final cover must have a permeability less than or equal to that of the bottom liner
- The facility must have run-on and runoff management systems to collect and control the water resulting from a 24-hour 25-year storm
- Run-on control should be designed to prevent flow of a 25-year storm from entering the active portion of the landfill
- Design should also include safeguards against side slope failure, excessive settlement that may lead to cracking, and the like

After the passage of the last phase of the Hazardous and Solid Waste Amendments (HSWA, 1984), the rule known as the Land Disposal Restrictions (LDR) banned hazardous wastes from landfills as of May 8, 1990. This rule, besides banning certain hazardous wastes from land disposal altogether, requires that no hazardous waste shall be sent to a land disposal facility unless it has been suitably treated and proper notification and certification attesting to the fact have been issued to the appropriate authorities (U.S. EPA, 1991). Details of LDR and the criteria to be used for design of a secure landfill are discussed in Chapter 14.

ABANDONED HAZARDOUS WASTE SITES

General Aspects

Before the passage of RCRA in 1976, large quantities of hazardous wastes were dumped on the land and into water all over the United States. Some were containerized, but the majority were disposed of directly into the environment in an "as is" condition. It was only a matter of time before the adverse effects of such activities began to manifest themselves, posing serious threat to the environment.

The problem of abandoned hazardous waste sites became so serious that legislation had to be enacted at the federal level to set aside funds and set up a mechanism to correct the situation. The Comprehensive Environmental Response, Compensation, and Liability Act (CERCLA) was first passed in December 1980, and had a budget of $1.6 billion. CERCLA, commonly known as

Superfund, was reauthorized in October, 1986, and came to be known as SARA (Superfund Reauthorization and Amendment Act), with a 5-year budget of $8.5 billion. It expired in 1991, but was extended for four years to September 30, 1994, with a budget of $5.1 billion. As of this writing, the reauthorization bill was under consideration in the U.S. Congress.

The EPA maintains a computerized data bank of the nation's abandoned hazardous waste sites; this is called CERCLIS (CERCLA Information System). Based on the degree of severity of the hazard and its potential impact on human health and environment, a priority list—the National Priority List (NPL)—for cleanup was developed and is maintained by the EPA. Funds are allocated to clean up the most critical sites first, and so on. The NPL is constantly updated, and priorities are reassigned if new information or preliminary investigations indicate that the level and type of contamination is greater or lesser than what was earlier determined. The following statistics are noteworthy:

- As of April 4, 1991, there were 34,225 sites on CERCLIS, of which 1189 were on the NPL.
- As of March 31, 1992, there were 35,984 sites on CERCLIS, and as of October 14, 1992, there were 1236 sites on the NPL.
- In December, 1993, there were about 38,000 sites on CERCLIS. As of November 2, 1994, there were 1227 sites on the NPL (65 sites had been cleaned up).

Current updates of this information can be obtained from the NPL/Superfund Hot Line.

Special Considerations

Investigation of an abandoned hazardous waste site includes three important but related aspects:

1. Assessment of the *nature* of the contaminants, i.e., type(s) of chemical(s) and concentration.
2. Assessment of the *extent* of contamination; this includes investigation of the contaminated medium or media not only at the surface, but also in the subsurface, to get a three-dimensional perspective.
3. Assessment of the *risk* to humans, other life forms, and the environment.

These considerations are key elements in any abandoned hazardous waste site evaluation program, because the ultimate choice of the most suitable remediation method depends on them. In addition, the degree of success of the remediation program is a function of how accurately considerations 1 and 2 are determined.

Topographic and geologic maps and reports, along with hydrological data, are very useful; they provide valuable background information for site investigations and development of a remediation plan.

Topography

The location of an abandoned hazardous waste site in relation to the topographic position—upland, bottomland, or depression—helps in understanding contaminant migration paths. Evaluation of flood hazard, wetland considerations, and aquifer contamination are all related to the topography.

Geology and Hydrology

The nature and characteristics of the earth materials provide information on potential migration paths. In general, loose earth materials and porous and permeable rocks allow for relatively easy movement of fluids through them. Consolidated sediments and impermeable rocks generally impede fluid movement, unless they have developed secondary porosity and permeability from fracturing, expansion cracking, and related processes. The location of permeable/impermeable layers with respect to the groundwater table also plays a critical role in the contamination of aquifers. As can be seen from Figure 11–4, in the first case (A), hazardous constituents will rapidly move into the groundwater and contaminate it as a result of direct vertical migration of contaminants. In the second case (B), the aquifer below the shale is protected and most migration will be lateral. In the third case (C), the first aquifer will be contaminated by vertical movement, but the second aquifer will remain uncontaminated. These illustrations explain the relationship of earth materials and aquifer locations to contaminant transport.

The possible modes of migration of hazardous constituents include

- Mobilization with groundwater—depending upon the nature of contaminant, it can be called a reactive contaminant (chemically reacts with groundwater) or non-reactive contaminant (does not react with groundwater)
- Flushing of contaminated materials with surface runoff
- Movement of soil particles, with adsorbed contaminants, by wind action
- Ingestion by biota—birds, rodents, and other animals
- Conveyance by biota

Moylan (1991) discusses the importance of geologic and related features in studies involving assessment and remediation of hazardous waste sites. Table 11–2 gives a summary of these features.

The determination of the nature, concentration, and penetration of the contaminants into the environment is typically done by sampling all suspected media, including air, water, soil, biota, and artifacts.

After objectives have been set and before starting field investigations, it is essential that a thorough review of the historical records relevant to the abandoned hazardous waste site be done. Every effort should be made to seek information on

- Nature of operation conducted at the site
- Materials used in the process
- Nature of wastes and/or the by-products
- Handling, storage, treatment, and disposal of hazardous materials and hazardous wastes
- Past land uses

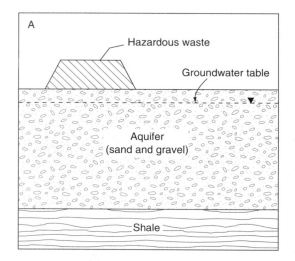

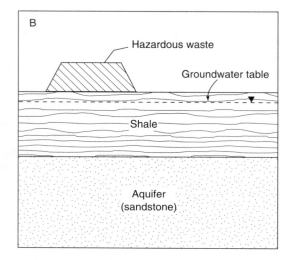

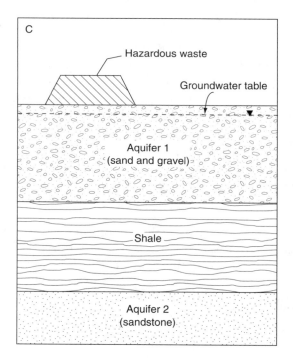

FIGURE 11–4   Geologic control on aquifer contamination

Information on past land use can be obtained from corporate and municipal records, which contain information on change in ownership and land use. Names of the potentially responsible parties (PRPs) may also be available from these records. Personal interviews with area residents may provide detailed information on past land uses and on the history of the site.

TABLE 11–2 Remediation Features (adopted from Moylan, 1991)

		Withdrawal & Injection Wells	Internal Drains	Slurry Walls	Slurry Wall Key Layer	Caps	Chemical Stabilization	Groundwater Treatment	Landfills	Thermal Treatment	Soil Washing	Excavation	Dredging	Vapor Extraction
Site Data	Topographic Surveys	1	1,3	1,3	1,3	1,3		1	1,3	1	1	1,3	1,3,4	1
	Utility Availability	1	1	1				1		1	1			1
	Borrow Availability		2	2		1			1					
	Transportation Network			1	1	1	1	1	1	1	1	1	1	1
Geochemical Data	Multiple Sampling Rounds	1,3,4	1,4	1,4	1			1,3,4						1
	Anion/Cation Analysis	1,3,4	1,4	1	1		1	1,3,4					1,3	
Geotechnical Data	Soil Moisture Content		1	1,3	1	2,3	1,3		1,3	1,3	1,3	1,3	1,3	1,3,4
	Atterberg Limits	1	1,3	1,3	1,3	2,3	1,3		1,3	1	1	1	1	1
	Soil Strength Parameters		2	2		2,3		2	2	2	2	2,3	2	
	Gradations	1,3	1,3	1,3	1,3	1,3	1		2,3	1	1	1	1	1
	Excavatability	1	1	1	1		1			1	1	1	1	
	Landfill Settlement		1,3	1,3		1,3,4								
Hydrogeological Data	Multiple Water Level	1,3,4	1,3	1,3,4	1,4	1	1,3		1,3,4	1	1	1,3		1,3,4
	Detailed Statigraphy	1	1,3	1	1,3		1		1			1,3	1	1
	Secondary Porosity Features	1,3	1,3	1,3	1,3				1,3			1,3		1

*Recommended Times to Collect Data: 1 Limited data in RI/FS phase, greater amount in RD; 2 Data collection begins in RC; 3 During RA; 4 During opperation and maintenance.

Historical Maps and Aerial Photos

Historical maps and aerial photos are very useful in determining land use changes. In addition to historical stereo pairs of aerial photos of a site, the relatively lesser known Sanborn fire insurance maps, first published during the 1860s, are very valuable in determining the history of past land use of a given parcel of land (Colten, 1991). According to Hatheway (1992), the original maps were prepared at scales of 1:600 and 1:1200, and are available from the Sanborn Maps and Information Service. The company charged (in 1992) a fee of $35 to search for the availability of the Sanborn map for any town or city in the United States; for an $85 fee the company will do both the search and, if available, print and supply individual Sanborn map sheets (maximum 10) for the given location. Many public and university libraries also carry the Sanborn fire insurance maps on microfilm. The following case study (Hatheway, 1992) is a good illustration of the use of such maps in tracing the site history in relation to use, storage, treatment, and disposal of hazardous materials.

Case History: *Sanborn Fire Insurance Maps in Site Assessment*

The three maps in Figure 11–5 show the layout of a gas works located in a major city in the U.S. Midwest. These three maps record the history of changes and expansion that took place from the time the plant was set up to its shutdown; in this case, a period of 96 years. The business was established in 1856; Figure 11–5A shows the initial setup and locations of gas tanks and related structures. Figure 11–5B shows the expansion and layout 44 years after the beginning of operations. Finally, two years before closing of operations in 1952 (Figure 11–5C) and after the elapse of half a century, nearly all land within the premises had been developed for storage of hydrocarbons and related materials. To anyone involved in transactions affecting this kind of property, the immense value of such maps cannot be overemphasized. The importance of such maps has been aptly amplified by Hatheway (1992, p. 25): "The maps provide us with locations and outlines of various stand-alone components of industrial facilities, as well as some of the underground utilities servicing the plant site. How fortunate we are to have such fine records of the past nature of facilities that now occupy our efforts in environmental assessments and hazardous waste site remediation."

Sanborn maps should *always* be sought in any study involving environmental audit of sites, evaluation of abandoned hazardous waste sites, or remediation planning.

Historical records usually provide information on

- How, when, and where raw materials were used and stored
- How, when, and where wastes were stored, treated, or disposed

Of course, availability of data depends on the adequacy of historic records. Based on the completeness and accuracy of historic information, a sampling program may be initiated to look for specific chemical elements/compounds in the waste. For example, if the historic records indicate that only electric capacitors and transformers were discarded at the site, the sampling and analytical program

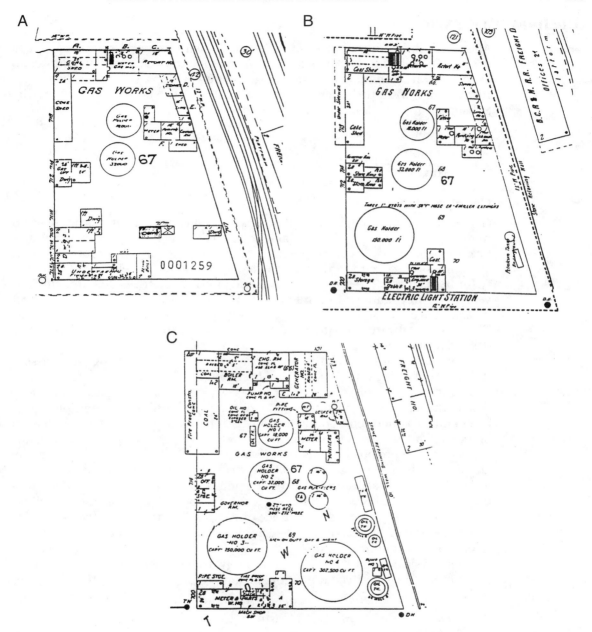

FIGURE 11–5 Sanborn maps of an industrial site in the Midwestern United States (adapted from Hatheway, 1992)

would be designed specifically to determine contamination by PCBs. On the other hand, if historical records indicate that the site was a landfill for industrial wastes, a comprehensive and elaborate sampling and analytical program would be mandatory. This might entail the collection of a large number of soil and water samples and the implementation of hundreds of chemical analyses.

SUBSURFACE INVESTIGATIONS

After the office study and the initial field investigations have been completed, a program of subsurface investigation is developed to determine the nature and extent of contamination. In conjunction with areal maps, subsurface investigations help define the shape, size, and nature of the contaminated media. Subsurface investigations usually include

- Geophysical surveys
- Drilling
- Sampling

Geophysical Surveys

Geophysics relates to the study of the physical and chemical properties of earth materials to interpret the nature and geometry of the geologic formations and hazardous materials occurring in the subsurface. The physical properties that are of importance in hazardous waste management include electric conductivity (or its reciprocal, resistivity), electromagnetic properties, magnetism, and velocity of propagation of seismic waves. All geophysical methods seek to detect measurable contrast in the physical property of the target and its surroundings. These methods are non-invasive and provide in situ measurements of the desired properties.

It must be emphasized that these are indirect methods, designed to assess the extent of contamination, and are based on the interpretation of numerical data obtained from surveys. In order to be certain, physical samples of the materials in the subsurface must be obtained, usually by drilling. The samples must be chemically analyzed to provide information on the extent of contamination of the media. The use of geophysical methods without actual probes of the subsurface cannot provide conclusive information on the nature of the subsurface. Despite this limitation, geophysical surveys afford a relatively inexpensive tool to obtain generalized information on the nature and extent of contamination in the subsurface, and help in selecting locations for final drilling and sampling. Before planning systematic subsurface drilling and sampling at large hazardous waste sites—tens of acres or more—geophysical surveys should always be considered as an economic option.

Geophysical methods that are commonly used in hazardous waste management, either singly or in combination, include

- *Electrical resistivity method.* Measures electrical resistivity
- *Electromagnetic method.* Measures electrical conductivity
- *Ground-penetrating radar (GPR) method.* Measures dielectric constant and conductivity
- *Magnetic method.* Measures magnetic susceptibility
- *Seismic refraction method.* Measures seismic wave velocity

How successfully geophysical methods can be used to get the needed information on hazardous materials depends on (a) the depth at which hazardous materials occur, (b) the degree of physical contrast between the hazardous material and its surroundings, and (c) background interference, ("noise")—a high signal-to-noise ratio produces better data. Geophysical methods have been used to

- Locate buried objects, such as drums, trenches, landfills, and utility lines
- Map contaminant plumes
- Map geologic and hydrogeologic features at sites, including stratigraphy, geologic structures, and related features

A brief discussion of geophysical methods, emphasizing the application rather than the theoretical considerations, is presented in the following sections.

Electrical Resistivity Method

The electrical resistivity method measures the resistance offered by a medium to the flow of electric current (resistivity is the inverse of conductivity). Various factors, such as mineralogy of the earth materials, layer thicknesses, nature of contained fluid, and concentration of ions, control the resistivity values. Temperature, porosity, permeability, water content, and salinity (or ion content) of the geologic formation are physical manifestations of the above factors. In general, dry sands, gravel, and massive, unweathered and nonfractured rocks give high resistivity values; clays, saturated sediments, and weathered and/or porous and permeable rocks yield low resistivity values.

Resistivity survey involves placing electrodes on the ground, sending electric current through them, and then measuring the potential drop between them. This makes the actual survey time demanding.

Electromagnetic Method

The electromagnetic (EM) method measures the electrical conductivity of a subsurface medium. Because conductivity is the inverse of resistivity, it is controlled by the same factors that affect resistivity; for example, low conductivity is generally associated with dry sand, gravel, and massive, unweathered rocks and, high conductivity values are generally associated with clays and weathered and/or fractured rocks. Unlike the resistivity method, the EM method does not require ground contact, which means that rapid data collection is possible—a distinct advantage when a large area is to be covered.

Both the resistivity and the EM methods have been successfully used to (a) detect and map contaminant plumes, (b) estimate vertical and horizontal distribution of earth materials below the surface, (c) evaluate groundwater characteristics, and (d) determine the flow paths of contaminated plumes (Benson et al., 1988).

Ground-Penetrating Radar

Ground-penetrating radar (GPR) is a relatively new method that has been used with some success in hazardous waste detection. GPR uses high-frequency electromagnetic (em) waves to obtain information on subsurface conditions. It involves sending a pulse of high-frequency em waves from the surface and recording the time it takes for them to be reflected back from the subsurface to the receiving antenna. Both the transmitting and receiving antennas can either be pulled behind a vehicle or dragged across the surface to record the data.

GPR surveys provide continuous recording of subsurface conditions producing a profile view of the subsurface with high resolution. GPR's applications in the hazardous waste field can be grouped into two categories: (a) site characterization, and (b) direct detection of subsurface features.

Site characterization provides useful information for mapping geologic and hydrogeologic features, such as occurrence and distribution of layers of sediments (engineering soils), depth to bedrock, and the groundwater table. The GPR method is, however, more commonly utilized for detection of hazardous materials and buried objects. Thus, using the GPR methods, it is possible to map the depths and locations of buried drums and storage tanks containing hazardous substances. The GPR method has also been used to detect pipes and vaults (Fenner and Vendi, 1992) and contaminant plumes in coarse-grained soils (Olhoeft, 1986). Unlike other geophysical methods, GPR is the *only* method that can also detect nonmetallic materials—tanks and drums made of fiberglass for example.

Two variations of GPR are available: (1) the continuous measurement system, and (2) the station measurement system. The continuous system provides a continuous profile with high resolution along level and smooth surfaces (pavement, level ground) of the traverse line, producing a real-time graphic record of the subsurface features that can be displayed on a CRT terminal or on a hardcopy printout.

The station measurement system is suitable for rough terrain and allows for deeper penetration—up to 150 m, compared to the 30-m maximum for the continuous system. GPR techniques have also been successful in locating karst features in carbonate rocks (Benson and La Fountain, 1984).

Seismic Refraction Method

The seismic refraction method involves measurement of the velocity of elastic waves through different media. It measures the travel time of seismic waves refracted from a geologic interface back to the source of seismic energy at the surface. Because P-waves (one of the several seismic waves) travel at different speeds in different geologic media, by determining the travel time of a P-wave from the surface to the geologic interface and back to the surface, the nature of the geologic material and conditions in the subsurface can be interpreted. The two main factors that control seismic wave velocity are the density and the elastic properties of the geologic materials.

In a typical seismic refraction survey, seismic waves are generated by impact, explosives, or projectiles (gun)—the waves are detected by geophones laid on the surface, and are either processed and recorded or displayed on seismographs.

In the hazardous waste field, the seismic refraction method has been used to estimate the depth to bedrock and water table, the thickness of various geologic formations, and the geologic structures, as well as to locate buried wastes (Benson and Yuhr, 1993).

Magnetic Method

The ambient magnetic field of the Earth is affected by presence of ferrous materials. A magnetometer is used to record changes in the intensity of the Earth's magnetic field. In hazardous waste management, the magnetic method can be used to locate buried drums, tanks, and the outlines of pits and trenches containing steel drums. The method has been used to locate large masses of steel/iron drums located at depths of 6 m to 20 m (Gretsky et al., 1990).

All geophysical methods have limitations that must be borne in mind before planning their use. However, by using a combination of more than one method, the results can be optimized. Table 11–3 summarizes the advantages and limitations of various geophysical methods in relation to their use in hazardous waste management.

Drilling

Drilling, or boring, is done to obtain information about the nature of earth materials and groundwater in the subsurface and to obtain samples for laboratory testing. It provides information on (a) the thickness of strata in the subsurface, (b) the nature of earth materials, and (c) the groundwater table. Although the location, spacing, and depth of bore holes vary with the local geology, topography, and size of the proposed project, the general guidelines given in Table 11–4 are helpful in designing an initial drilling program.

Because boring is expensive, it may be more economical to run geophysical surveys to correlate the subsurface features *between* the bore holes at projects covering large areas. In other cases, it is more convenient to excavate exploratory pits or trenches to get information on strata thickness and other characteristics. Pitting and trenching are relatively inexpensive, and should be considered whenever possible.

Data on the groundwater table and aquifer characteristics are collected during the drilling phase. Groundwater characteristics do vary with the season. Therefore, it is very important that more than one groundwater sampling and measurement operation be performed. These should be spread over a period of time to cover seasonal changes in both water levels and groundwater geochemistry. Since protection of groundwater is of paramount importance, long-term monitoring of groundwater table fluctuations, discharges, and chemical characteristics must be carried out.

After all data have been gathered, the next task involves data evaluation. It is very important that all key personnel—geologists, geotechnical engineers, chemists, and others—be present to consider every aspect of the information in order to come up with the best possible interpretation.

Sampling

Sampling of various media—soil/sediments, surface water and groundwater, air, biota, and artifacts—is very common at sites containing, or suspected of containing, hazardous wastes. The objectives of sampling are (a) estimation of the nature, extent, and concentration of contamination in all affected media in all three spatial directions, and (b) determination of the migration path(s)—whether through water (surface or subsurface) or by wind (adsorbed solids). Before commencement of actual sampling, a sampling plan must be developed. Some useful considerations in developing a sampling program include

- Objective(s) of investigation
- Background information on the site

TABLE 11-3 Relative Advantages and Limitations of Geophysical Methods (adapted from Benson and Yuhr, 1993)

| METHOD | PROPERTY MEASURED | Resolution | | Depth (m) | Limitations | |
		Vertical	Horizontal		Geologic	Cultural
Electromagnetic	Electrical conductivity	Fair	Excellent	1.5–61	Low conductivity materials	Metallic objects on surface & buried
Electric Resistivity	Resistivity	Good	Good to Fair	100s of meters	Low resistivity materials	Grounded metallic objects on surface and buried metals
Magnetic	Magnetism	—	Excellent	1.5–10	—	Magnetic objects
Radar	Dielectric constant & conductivity	Excellent	Excellent	1.5–15	Sediments	Surface objects road salts
Seismic	P-wave velocity	Good	Good to Fair	<30	Thin beds overlying sediments	Ground vibrations; concrete, asphalt

TABLE 11–4   General Guidelines for Initial Boring Plan
(Minimum Number)

Size of Facility		Total Number of Borings	Deep Borings
(ha)	**(ac)**		
<4.1	<10	4	1
4.1–19.8	10–49	8	2
20.3–40.1	50–99	14	4
40.5–81.0	100–200	20	5
>81.0	>200	24, plus 1 for each additional 10 acres	6, plus 1 for each additional 10 acres

- Analyses of existing data
- Sampling locations
- Type of sampling device(s)
- Analytes of interest
- Analytical procedures
- Operational plan and work schedule

Sampling Methods

A sample, by definition, is a part of a large mass; therefore, it is extremely important that the sample be representative of the hazardous waste being investigated. For sampling of solid hazardous wastes, three general methods of sampling are available:

1. *Simple random sampling.* Good when hazardous waste is heterogeneous in nature and no information on its chemical makeup is available. Random samples should be taken from several locations within a hazardous waste mass.

2. *Systematic random sampling.* The first batch of samples are chosen randomly, but subsequent samples are taken at fixed space and time intervals (e.g., grid sampling).

3. *Stratified sampling.* When a hazardous waste comprises several piles, each with one general type of waste, separate sampling from each pile is called stratified sampling. This method is useful when more information about the hazardous waste site is available and when waste is not randomly heterogeneous in its chemical nature. If the waste can be divided into different categories based on its chemical constituents (soils with hydrocarbons, soil with heavy metal sludges, etc.), then random or systematic samples can be taken from each stratified (segregated) part of the waste. LNAPL and DNAPL are good examples of stratification in groundwater sampling.

Knowledge of site geology is of critical importance when designing a successful sampling program. The following case history illustrates the importance of geologic factors in determining whether the observed concentrations of inorganic constituents are the result of site contamination or whether they represent the ambient geochemical conditions, as well as their significance in hazardous waste site investigation in general.

Case History: *Sampling Program at Vandenberg Air Force Base*

The Vandenberg Air Force Base (VAFB) occupies nearly 399 km^2 (154 mi^2) of coastal California, and is the site of a variety of military and industrial operations (Figure 11–6). VAFB was particularly active in the development of missile technology, and numerous missiles were launched from the base. Large volumes of water were released during launches to dissipate heat and the destructive forces associated with the launch blast; this deluge water contained metals which also occur naturally in the soil and the groundwater at VAFB. Wastes from plating shops and other facilities at VAFB may also have introduced metals into the soil and groundwater. Approximately 50 potentially contaminated sites were investigated as part of the Air Force's Installation Restoration Program (IRP). Because metals occur naturally, as well as in the waste streams, it was extremely important to establish the background values, reflecting the ambient concentration, for comparison with site data to determine if the sites contained elevated concentrations of inorganic constituents.

The design of the background sampling program was based on the geologic and physiographic settings at VAFB. The main geologic units include unconsolidated sand and alluvium of Quaternary age, and siliceous shales of Miocene age (Figure 11–7). The base comprises three distinct physiographic regions: the Lompoc Terrace, the Burton Mesa, and the San Antonio Terrace (Figure 11–6). In addition, the base contains seven drainage basins that roughly coincide with the groundwater basins. The shallowest groundwater generally occurs in unconsolidated sands of alluvial or eolian origin. The sampling program was designed to provide adequate numbers of soil and groundwater samples from each geologic unit within each physiographic region.

The following factors were considered for obtaining representative background samples:

- Air photos and topographic maps were studied to determine any past land use that might have contributed to contamination. Field visits were made to confirm that the planned sampling sites were in uncontaminated areas. Areas of known contamination were avoided; sampling points were located uphill or downhill from such areas.

- Adequate distance between sampling locations and roadways was maintained to avoid potential contamination from automobile exhaust, hydrocarbons, lead, and other substances. A distance of 61 m (200 ft) was found to be adequate.

- Sampling locations that were evenly distributed throughout the three main physiographic regions where all three geologic units are present were selected.

Soil samples were collected from the surface down to depths of about 30 m (100 ft) using hollow-stem auger and cone penetrometer rigs. At least 10 borings

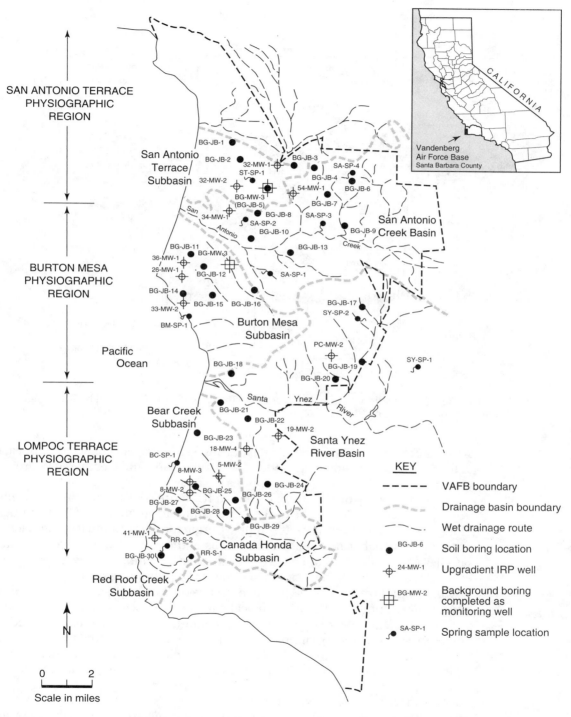

FIGURE 11–6 Groundwater and soil sampling locations, Vandenberg Air Force Base, California (provided by Todd Beatty, Jacobs Engineering Group, Inc., Santa Barbara, CA)

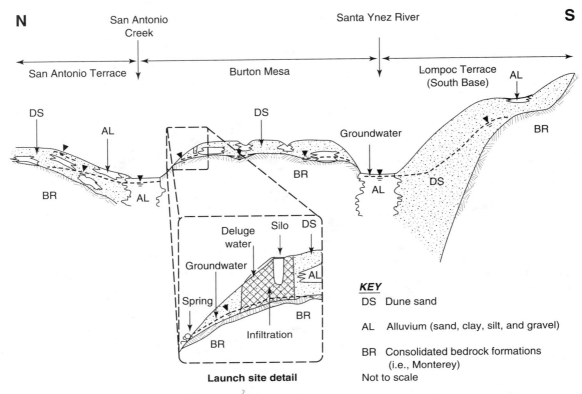

FIGURE 11–7 Geologic cross section, Vandenberg Air Force Base, California (provided by Todd Beatty, Jacobs Engineering Group, Inc., Santa Barbara, CA)

were made in each physiographic region, and samples were collected from each of the three main geologic units (dune sand, alluvium, and weathered bedrock), yielding 104 samples (sampling locations are shown on Figure 11–6).

Groundwater samples were collected from preexisting monitoring wells, from wells installed during the background sampling program, and from springs. The shallowest groundwater was sampled during the background level characterization program, because that was the focus of the site sampling. Groundwater samples were collected from each drainage basin, resulting in 27 water samples. Figure 11–6 also shows the location of groundwater sampling sites.

All samples were analyzed for the presence of aluminum, arsenic, antimony, barium, beryllium, boron, cadmium, calcium, chloride, chromium, cobalt, copper, fluoride, lead, silver, sodium, sulfate, thallium, vanadium, and zinc. In addition to these inorganics, groundwater samples were also analyzed for petroleum hydrocarbons, VOCs, and synthetic organics, including various pesticides and hydrazine. Table 11–5 gives the analytes for surface soil, subsurface soil, and groundwater.

The analytical results were subjected to rigorous statistical analyses, which showed that (a) the background concentration levels were generally lower for surface soil than for subsurface units, and (b) the concentrations of the majority of metals are higher in bedrock than in the unconsolidated alluvium and sand

TABLE 11–5 Background Analytical Program (Data provided by Todd Beatty, Jacobs Engineering Group, Inc., Santa Barbara, CA.)

Sample Type*	Metals**	Common Anions[†]	pH (SW9045)	TOC (SW9060)	TDS (E160.1)	SVOCs (SW8270)	TPH (E418.1)	Other Organic Analyses[‡]
Surface Soil	X	X	X	X		X		
Subsurface Soil	X	X	X					
Groundwater	X	X			X		X	X

*TOC = total organic carbon; SVOCs = semivolatile organic compounds; TDS = total dissolved solids; TPH = total petroleum hydrocarbons.

**In addition to SW6010, samples were analyzed by graphite furnace atomic absorption for Antimony (SW7041), Arsenic (SW7060), Beryllium (SW7091), Total Lead (SW7421), Selenium (SW7740), Silver (SW7761), and Thallium (SW7841). Samples were also analyzed for mercury by method SW7471.

[†]Soil and groundwater samples were analyzed by method E300 for chloride, nitrate-nitrite, and sulfate. Samples were also analyzed by method E340.2 for fluoride.

[‡]Groundwater samples were analyzed for chemicals of concern associated with neighboring sites, which included volatile organic compounds (SW8260), semivolatile organic compounds (SW8270), pesticides (SW8080, SW8140), and hydrazine (ASTM D1385-88).

deposits. This implies that observed concentration should always be compared with the background concentration found in similar geologic units.

The study established background threshold values for the various metals and non-metals present in the geologic units at the site. These values were used to screen analytical data for evidence of contamination. When the concentration of inorganic constituents at an IRP site was found to be lower than the threshold value, the site was considered uncontaminated. On the other hand, if the concentrations detected at a site were greater than the threshold values, the site was suspected of being contaminated and was included in further investigations for remedial action. This approach resulted in elimination of many sites from further investigations for remedial actions, providing significant cost savings.

(Materials for the case history were provided by Todd Beatty, Jacobs Engineering Group, Inc., Santa Barbara, CA.)

Error Minimization and Quality Control During Sampling

Sampling at an abandoned hazardous waste site usually generates a large volume of data. A quality control procedure (QCP) should be adopted to ensure that errors are minimized. The QCP should include information on

- Sampling method(s) to be used
- Sample handling
- Calibration of field equipment
- Decontamination of equipment
- Proper documentation (chain of custody document)
- Accurate recordkeeping
- Proper training of field crew

A standard form can be designed to maintain proper records and to establish the chain of custody of samples. This form should also contain: required analyses; type of media sampled—water, air, soil, biota, artifact; sampling method used; sampling location; date and time of sampling; and name of responsible persons. Proper documentation is vital, as the infomation may be used in legal disputes. Therefore, the chain of custody must be clearly established. Signatures should be obtained whenever samples pass hands. A sample chain of custody form is shown in Figure 11–8.

Details of quality control and quality assurance are discussed in Chapter 10.

SOURCES OF GEOLOGIC AND RELATED INFORMATION

Geologic Information

Maps and reports published by the U.S. Geological Survey (USGS) and state geological surveys (see Appendix J for names and addresses of federal and state geological surveys) are good sources of geologic information. If a published map or report is not available, chances are that the information will be available in open-file reports and may still be accessible. Write or call these agencies for information.

Project				Sampled by:						
Station Number	Station Location	Date	Time	Sample Type					Number of Containers	Analysis Required
				Water		Air	Soil			
				Comp (C)	Grab (G)		C	G		

Relinquished by: (Signature)		Received by: (Signature)		Date	Time
Relinquished by: (Signature)		Received by: (Signature)			
Relinquished by: (Signature)		Received by: (Signature)			
Relinquished by: (Signature)		Received by: (Signature)			
Dispatched by: (Signature)	Date Time	Received for Laboratory by:			
Method of Shipment					

FIGURE 11–8   Sampling record and chain-of-custody document

Geology/Geosciences departments at local universities are also valuable sources of geologic information (generally available at no cost).

Topographic Maps

The USGS is responsible for the preparation and updating of topographic maps of the country. These maps are generally drawn on fairly large scale, 1 in. = 2000 ft

(1:24,000), and are available from the USGS and their authorized agents. Many bookstores and government document sale outlets carry these maps. They are inexpensive and contain valuable information on topography, slope, and cultural features.

Aerial Photos

The Soil Conservation Service (SCS) of the U.S. Department of Agriculture (USDA) and the U.S. Forest Service have been making aerial photos of the country for a long time. Black-and-white aerial photos dating back to the late 1930s (in 9-inch format and at scales of 1:12,000 or 1:80,000) are available from these agencies at a reasonable cost. Study of historical photographs provides information on land use changes over time, which, in many cases, gives clues to abandoned hazardous waste dumps (Lyon, 1982). Older aerial photos can be obtained from the National Archives, GSA, Cartographic Division.

Satellite Imagery

Remote sensing data include high-altitude aerial photography and satellite imagery. These are available at scales varying between 1:1,000,000 and 1:250,000. They provide a synoptic view of the landscape and are very useful in delineating large-scale features. Satellite images and high-altitude aerial photos can be obtained from the EROS Data Center.

County Soil Survey Reports

In 1896, the U.S. Congress authorized funds to carry out soil surveys of the country. The first soil survey report was published in 1899. Since that time, several hundred soil survey reports have been published. Soil surveys for other counties are underway, and reports are being prepared.

Soil survey reports are produced by the Soil Conservation Service (SCS) of the U.S. Department of Agriculture. Each report generally covers one county, and can be obtained from the county office of the SCS. These reports contain useful information on physical, chemical, hydrologic, and engineering properties of soils, such as

- Thickness of soil profile and nature of parent material
- Topographic position
- Percent of slope
- Grain size (USCS classification)
- Atterberg limits
- pH characteristics and corrosion potential
- Frost-heave potential
- Shrink-swell potential
- Flooding potential
- Seasonal high-water tables
- Permeability
- Erodibility

Soil survey reports are very useful in the initial phase of project planning and in the design of the field investigation program. Prokopovich (1984) and Hasan (1994) have discussed the importance of soil survey reports in geotechnical projects. The information is also applicable to hazardous waste site evaluation and remediation studies.

Other Sources

Many national and state agencies may be contacted for procurement of geologic and other information. Appendix J contains addresses and telephone numbers of organizations that can provide geologic and related information.

STUDY QUESTIONS

1. What is a geologic model? What aspects of site geology should be incorporated into the model? Why is it important to relate the model to the nature of the project?
2. What are the two main phases of the geologic site-selection process? What are the sources of geologic and hydrogeologic information?
3. Why is it important to define the purpose and scope of a field investigation project? Illustrate your answer by using the example of site selection for a secure landfill.
4. What factors make the process of site selection for a hazardous waste facility a complex undertaking? Besides the technical factors, what other factors should be included in site selection for such facilities?
5. What site-selection criteria have been set by the EPA for hazardous waste sites?
6. What special considerations should be included in investigation of an abandoned hazardous waste site? How can the review of historical records help in such investigations?

7. What is the importance of aerial photos and the Sanborn fire insurance maps in planning an investigation program for an abandoned hazardous waste site?
8. What subsurface methods may be used for site investigations? In what ways are geophysical surveys useful in investigations of (abandoned) hazardous waste sites?
9. Briefly explain the difference between the various geophysical surveys. Which method(s) would you use for a hazardous waste site where a large number of 55-gallon steel drums are known to have been buried for long a time? Why?
10. What special care should be taken when designing a sampling program for an abandoned hazardous waste site? What is the chain-of-custody concept in sample handling?
11. What features should be included in ensuring quality control for sampling at a hazardous waste site?

REFERENCES CITED

40 Code of Federal Regulations, Part 264, July 1, 1990, Appendix IX: Washington, D.C., U.S. Government Printing Office, p. 302–308.

Benson, R.C., and L. La Fountain, 1984, Evaluation of subsidence or collapse potential due to subsurface cavities, *in* B.F. Beck, ed., Proceedings of the First Multidisciplinary Conference on Sinkholes, Orlando, Florida, October, 1984: Rotterdam/Boston, A. Balkema, Publishers, p. 201–215.

Benson, R.C., and L.M. Yuhr, 1993, Application of Surface Geophysics to Site Characterization (Short Course Notes), March 23–26, Tampa, FL: Ohio Environmental Education Enterprises, 715 p.

Benson, R.C., M. Turner, P. Turner, and W. Vogelson, 1988, In situ time-series measurements for long-term ground-water monitoring, *in* A. G. Collins, and A.I. Johnson, eds., Ground Water Contamination: Field Methods,

Philadelphia, American Society for Testing and Materials, ASTM STP-963, p. 58–72

Colten, C.C., 1991, The Illinois Sanborn Geographic Information System, Chicago, Proceedings, 34th Annual Meeting, Association of Engineering Geologists, October, 1991, p. 287–295.

Fenner, T.J., and M.A. Vendi, 1992, Ground penetrating radar for hazardous waste site investigations, University of Edinburgh, U.K., Proceedings: Polluted and Marginal Lands Conference, p. 107–114.

Gretsky, P., R. Barbour, and G.S. Asimenios, 1990, Geophysics, pit surveys reduce uncertainty: Pollution Engineering, vol. 22, p. 102–108.

Hasan, S.E., 1994, Use of soil survey reports in geotechnical projects: Bulletin of the Association of Engineering Geologists, vol. 32, no. 2, p. 367–376.

Hasan, S.E., R.L. Moberly, and J.A. Caoile, 1988, Geology of Greater Kansas City, Missouri and Kansas, United States of America: Bulletin of the Association of Engineering Geologists, vol. 25, no. 3, p. 322–323.

Hatheway, A.W., 1992, Perspectives Number 13, Don't Forget the Sanborn Maps, carefully and artfully-compiled historic plat maps of America's cities, 1867–1992: AEG News, v. 35, no. 4, p. 25–27.

Lyon, J.G., 1982, Use of aerial photography and remote sensing in the management of hazardous wastes, *in* T.L. Sweeney, H.G. Bhatt, R.M. Sykes, and O.J. Sproul, eds., Hazardous Waste Management for the 80s: Ann Arbor, MI, Ann Arbor Science, p. 163–171.

Moylan, J.E., 1991, Site characterization data needs for effective RD and RA: Proceedings, Conference on Design and Construction Issues at Hazardous Waste Sites, Dallas, TX: U.S. EPA Report No. EPA/540/8-91/012, May 1991, p. 1103–1109.

Olhoeft, G.R., 1986, Direct detection of hydrocarbon and organic chemicals with ground penetrating radar and complex resistivity: Proceedings, NWWA/API Conference on Petroleum Hydrocarbons and Organic Chemicals in Groundwater-Prevention, Detection and Restoration, November, 1986: Dublin, OH, National Water Well Association, p. 284–305.

O'Reilly, E.O., 1992, Effective use of geotechnical expertise in superfund sites: case studies and a proposal for expanding the role of the "Special Master"; M.S. thesis; Department of Geosciences, University of Missouri, Kansas City, Missouri, 58 p.

Prokopovich, N.P., 1984, Use of agricultural soil survey maps for engineering geological mapping: Bulletin of the Association of Engineering Geologists, vol. 21, no. 4, p. 437–447.

Simms, J.J., 1975, A study of the bedrock valleys of the Kansas and Missouri Rivers in the vicinity of Kansas City: Unpublished M.S. thesis: Department of Geology, University of Kansas, Lawrence, Kansas, 78 p.

U.S. Environmental Protection Agency, 1987, Second Record of Decision: CCC Superfund Site; Washington, D.C., 88 p.

U.S. Environmental Protection Agency, 1991, Land disposal restrictions, summary of requirements; Report No. OSWER 9934.0-1A, 26 p. plus appendices.

SUPPLEMENTAL READING

Freeman, H.M. ed., 1989, Standard Handbook of Hazardous Waste Treatment and Disposal: New York, McGraw-Hill, Inc., 14 chapters plus index.

Johnson, R.B., and J.V. DeGraff, 1988, Principles of Engineering Geology: New York, John Wiley & Sons, Inc., 497 p.

West, T.R., 1995, Geology Applied to Engineering: Englewood Cliffs, NJ, Prentice Hall, Inc., 560 p.

Wentz, C.A., 1989, Hazardous Waste Management: New York, McGraw-Hill Publishing Co., 461 p.

12

Personnel Protection and Safety at Hazardous Waste Sites

INTRODUCTION

The fact that the nature of chemicals and other toxic substances at an uncontrolled hazardous waste site is unknown, and therefore full of potential hazards that pose health and safety risks to the workers, makes the task of site evaluation extremely important. The hazards range from minor to deadly—from a simple fall on uneven ground covered with debris to exposure to phosgene gas that could be life threatening. In order to minimize the risks to personnel, and to ensure that all reasonable precautions are taken, activities at hazardous waste sites are regulated by federal and local agencies. The goal of all regulations is to afford maximum protection to workers by minimizing health and safety risks. Investigation at an uncontrolled hazardous waste site is an extremely risky operation, and all possible precautions must be taken to ensure the safety of personnel at all times.

Potential Hazards

Hazardous waste sites pose a variety of risks that could adversely impact the health and safety of workers. These risks are associated both with the nature of the site and with the work being carried out, and include

- Biologic hazards
- Chemical exposure
- Fire and explosion
- Heat stress
- Ionizing radiation

- Oxygen deficiency
- Noise
- Other safety hazards

The uncontrolled nature of hazardous waste sites and the possibility that a single location may contain hundreds or even thousands of chemicals make hazardous waste site investigations a very risky venture. However, if the guidelines are properly followed and adequate care is taken to ensure personnel safety, even extremely hazardous substances will not endanger the health and safety of the workers. Problems arise when the guidelines are not followed properly and details are overlooked or bypassed. Robotics offers a great potential for reducing risk at hazardous waste sites.

Personnel protection requires use of specialized clothing and equipment, along with elaborate training in the use of equipment and work-site safety, all of which make it a costly item in hazardous waste site investigations. Nevertheless, the high cost is fully justified in view of the risks involved. No shortcuts should ever be taken, and the safety aspects should never be compromised. Compared to the overall cost of cleanup of a contaminated site, the cost of personnel protection is small: generally 5 to 10 percent of the total cost.

A Personnel Safety and Health Plan (PSHP) is the key to ensuring the health and safety of workers at hazardous waste sites. Design of a PSHP is usually done by an industrial hygienist with input from other experts. While the details of how to develop a PSHP are beyond the scope of this book, the main features of such plans are presented in the following sections to provide an overview of this important aspect of hazardous waste management; the regulations that apply to personnel safety, plan development, site characterization, personnel protection equipment (PPE), site control, decontamination procedures, and medical surveillance are discussed.

REGULATIONS

Several federal agencies have developed regulations that relate to the health and safety of personnel likely to be exposed to hazardous substances. Many states have their own laws to regulate activities at hazardous waste sites. Generally, federal laws take precedence over state laws, unless the latter set more stringent standards.

The EPA, under the Superfund Amendments and Reauthorization Act (SARA) of 1986, set forth in Section 126 a revision to and amendment of 29 CFR Part 1910.120 to address hazardous waste operations and emergency response regulations. It is a comprehensive regulation that deals with personnel health and safety, medical surveillance, training, and site control. Use of protective equipment is addressed by OSHA in 29 CFR Sections 1910.132, 133, 135, and 136. Other agencies, such as the U.S. Coast Guard and the Department of Transportation, have also set regulations to control storage, handling, and transportation of hazardous substances.

Applicability

Hazardous waste operations and emergency response regulations are applicable to anyone involved in management of hazardous materials and hazardous wastes. The rules specifically relate to workers who are likely to be exposed to such substances in a way that could pose significant threat to their health and safety. Four operations are covered by these regulations:

- All corrective actions under RCRA
- Activities associated with storage, treatment, or disposal of hazardous waste
- Government (all levels—federal to local) mandated or voluntary cleanup operations at hazardous waste sites
- All emergency response activities following release or impending release of hazardous substances regardless of location

Various aspects of personnel safety at hazardous waste sites can be grouped into three categories:

1. Preliminary procedures
2. Onsite activities
3. Emergency response

PRELIMINARY PROCEDURES

Proper planning is the key to successful completion of investigations at a hazardous waste site. Available information on the uncontrolled site (which is usually very scanty), along with adequate anticipation of potential hazards, can help in formulating a plan that will ensure that work at the site will proceed with minimum risk to personnel and the public. Formulation of a health and safety plan involves four steps:

1. Developing an organizational structure for the site operations
2. Preparing a comprehensive work plan that addresses each phase of the operation
3. Implementing a site safety and health plan
4. Coordinating the operations with local officials

Figure 12–1 is a generalized organizational chart showing the responsibilities and chain of command.

Work Plan

A work plan describing the cleanup activities is an essential component of the PSHP. Like any other planning document, the work plan should be periodically updated and modified as new information on the site becomes available. A comprehensive work plan should be based on and include the following:

- Review of all available information about the site, such as

    Site location, topography, ownership, and access

    Waste inventories and generator's and transporters' manifests (if available)

    Results of any prior sampling and monitoring

    Site maps and photographs

    Records available at local environmental and health agencies

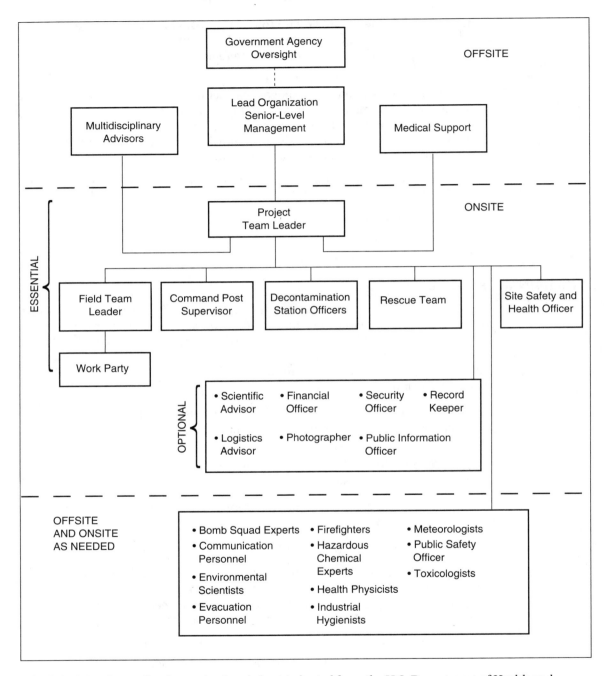

FIGURE 12–1 Generalized organizational chart (adapted from the U.S. Department of Health and Human Services, 1985)

- Work objectives
- Procedures for achieving the objectives—this should include details of the sampling plan, inventory preparation, and disposal methods
- Personnel requirements, including any additional training needed
- Equipment requirements, including special equipment or services, such as drilling equipment, geophysical surveying equipment, and operators required

Formulation of the work plan requires a multidisciplinary approach with input from many experts, all of whom should be consulted in the development process.

Site Safety Plan

Before commencement of activities at a hazardous waste site, a site safety plan must be in place. The plan should establish procedures and policies to safeguard the workers and the public from potential hazards at the site. Accident minimization during normal activities at the site should be one of the main objectives of the site safety plan. A site safety plan that includes the following considerations usually provides the desired results:

- Description of risk associated with each operation to be conducted at the site
- List of names of key personnel who will be responsible for site safety
- Protective clothing and equipment requirements for various personnel
- Decontamination procedures for workers and equipment
- Description of site control measures
- Medical surveillance program
- Environmental monitoring and sampling at the site
- Contingency plan for safe and effective response to emergency situations developing at the site

Personnel Training

Proper training of personnel is essential for successful completion of operations at a hazardous waste site. Training enables the workers to recognize safety, health, and related site hazards; familiarizes them with use of personal protective equipment; and provides work practices that minimize risk. It also emphasizes the individual's responsibilities for implementing site health and safety procedures.

Initial training comprises 40 hours of classroom instruction and 24 hours of field training. Supervisory and managerial personnel must complete an additional 8 hours of training over and above the required initial training. The trainers have to be academically qualified to impart the training or must have completed training programs themselves. A certificate is an acceptable documentary proof of completion of the training program. All employees, including the supervisors, are required to receive 8 hours of refresher training annually. Table 12–1 gives a summary of training required for various personnel involved in hazardous waste site activities.

TABLE 12–1 Training Requirements for Hazardous Waste Site Workers (adapted from U.S. Department of Health and Human Services, 1985)

Training Topic	Emphasis of Training	General Site Worker*	Onsite Management and Supervisors*	Health and Safety Staff*
Biology, Chemistry, and Physics of Hazardous Materials	Chemical and physical properties; chemical reactions; chemical compatibilities	R	R	R
Toxicology	Dosage, routes of exposure, toxic effects, immediately dangerous to life or health (IDLH) values, permissible exposure limits (PELs), recommended exposure limits (RELs), threshold limit values (TLVs)	R	R	R
Industrial Hygiene	Selection and monitoring of personnel protective clothing and equipment	R	R	R
	Calculation of doses and exposure levels; evaluation of hazards; selection of worker health and safety protective measures	R	R	R
Rights and Responsibilities of Workers Under OSHA	Applicable provisions of Title 29 of the Code of Federal Regulations (OSHA))	R	R	R
Monitoring Equipment	Functions, capabilities, selection, use limitations, and maintenance	R	R	R
Hazard Evaluation	Techniques of sampling and assessment		R	R
	Evaluation of field and lab results		R	R
	Risk assessment		O	R
Site Safety Plan	Safe practices, safety briefings and meetings, Standard Operating Procedures, site safety map	R	R	R
Standard Operating Procedures (SOPs)	Hands-on practice	R	R	R
	Development and compliance		R	R
Engineering Controls	The use of barriers, isolation, and distance to minimize hazards	R	R	R

* R = Recommended; O = Optional.

TABLE 12–1 (continued)

Training Topic	Emphasis of Training	General Site Worker	Onsite Management and Supervisors	Health and Safety Staff
Personnel Protective Clothing and Equipment (PPE)	Assignment, sizing, fitting, maintenance, use, limitations, and hands-on training	R	R	R
	Selection of PPE		O	R
	Ergonomics			R
Medical Program	Medical monitoring, first aid, stress recognition	R	R	R
	Advanced first aid, cardiopulmonary resuscitation (CPR), emergency drills	O	R	R
	Design, planning, and implementation			R
Decontamination	Hands-on training using simulated field conditions	R	R	R
	Design and maintenance	R	R	R
Legal and Regulatory Aspects	Applicable safety and health regulations (OSHA, EPA, etc.)	O	R	R
Emergencies	Emergency help and self-rescue; emergency drills	R	R	R
	Response to emergencies; follow-up investigation and documentation		R	R

Medical Surveillance

Employers are required to institute a medical surveillance program. This applies to employees who (a) wear a respirator for a period of 30 days or more in a year, (b) employees who will be exposed to hazardous substances or health hazards at or above the permissible exposure limits for 30 days or more in a year, (c) employees injured due to overexposure from an emergency incident involving hazardous substances or health hazards, or (d) employees serving on a hazardous materials (HAZMAT) team.

Many chemicals found at hazardous waste sites are highly toxic, and the medical surveillance program should be designed to detect workers' exposure to these chemicals and associated toxic substances. Table 12–2 lists some of the common chemicals found at hazardous waste sites and their effects on health.

The employer has to arrange for medical examination of the employees and pay for such examinations. Medical examinations are required on a preemployment and annual basis. Medical examination is mandatory for employees who are exposed to hazardous materials or who develop symptoms of overexposure. The law has established guidelines that are to be followed by health professionals. Table 12–3 lists medical surveillance program for hazardous waste site workers.

ONSITE ACTIVITIES

Onsite activities represent the implementation part of the work plan, and include the following five components:

- Site characterization
- Site control
- Use of personnel protective equipment (PPE)
- Decontamination of workers and equipment
- Emergency plan

Site Characterization

The purpose of site characterization is to identify hazards present at the site and to select suitable methods to protect the workers. This is accomplished by obtaining all background information about the site with as much accuracy and detail as possible. Thoroughness and accuracy form the basis for adoption of the most suitable protective measures in relation to the actual hazards to which the workers might be exposed.

Site characterization involves three phases:

1. Preentry offsite characterization: Gathering relevant information away from the site and carrying out reconnaissance from the site perimeter.

2. Onsite survey: Conducting investigations at the site. This requires the workers to have adequate PPE.

3. Other activities: Sampling, collection, removal and/or cleanup operations. Such activities have to be deferred until all information on the nature of hazards present at the site has been evaluated and suitable PPE have been selected. It is essential that conditions be monitored on an ongoing basis throughout the entire period of site activities.

TABLE 12–2 Common Chemicals Found at Hazardous Waste Sites, Their Health Effects, and Medical Monitoring (adapted from U.S. Department of Health and Human Services, 1985)

Hazardous Substance/Chemical Group	Compounds	Uses	Target Organs	Potential Health Effects**	Medical Monitoring
Aromatic Hydrocarbons	Benzene Ethylbenzene Toluene Xylene	Commercial solvents and intermediates for synthesis in the chemical and pharmaceutical industries	Blood Bone marrow CNS* Eyes Respiratory system Skin Liver Kidney	All cause: CNS depression: decreased alertness, headache, sleepiness, loss of consciousness Defatting dermatitis Benzene suppresses bone-marrow function, causing blood changes. Chronic exposure can cause leukemia. *Note:* Because other aromatic hydrocarbons may be contaminated with benzene during distillation, benzene-related health effects should be considered when exposure to any of these agents is suspected.	Occupational/general medical history emphasizing prior exposure to these or other toxic agents Medical examination with focus on liver, kidney, nervous system, and skin Laboratory testing: CBC† Platelet count Measurement of kidney and liver function
Asbestos (or asbestiform particles)		A variety of industrial uses, including: Building Construction Cement work Insulation Fireproofing Pipes and ducts for water, air, and chemicals Automobile brake pads and linings	Lungs Gastrointestinal system	Chronic effects: Lung cancer Mesothelioma Asbestosis Gastrointestinal malignancies Asbestos exposure coupled with cigarette smoking has been shown to have a synergistic effect in the development of lung cancer.	History and physical examination should focus on the lungs and gastrointestinal system Laboratory tests should include: a stool test for occult blood evaluation as a check for possible hidden gastrointestinal malignancy A high quality chest x-ray and pulmonary function test may help to identify long-term changes associated with asbestos diseases; however, early identification of low-dose exposure is unlikely

*CNS = Central nervous system.
**Long-Term effects generally manifest in 10 to 30 years.
†CBC = Complete blood count; RBC = Red blood count.

TABLE 12-2 (continued)

Hazardous Substance/Chemical Group	Compounds	Uses	Target Organs	Potential Health Effects	Medical Monitoring
Halogenated Aliphatic Hydrocarbons	Carbon tetrachloride Chloroform Ethyl bromide Ethyl chloride Ethylene dibromide Ethylene dichloride Methyl chloride Methyl chloroform Methylene chloride Tetrachloroethane Tetrachloroethylene (perchloroethylene) Trichloroethylene Vinyl chloride	Commercial solvents and intermediates in organic synthesis	CNS Kidney Liver Skin	All cause: CNS depression: decreased alertness, headaches, sleepiness, loss of consciousness Kidney changes: decreased urine flow, swelling (especially around eyes), anemia Liver changes: fatigue, malaise, dark urine, liver enlargement, jaundice Vinyl chloride is a known carcinogen; several others in this group are potential carcinogens.	Occupational/general medical history emphasizing prior exposure to these or other toxic agents Medical examination with focus on liver, kidney, nervous system, and skin Laboratory testing for liver and kidney function; carboxyhemoglobin where relevant
Heavy Metals	Arsenic Beryllium Cadmium Chromium Lead Mercury	Wide variety of industrial and commercial uses	Multiple organs and systems, including: Blood Cardiopulmonary Gastrointestinal Kidney Liver Lung CNS Skin	All are toxic to the kidneys. Each heavy metal has its own characteristic symptom cluster; for example, lead causes decreased mental ability, weakness (especially hands), headache, abdominal cramps, diarrhea, and anemia. Lead can also affect the blood-forming mechanism, kidneys, and the peripheral nervous system. Long-term effects also vary. Lead toxicity can cause permanent kidney and brain damage; cadmium can cause kidney or lung disease. Chromium, beryllium, arsenic, and cadmium have been implicated as human carcinogens.	History-taking and physical exam: search for symptom clusters associated with specific metal exposure; e.g., for lead, look for neurological deficit, anemia, and gastrointestinal symptoms Laboratory testing: Measurements of metallic content in blood, urine, and tissues (e.g., blood lead level; urine screen for arsenic, mercury, chromium, and cadmium) CBC Measurement of kidney function, and liver function where relevant Chest x-ray or pulmonary function testing where relevant

TABLE 12–2 *(continued)*

Hazardous Substance/Chemical Group	Compounds	Uses	Target Organs	Potential Health Effects	Medical Monitoring
Herbicides	Chlorophenoxy compounds: 2,4-dichlorophen-oxyacetic acid (2,4-D), 2,4,5-tri-chlorophenoxyacetic acid (2,4,5-T) Dioxin (tetracholoro-dibenzo-p-dioxin, TCDD) occurs as a trace contami-nant in these compounds.	Vegetation control	Kidney Liver CNS Skin	Chlorophenoxy compounds can cause chloracne, weakness or numbness of the arms and legs, and may result in long-term nerve damage. Dioxin causes chloracne and may aggravate preexisting liver and kidney diseases.	History and physical exam should focus on the skin and nervous system. Laboratory tests include: Measurement of liver and kidney function, where relevant Urinalysis
Organochlorine Insecticides	Chlorinated ethanes: DDT Cyclodienes: Aldrin Chlordane Dieldrin Endrin Chlorocyclohexanes: Lindane	Pest control	Kidney Liver CNS	All cause acute symptoms of apprehension, irritability, dizziness, disturbed equili-brium, tremor, and convulsions Cyclodienes may cause convulsions without any other initial symptoms. Chlorocyclohexanes can cause liver toxicity and can cause permanent kidney damage.	History and physical exam should focus on the nervous system Laboratory tests include: Measurement of kidney and liver function CBC for exposure to chlorocyclohexanes
Organo-phosphate and Carbamate Insecticides	Organophosphate: Diazinon Dichlorovos Dimethoate Trichlorfon Malathion Methyl parathion Parathion Carbamate: Aldicarb Baygon Zectran	Pest control	CNS Liver Kidney	All cause a chain of internal reactions leading to neuro-muscular blockage. Depending on the extent of poisoning, acute symptoms range from headaches, fatigue, dizziness, increased salivation and crying, profuse sweating, nausea, vomiting, cramps, and diarrhea to tightness in the chest, muscle twitching, and slowing of the heartbeat	Physical exam should focus on the nervous system. Laboratory tests should include: RBC cholinesterase levels for recent exposure (plasma cholinesterase for acute exposures) Measurement of delayed neuro-toxicity and other effects

TABLE 12–2 *(continued)*

Hazardous Substance/Chemical Group	Compounds	Uses	Target Organs	Potential Health Effects	Medical Monitoring
Organophosphate and Carbamate Insecticides (continued)				Severe cases may result in rapid onset of unconsciousness and seizures. A delayed effect may be weakness and numbness in the feet and hands. Long-term, permanent nerve damage is possible.	
Polychlorinated Biphenyls (PCBs)		Wide variety of industrial uses	Liver CNS (speculative) Respiratory system (speculative) Skin	Various skin ailments, including chloracne; may cause liver toxicity; carcinogenic to animals.	Physical exam should focus on the skin and liver. Laboratory tests include: Serum PCB levels Triglycerides and cholesterol Measurement of liver function

TABLE 12-3 Recommended Medical Surveillance Program for Hazardous Waste Site Workers (adapted from U.S. Department of Health and Human Services, 1985)

Component	Recommended	Optional
Preemployment Screening	Medical history Occupational history Physical examination Determination of fitness to work wearing protective equipment Baseline monitoring for specific exposures	Freezing pre-employment serum specimen for later testing (limited to specific situations)
Periodic Medical Examinations	Yearly update of medical and occupational history; yearly physical examination; testing based on (1) examination results, (2) exposures, and (3) job class and task More frequent testing based on specific exposures	Yearly testing with routine medical tests
Emergency Treatment	Provide emergency first aid on site Develop liaison with local hospital and medical specialists Arrange for decontamination of victims Arrange in advance for transport of victims Transfer medical records; give details of incident and medical history to next care provider	
Nonemergency Treatment	Develop mechanism for non-emergency health care	
Recordkeeping and Review	Maintain and provide access to medical records in accordance with OSHA and state regulations Report and record occupational injuries and illnesses Review site safety plan regularly to determine if additional testing is needed Review program periodically; focus on current site hazards, exposures, and industrial hygiene standards	

Preentry Offsite Characterization

The goal during this phase of the investigation should be to gather as much information as possible before actual entry into the site. This will provide a sound basis for evaluation of site hazards and for adoption of proper control measures to protect the workers. Inhalation hazards and other conditions that may be immediately dangerous to life or health (IDLH) should be specifically targeted for identification. There are many visible signs of IDLH and related conditions that serve as good indicators. Common visible indicators of potentially dangerous or life-threatening conditions (U.S. Department of Health, 1985) include

- Large containers or tanks that must be entered
- Enclosed spaces, such as buildings or trenches, that must be entered
- Potentially explosive or flammable situations (indicated by bulging drums, effervescence, gas generation, or instrument readings)
- Extremely hazardous materials (such as cyanide, phosgene, or radiation sources)
- Visible vapor clouds
- Areas where biological indicators (such as dead animals or vegetation) are located

Background research involving review of the company's operation records, local fire and police departments records, court records, media reports, and previous survey records (soil, geologic, hydrogeologic, and geophysical); sampling, chemical analyses, and monitoring data may prove very useful in preliminary assessment of potential hazards at the site.

Perimeter reconnaissance should be conducted by trained personnel with proper PPE. Generally, level B personnel protective equipment (Table 12–4) is recommended for most situations. Visual observations, air-quality monitoring at the perimeter, sampling of soil, surface water and groundwater, and any runoff from the site should be routinely carried out. Biologic indicators, such as dead vegetation or animals, presence of vapor clouds, discolored liquids, oil slicks, and vapors are good indicators of potential hazards and should always be looked for during perimeter reconnaissance. A sketch map of the site should be prepared showing the location of buildings, pits, ponds, containers of all kinds, tanks, dumps, and access roads.

Onsite Survey

Data and information obtained from the offsite survey should be used to develop a site safety plan to be used during entry into the site. Based on available information, recommendations on use of suitable PPE and precautions to be taken by workers should be included in the site safety plan. Actual conditions will be known only after entry into the site; therefore, prudence and caution should be the guiding principles, and a conservative approach should always be taken. It is important that in addition to persons actually entering the site—a minimum of two—at least two properly equipped outside support persons must be present and ready to enter the site in case of an emergency.

Upon entering the site, the worker should monitor the air for IDLH and other conditions, such as oxygen deficiency, explosive air, and toxic substances. Suitable devices should be used to monitor ionizing radiation. The presence of

TABLE 12–4 Personnel Protective Equipment (adapted from U.S. Department of Health and Human Services, 1985)

Level of Protection	Equipment	Protection Provided	Use	Limiting Criteria

A	RECOMMENDED: Pressure-demand, full-facepiece SCBA or pressure-demand supplied-air respirator with escape SCBA
	Fully encapsulating chemical-resistant suit
	Inner chemical-resistant gloves
	Chemical-resistant safety boots/shoes
	Two-way radio communications
	OPTIONAL:
	Cooling unit
	Coveralls
	Long cotton underwear
	Hard hat
	Disposable gloves and boot covers
		The highest available level of respiratory, skin, and eye protection	The chemical substance has been identified and requires the highest level of protection for skin, eyes, and the respiratory system based on either:
			
			  measured (or potential for) high concentration of atmospheric vapors, gases, or particulates
			
			  or
			
			  site operations and work functions involving a high potential for splash, immersion, or exposure to unexpected vapors, gases, or particulate matter that are harmful to skin or capable of being absorbed through the intact skin
			Substances with a high degree of hazard to the skin are known or suspected to be present, and skin contact is possible
			Operations must be conducted in confined, poorly ventilated areas until the absence of conditions requiring Level A protection is determined
				Fully encapsulating suit material must be compatible with the substances involved

TABLE 12–4 *(continued)*

Level of Protection	Equipment	Protection Provided	Use	Limiting Criteria
B	RECOMMENDED: Pressure-demand, full-facepiece SCBA or pressure-demand supplied-air respirator with escape SCBA Chemical-resistant clothing (overalls and long-sleeved jacket; hooded, one- or two-piece chemical splash suit; disposable chemical-resistant one-piece suit) Inner and outer chemical-resistant gloves Chemical-resistant safety boots/shoes Hard hat Two-way radio communications OPTIONAL: Coveralls Disposable boot covers Face shield Long cotton underwear	The same level of respiratory protection but less skin protection than Level A It is the minimum level recommended for initial site entries until the hazards have been further identified.	The type and atmospheric concentration of substances have been identified and require a high level of respiratory protection, but less skin protection. This involves atmospheres: with IDLH concentrations of specific substances that do not represent a severe skin hazard or that do not meet the criteria for use of air-purifying respirators. Atmosphere contains less than 19.5 percent oxygen. Presence of incompletely identified vapors or gases is indicated by direct-reading organic vapor detection instrument, but vapors and gases are not suspected of containing high levels of chemicals harmful to skin or capable of being absorbed through the intact skin.	Use only when the vapor or gases present are not suspected of containing high concentrations of chemicals that are harmful to skin or capable of being absorbed through the intact skin. Use only when it is highly unlikely that the work being done will generate either high concentrations of vapors, gases, or particulates or splashes of material that will affect exposed skin.

TABLE 12–4 *(continued)*

Level of Protection	Equipment	Protection Provided	Use	Limiting Criteria
C	RECOMMENDED: Full-facepiece, air-purifying canister-equipped respirator Chemical-resistant clothing (overalls and long-sleeved jacket; hooded, one- or two-piece chemical splash suit; disposable chemical-resistant one-piece suit) Inner and outer chemical-resistant gloves Chemical-resistant safety boots/shoes Hard hat Two-way radio communications OPTIONAL: Coveralls Disposable boot covers Face shield Escape mask Long cotton underwear	The same level of skin protection as Level B, but a lower level of respiratory protection	The atmospheric contaminants, liquid splashes, or other direct contact will not adversely affect any exposed skin The types of air contaminants have been identified, concentrations measured, and a canister is available that can remove the contaminant All criteria for the use of air-purifying respirators are met	Atmospheric concentration of chemicals must not exceed IDLH levels The atmosphere must contain at least 19.5 percent oxygen
D	RECOMMENDED: Coveralls Safety boots/shoes Safety glasses or chemical splash goggles Hard hat OPTIONAL: Gloves Escape mask Face shield	No respiratory protection Minimal skin protection	The atmosphere contains no known hazard Work functions preclude splashes, immersion, or the potential for unexpected inhalation of or contact with hazardous levels of any chemicals	This level should not be worn in the Exclusion Zone The atmosphere must contain at least 19.5 percent oxygen

plants causing allergic reactions, such as poison oak, poison ivy, or poison sumac, should be noted.

SITE CONTROL

Once the nature of hazards associated with the site has been determined and proper PPE and safety plan put in place, various zones should be delineated to facilitate removal/cleanup operations and to prevent unauthorized persons from entering the site. This is usually done by demarcating three zones as shown in Figure 12–2, along with a fence having controlled entry and exit points and clear warning signs. The three most commonly demarcated zones are:

1. Exclusion Zone (EZ), which includes the contaminated area.
2. Contaminant Reduction Zone (CRZ), which is the area set aside for personnel and equipment decontamination. The CRZ also serves as a transition zone between the EZ and the support zone.
3. Support Zone (SZ), which is the area free from contamination; workers in the SZ are not exposed to hazardous conditions.

The boundary between the EZ and CRZ is called the *hotline*. The hotline can be established by using the following criteria (U.S. Department of Health, 1985):

- Visually survey immediate site environs
- Determine the locations of
  - Hazardous substances
  - Drainage, leachate, and spilled material
  - Visible discolorations
- Evaluate data from initial site survey indicating the presence of
  - Combustible gases
  - Organic and inorganic gases, particulates, or vapors
  - Ionizing radiation
- Evaluate results of soil and water sampling
- Consider the distances needed to prevent an explosion or fire from affecting personnel outside the Exclusion Zone
- Consider meteorological conditions and the potential for contaminants to be blown from the area
- Secure or mark the hotline
- Modify location of the hotline if necessary, as more information becomes available

Depending upon the degree of hazards or on the incompatibility of waste streams at the site, the EZ may be subdivided into different areas.

The level of PPE in the exclusion zone varies with specific job assignments. For example, a person responsible for sample collection from open containers may require Level B PPE, but another worker who performs air monitoring may require Level C PPE.

The Buddy System

The buddy system involves organizing work groups so that each employee of the group is designated to be observed by at least one other employee in the same work group. The buddy system allows for

- Rendering assistance to partner as and when needed
- Observing partner for signs of chemical or heat exposure
- Checking the functioning and overall integrity of partner's PPE
- Notifying the command post supervisor or others if emergency help is needed

The buddy system in itself may not be the solution for all emergency situations that may arise at a hazardous waste site. It is therefore important that workers in

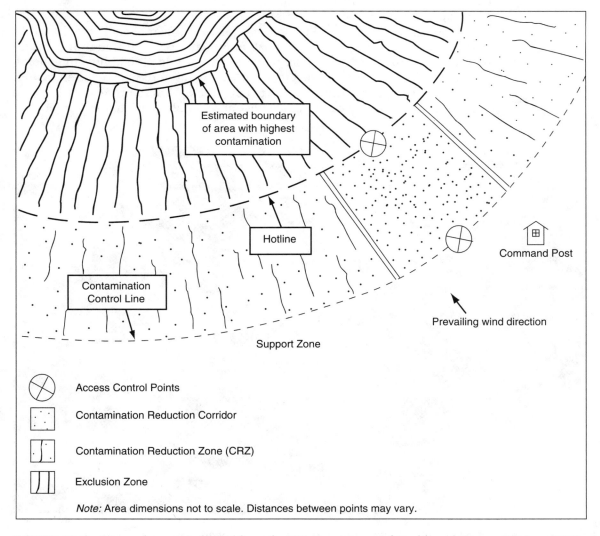

FIGURE 12–2 Site work zones (adopted from the U.S. Department of Health and Human Services, 1985)

the exclusion zone should be in direct line-of-sight or communication contact with command post personnel or with the backup person in the support zone.

PERSONNEL PROTECTIVE EQUIPMENT (PPE)

The purpose of PPE is to isolate or shield the worker from physical, chemical, and biologic hazards that may occur at a hazardous waste site. Properly selected PPE allows protection of the worker's eyes, face, hands, feet, head, body, skin, and ears against dangerous chemicals and other hazards at the site.

Personal protective equipment is intended to reduce the anticipated exposure levels to below permissible exposure limits (PEL). If sufficient information is not available to identify the hazards at a site, an ensemble providing protection equivalent to Level B must be used. Once the hazards at the site have been identified, appropriate PPE should be used.

It should be noted that PPE itself may pose significant hazards to the workers, including physical, psychological, and heat stress; limited mobility and restricted vision; and reduced hearing and communication ability. In general, the higher the level of PPE, the greater the associated risks. Nonetheless, PPE should be selected with the goal of providing the optimum protection, because both over- and underprotection may pose additional hazard to the worker.

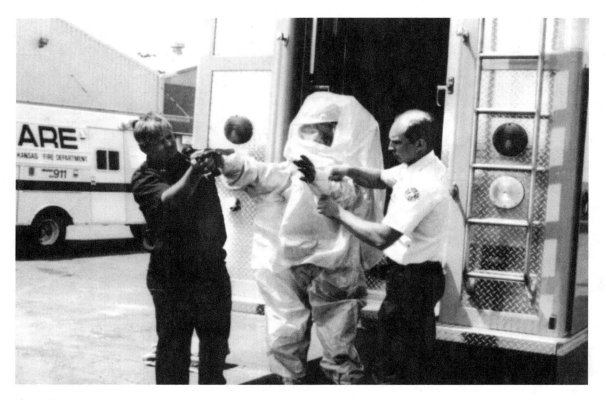

FIGURE 12–3   A worker in fully encapsulating PPE getting ready for entry into the EZ (photo courtesy of Bill Keffer, EPA)

Types of PPE

PPEs are classified into four levels, A, B, C, and D. Level A is the highest level of PPE; it features a fully encapsulated chemical-resistant suit that is covered by a disposable protective suit, gloves, boots, and a hard hat. Level A PPE also includes a pressure-demand, self-contained breathing apparatus (SCBA). Conditions that warrant use of Level A PPE include high concentrations of atmospheric vapors and gases, or an operation with a high potential for splash, immersion, or exposure to unexpected vapors or gases. Level A ensemble is also the costliest.

Level B PPE includes everything in level A except the encapsulating suit. Level C substitutes an air-purifying respirator for the SCBA of level B. Level D, comprising boots, gloves, and hard hat, is the least expensive. Figure 12–3 shows a worker getting ready for entry into the EZ in a fully encapsulating PPE. Table 12–4 summarizes the four levels of personnel protection and their limitations.

DECONTAMINATION

Use of suitable protective equipment affords adequate body protection to the person entering or working at the contaminated site. However, the protective ensemble itself becomes contaminated with hazardous substances present in the exclusion zone. In order to prevent contaminants from getting transferred outside the EZ and into the clean areas, the contaminants that accumulate on the workers' PPE and on the tools and equipment used must be physically removed or chemically neutralized (Figure 12–4). Decontamination protects

FIGURE 12–4 Decontamination of boot covers in a series of rinsates (photo courtesy of Bill Keffer, EPA)

workers from hazardous substances that could contaminate or permeate clothing, breathing equipment, tools, and vehicles used at the site. Decontamination also prevents mixing of incompatible materials, which can create a new hazard.

Decontamination Plan

The decontamination plan, which is a part of the site safety plan, should include

- Number and layout of decontamination stations
- Decontamination equipment and supplies needed
- Decontamination method(s) to be used
- Procedure(s) to prevent contamination of the clean areas
- Disposal of contaminated clothing, fluids, and/or equipment

The decontamination plan should be modified whenever a change in site conditions, PPE, personnel, or site hazards occurs.

Figure 12–5(A)–(C) shows the layout for achieving maximum decontamination for levels A, B, and C, respectively.

Methods

Decontamination may be performed by (a) physically removing the contaminants from clothing and equipment, (b) chemically neutralizing or detoxifying the contaminants, or (c) using a combination of physical and chemical methods. Table 12–5 lists some of the common decontamination methods.

Testing Efficacy

The effectiveness of a particular decontamination method should be assessed at an early stage, and should be periodically reevaluated during the lifetime of the removal/cleanup operations. If the contaminated materials are not being removed from the PPE or are permeating through workers' protective clothing, the plan must be revised. The following methods are useful in ascertaining the efficacy of decontamination methods:

- Visual observation. Discolorations, stains, presence of dust, and similar substances can be generally detected using natural light. The presence of polycyclic aromatic hydrocarbons can be detected by using UV light, which causes them to fluoresce. Use of UV light must be done under the supervision of a qualified health professional at the site, because exposure to UV light is known to increase the risk of skin cancer.
- Wipe sampling. A dry or wet cloth, glass fiber filter paper, or swab can be wiped over the surface of the contaminated object and then analyzed in the laboratory. Not only the outer and inner surfaces of protective clothing, but also the worker's skin, should be tested by wipe sampling.
- Cleaning solutions analysis. If chemical analysis of final rinse solution indicates elevated levels of contaminants, it is a positive indication that the method is not effective. In such cases, the contamination plan must be revised and/or additional cleaning and rinsing operations must be included.
- Permeation testing. If chemical analysis reveals that contaminants have permeated the protective clothing, the decontamination method should

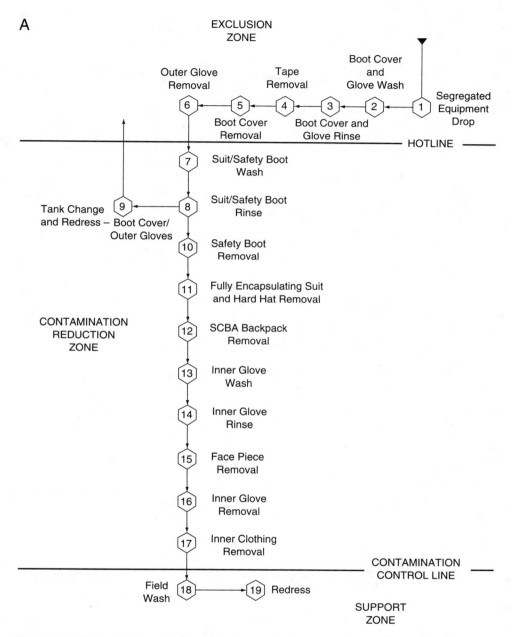

FIGURE 12–5(A) Maximum decontamination layout for Level A (adapted from the U.S. Department of Health and Human Services, 1985)

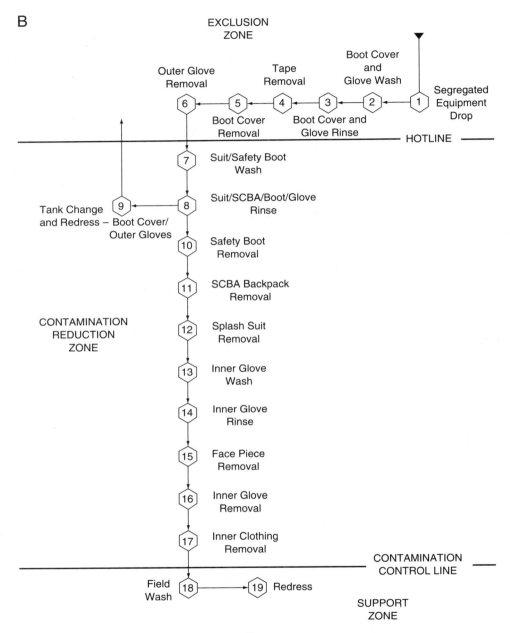

FIGURE 12–5(B)   Maximum decontamination layout for Level B (adapted from the U.S. Department of Health and Human Services, 1985)

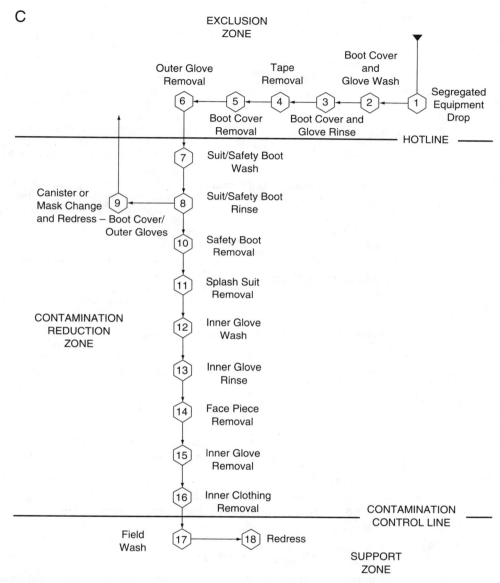

FIGURE 12–5(C) Maximum decontamination layout for Level C (adapted from the U. S. Department of Health and Human Services, 1985)

TABLE 12-5   Common Decontamination Methods (adapted from U.S. Department of Health and Human Services, 1985)

REMOVAL

Contaminant Removal	Water rinse, using pressurized or gravity flow
	Chemical leaching and extraction
	Evaporation/vaporization
	Pressurized air jets
	Scrubbing/scraping: Commonly done using brushes, scrapers, or sponges and water-compatible solvent cleaning solutions
	Steam jets
Removal of Contaminated Surfaces	Disposal of deeply permeated material, e.g., clothing, floor mats, and seats
	Disposal of protective coverings/coatings

INACTIVATION

Chemical Detoxification	Halogen stripping
	Neutralization
	Oxidation/reduction
	Thermal degradation
Disinfection/Sterilization	Chemical disinfection
	Dry heat sterilization
	Gas/vapor sterilization
	Irradiation
	Steam sterilization

be changed. Permeation may also suggest that the PPE is not compatible with the chemicals occurring at the site, and calls for a change in the PPE.

Decontamination is a very critical aspect of work at a hazardous waste site, and appropriate decisions must be made to determine the most suitable method. Figure 12–6 is a flow chart that aids in evaluating health and safety aspects of a decontamination plan.

Disposal of Contaminated Objects

All equipment and objects, including rinse water and other fluids used for decontamination, must be either decontaminated at the site or disposed of in an acceptable manner. Usually clothing, buckets, pans, brushes, and inexpensive tools are collected and placed in properly labeled containers for ultimate offsite disposal.

EMERGENCY PROCEDURES

Planning for an emergency situation is another critical element of the personnel safety and health plan. An emergency response plan must be developed for any unexpected emergency at the work site. The plan must identify personnel roles, line of authority, and communication, and must be made available to all

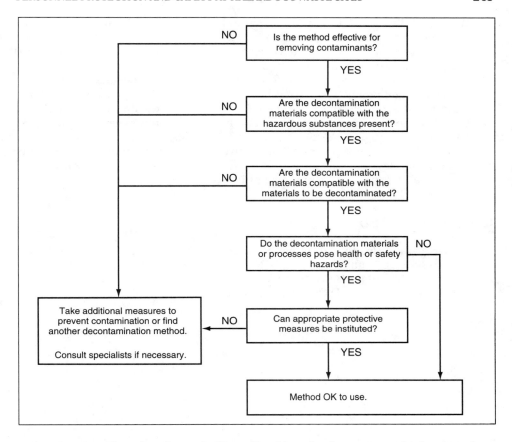

FIGURE 12–6 Flow chart for evaluation of health and safety aspects of a decontamination method (adopted from the U. S. Department of Health and Human Services, 1985)

employees. In order to recognize and prevent emergency situations, clearly marked places of refuge and anticipated safe distances must be included in the plan. Evacuation routes, emergency medical treatment, and decontamination procedures should be a part of the emergency plan. The plan should also contain information on the local authority responsible for emergency hazard response and who should be contacted if an emergency evacuation of employees from a hazardous material site is warranted.

STUDY QUESTIONS

1. Why is it important to provide adequate protection to personnel working at hazardous waste sites? What are the hazards to which these workers may be exposed?
2. Which federal agency has set the regulations regarding the health and safety of personnel working at hazardous waste sites? Does the same agency address personnel protective equipment? if not, name the agency that does this.
3. What are the specific situations where hazardous waste operations and emergency response regulations are applicable?
4. What information should be used in formulating a work plan for a hazardous waste site investigation?

5. Why is medical surveillance of personnel working at hazardous waste sites important? Describe the health effects of three chemicals commonly found at hazardous waste sites.

6. What are the three zones that are commonly demarcated at a hazardous waste site for ensuring proper site control? What is the hotline?

7. What is the buddy system? How does it help hazardous waste site workers?

8. What is the purpose of personnel protective equipment? How does the level A PPE differ from level B? What minimum level of PPE should be used at a hazardous waste for which sufficient information is not available? Discuss the hazards associated with the use of PPE.

9. What is decontamination? Discuss the various methods that can be used for decontamination of personnel and equipment. How would you determine if the decontamination process has been effective?

10. What essential information must be included in an emergency response plan for a hazardous waste site investigation?

REFERENCE CITED

U.S. Department of Health and Human Services, 1985, Occupational Safety and Health Guidance Manual for Hazardous Waste Site Activities: Washington, D.C., DHHS (NIOSH) Publication No. 85–115, Superintendent of Documents, U.S. Government Printing Office, unpaginated.

13

Treatment Technologies for Hazardous Wastes

INTRODUCTION

One of the major requirements of the Resource Conservation and Recovery Act of 1976 is proper control of hazardous waste after it is generated. Growing environmental awareness in the 1960s and 1970s prompted lawmakers to set up legislation to eliminate undesirable practices of hazardous waste disposal. Treatment methods, if selected properly, either eliminate the waste or drastically reduce the volume of the waste to be disposed of.

The nature and volume of hazardous waste depend on the manufacturing process, the waste minimization options used in the process, the nature of the feedstock, and the finished product(s). It is therefore very important that the waste streams for each process be fully characterized before selecting the most suitable treatment method.

Treatment technologies are evolving at a very rapid rate; innovative methods are being developed at a fast pace. The EPA, through its Superfund Innovative Technology Evaluation (SITE) initiative, has provided incentives to industries and research organizations to develop efficient, economical, and environmentally safe treatment technologies (EPA, 1990). The EPA spent $20 million between 1986 and 1991 to support research to develop new treatment technologies for hazardous waste.

The discussion of treatment of hazardous waste presented in this chapter is intended to provide an overview of the fundamental principles used in these technologies; it does not address engineering and other design details. A large amount of published literature dealing with treatment technologies is available.

In addition, proceedings of symposia addressing the subject of hazardous waste treatment that are held on a regular basis have been published. The four volumes in the series *Emerging Technologies in Hazardous Waste Management* (American Chemical Society, 1989–1992) are a very good source of information on newly evolving treatment technologies.

Treatment methods effectively reduce toxicity of hazardous wastes—making them suitable for disposal in a conventional manner. However, because of their unique chemical natures and economic considerations, many other kinds of hazardous wastes are disposed of without subjecting them to treatment, as in deep injection of liquid waste, for example. (Disposal technologies are discussed in Chapter 14.)

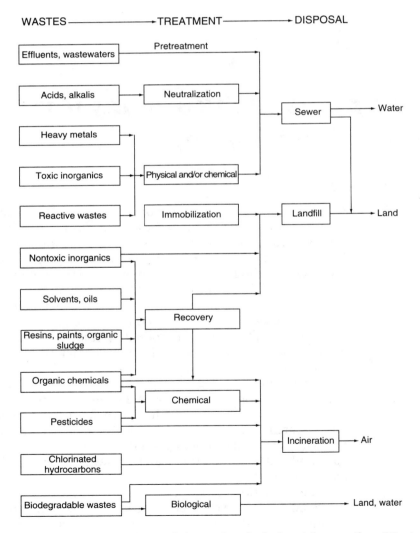

FIGURE 13–1   Treatment and disposal alternatives for industrial waste (from Wentz: *Hazardous Waste Management* ©1989, McGraw-Hill, Inc. reprinted wtih permission)

PLATE I

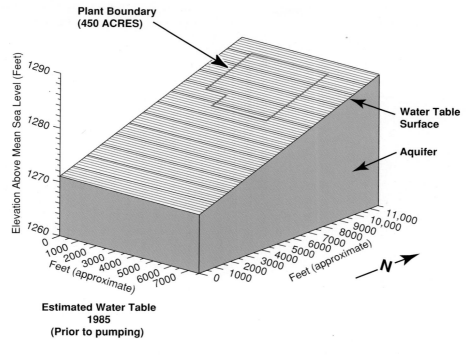

Estimated Water Table
1985
(Prior to pumping)

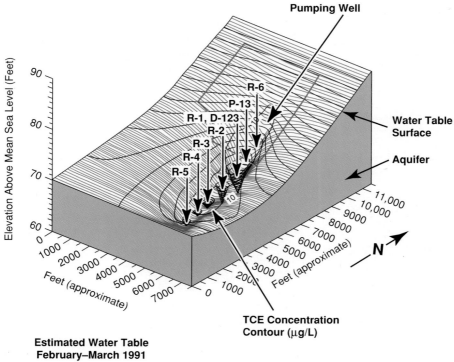

Estimated Water Table
February–March 1991

FIGURE 13–3　Cones of depression associated with pumping wells

PLATE II

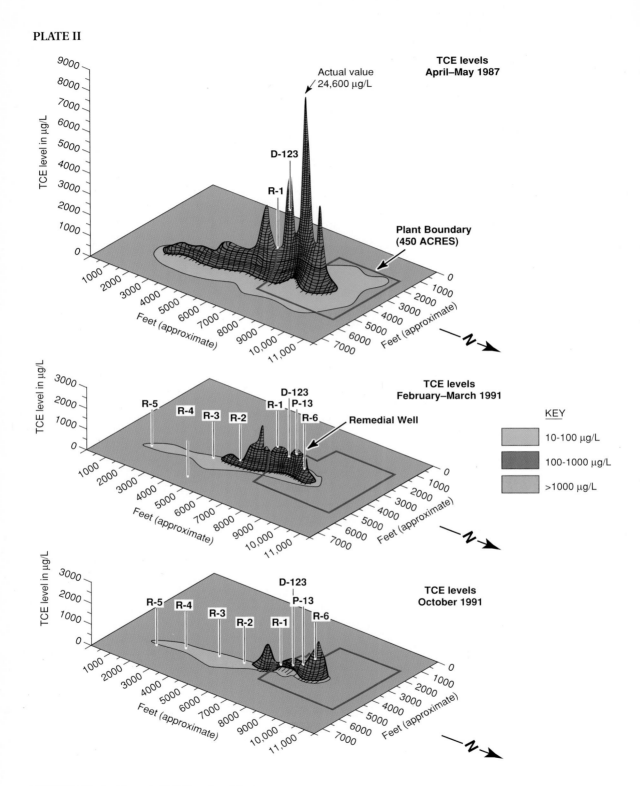

FIGURE 13–4 Drop in TCE levels with time

Existing treatment methods can be grouped into four categories:

- Physical
- Chemical
- Biological
- Miscellaneous

Most treatment methods use a combination of one or more of these methods. Figure 13–1 shows treatment and disposal alternatives for common types of industrial wastes.

PHYSICAL METHODS

Physical methods of hazardous waste treatment include processes as simple as separation using screens to highly complex processes such as reverse osmosis and ground freezing. Physical treatment should be a routine consideration in cases where the hazardous waste contains both liquid and solid fractions (mixed waste); it is a cost-effective process and is the least complicated solution to waste management problems. The following sections describe common physical treatment methods.

Screening

Bar racks, strainers, and screens can be used to separate large suspended solids from liquid waste. Screening is usually the first step in waste treatment and is followed by other methods.

Sedimentation

Gravitational settling, controlled by particle size, shape, density, and the viscosity of the liquid, causes separation of solids from liquid. Separation of waste fractions by sedimentation is done in settling tanks or ponds, where sufficient time is allowed for solids to settle down to the bottom of the system. Solids are periodically removed from the bottom for further treatment or (if nonhazardous) are thickened to reduce volume before disposal.

Centrifugation

Centrifuges are commonly used on sludges to separate solids from the liquid fraction of a mixed hazardous waste. Solid, high-strength cake having minimal water content—which allows for easy handling—can be conveniently obtained by subjecting the mixed waste to centrifugation. Sludge from industrial waste should be adequately dewatered before incineration, because drier sludge consumes less fuel during incineration.

Flotation

Air flotation can be used to separate low-density solids and hydrocarbon solids from liquids. Upon introduction of air into the waste, tiny air bubbles attach to the solids and rise to the surface, where they are removed by skimming.

Filtration

Filtration effectively removes suspended solids from liquid wastes. Instead of using a single layer of the filtration medium, two to three layers of minerals having different grain size and specific gravity are used to achieve a maximum level of filtration. Such systems are called multimedia filters. One multimedia filter, for example, uses a bed of ground anthracite (high-grade coal) having a specific gravity (SG) of 1.6 and grain diameter of 1.0 mm for the topmost layer, a layer of silica sand (SG = 2.6; grain diameter = 0.5 mm) in the middle, and a layer of garnet (a rock-forming silicate mineral; SG = 4.0; grain diameter = 0.3 mm) at the bottom. This arrangement allows for progressive entrapment of suspended solids from the waste stream as it passes through the layer of largest pore openings on top through the smallest openings in the bottom layer. Figure 13–2 shows the relationships among the mineralogy, SG, and grain diameter of a multimedia filtration system.

After initial separation, the backwash stream may be subjected to centrifugation to thicken the sludge. The sludge is then run through a belt press, filter press, or vacuum filter to remove excess water and to obtain a solid sludge with a low moisture content for further treatment or disposal.

After a period of time, pores in the various layers become clogged with solids, resulting in a decrease in efficiency of the system. Backflow washing is periodically done to clear the layers and open the pores. After washing, mineral grains settle down, with the least dense anthracite on top, silica sand at its base, and garnet again forming the bottommost layer.

Adsorption

The binding of solid particles to one another is called adsorption; there is no penetration of one solid into the other through chemical reactions. Adsorption is analogous to adhesion. Activated carbon is used to separate dilute organic compounds from an aqueous waste stream. Organic compounds having high molecular weights, high boiling points, and low water solubility (ppm range), such as polycyclic aromatic hydrocarbons (pyrene, phenantherene) and chlorinated hydrocarbons, are easily adsorbed by activated carbon.

Activated carbon, because of its extremely large surface area per unit weight (600 to 1000 m^2/g) and the presence of interconnected microscopic pores, serves as a very efficient adsorption medium. Two forms of activated carbon are generally used for waste treatment: granular activated carbon (GAC) and powdered activated carbon (PAC). A treatment system using granular carbon requires flow of liquids through fixed beds of granular carbon; PAC is usually

FIGURE 13–2 Multimedia filtration system utilizing beds of different minerals

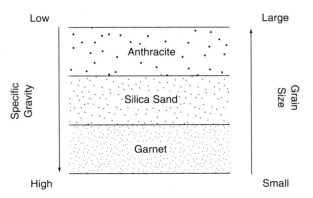

applied to liquid waste in a mechanically agitated tank or other type of contactor designed to allow maximum contact between the PAC and the liquid. After treatment, the PAC can be separated from the liquid either by filtration or sedimentation processes.

The efficiency of activated carbon goes down with use. For this reason, spent carbon has to be regenerated or replaced to maintain peak performance of the treatment system. In large treatment plants, activated carbon regeneration is done *in situ* by heating the bed of spent activated carbon above the boiling point of the adsorbed organic compounds to drive off (desorb) the volatiles. The volatiles can be either burned off in an incinerator or passed through condensers and collected as a usable liquid. Spent carbon that cannot be regenerated *in situ* can be regenerated by heating in an incinerator.

Evaporation

Evaporation results in volume reduction of the liquid component of a mixed hazardous waste. Evaporation is used to separate volatiles from nonvolatile components of a mixed waste stream, or a liquid from dissolved solids. In wastes with very high dissolved solids content, evaporation is carried out in specially designed equipment called a *dryer*. Commonly used dryers include drum dryers, pneumatic conveying dryers, tray and compartment dryers, and vacuum rotary dryers. Heat is applied to the waste stream and, as the liquid evaporates, the waste solution gets more concentrated and eventually saturated with dissolved solids. Evaporation is commonly used to reduce the volume of a liquid-rich hazardous waste.

Distillation

Two or more liquids can be separated by subjecting them to fractional distillation, which involves vaporization and subsequent condensation of the liquids. The process utilizes the difference in the vaporization temperatures of the liquids in the mixed waste stream. Distillation is used to separate contaminants from solvents to obtain purified solvents for reuse. The distilled liquids, generally of high purity, are recycled and marketed.

Treatment technologies using distillation are based on the difference in vapor pressures of the liquids in the waste stream. Vapor pressure controls the volatility of a liquid. Liquids with high vapor pressures evaporate easily; those with low vapor pressures evaporate slowly, and often need heat to accelerate the evaporation process. A pure liquid will attain its boiling temperature when its vapor pressure equals the atmospheric pressure. Dissolved salts and other contaminants cause a decrease in the vapor pressure, which in turn causes an increase in the boiling point.

Two common types of distillation used for hazardous waste treatment are batch distillation and continuous fractional distillation.

Batch Distillation

This process is suitable for waste streams that contain liquids of different compositions and varied vapor pressures. Batch distillation is carried out in a boiler or a heated evaporation chamber. Heat causes the liquid with high vapor pressure to volatilize first. The vapors pass through condensers; the condensate (liquid) is collected for reuse. Removal of the more volatile liquid results in increased concentrations of other liquids in the mixture. As the boiler temperature increases, the liquid with a lower vapor pressure will be the next one to volatilize, and so on. In this way, several fractions of the liquids, with different vapor pressures, may be recovered.

When all the desired fractions have been removed, a residue—called the still bottom—is left in the boiler. The still bottom has to be periodically removed for treatment and/or disposal.

Continuous Fractional Distillation

This method is suitable when the waste stream contains materials with similar vapor pressures. Upon heating, both materials will evaporate at nearly the same temperature. However, because of the slight difference in vapor pressures, the condensate will contain a larger quantity of the material that has the higher vapor pressure. In order to increase the purity of the condensate to the desired level, it is redistilled, which results in an increase in the concentration (purity) of the material with the higher vapor pressure. The process can be repeated until the final condensate meets the desired standard of purity. As with batch distillation, still bottom needs to be treated and/or disposed of in an appropriate manner.

Stripping

Stripping involves transfer of dissolved molecules from a liquid into a vapor stream or gas flow. Stripping is commonly done by passing air, heated nitrogen gas, or steam through the liquid waste. Stripping causes volatiles to be converted into the gaseous phase. *Air stripping* is used to remove volatile organic contaminants from water or aqueous waste streams. Groundwater contaminated with benzene, toluene, and TCE has been successfully remediated by air stripping (see the Case History on the remediation of a TCE-contaminated aquifer below). The process works best when the concentration of VOCs is less than 100 mg/L. It is possible to remove more than 99 percent of the VOCs from water by using a properly designed and efficiently operating air stripper. Air stripping, in which the flowing gas is air, is most efficiently accomplished in a packed tower where air and contaminated water flow in opposite directions; air is blown from the bottom of the tower and contaminated water enters from the top.

Air sparging involves introduction of air into conatminated media to remove the contaminants. Two variations of air sparging are used to clean up contaminated groundwater: physical air sparging and biological air sparging. In physical air sparging, air is introduced below the groundwater table to clean up groundwater contaminated with VOCs. The introduced air promotes volatilization of the VOCs present in the groundwater. The vapors purged from the groundwater move upward and accumulate in the vadose zone, where a vapor extraction system is used to remove the VOCs from the vadose zone. In biological air sparging, air is introduced into the contaminated groundwater to promote microbial growth for biodegradation.

In general, air/gas stripping is more suitable for removal of materials with high vapor pressures and low water solubilities, such as chlorinated hydrocarbons and aromatic compounds. *Steam stripping* is suitable for less volatile materials with low vapor pressures. These include high-boiling-point chlorinated aromatics and hydrocarbons, alcohols, and ketones (EPA, 1987). Air stripping can be used for removal of VOCs from materials having a hydraulic conductivity of as low as 10^{-5} cm/s.

Case History: *Remediation of an Aquifer Contaminated with Trichloroethylene (TCE)*

The following case history illustrates the pump-and-treat method for remediating TCE-contaminated groundwater at an industrial site.

At a 182-ha (450-ac) manufacturing plant, located in the U.S. Midwest in an alluvial geologic setting, TCE leaks and spills over a 20- to 30-year period resulted in groundwater contamination. The contamination plume, 2.4 km × 0.5 km (1.5 × 0.3 mi), was elongated by the groundwater flow down the regional hydraulic gradient.

The contamination was first discovered by the plant owners in 1985 during routine monitoring of onsite groundwater supply wells. Low levels (<10 µg/L) of TCE were discovered in these wells. These supply wells happened to be on the fringe of the contaminant plume, which was subsequently defined by drilling and sampling 62 monitoring wells and test borings along the plume length. The plume was found to contain approximately 5300 liters (1400 gallons) of TCE dissolved in groundwater. At some locations, the TCE concentration ranged as high as 20,000 µg/L; the MCL for TCE in groundwater is 5 µg/L. Unlike the commonly held belief that DNAPLs, because of their high density, will sink to the bottom of the aquifer to form pools, no such occurrence of TCE (density = 1.4 g/mL) was found at the site. This is due to the very slow rate of TCE loading and its high water solubility.

Average thickness of the alluvium is approximately 36.6 m (120 ft), but ranges from 31 m to 38 m (100–125 ft). It consists mainly of fine- to medium-grained sand, with some coarse sand and gravel layers, overlying shale bedrock at a depth of nearly 37 m (120 ft) below the ground surface [elevation about 400 m (1310 ft) above MSL]. The aquifer can be divided into an upper and lower layer defined by a series of clay interbeds occurring at a depth of approximately 14 m (45 ft). The clay beds are not continuous and were not encountered in some of the monitoring wells. This indicates that locally the upper and lower aquifer layers are in good hydraulic communication. The groundwater table occurs at depths of 4.6 to 6.1 m (15–20 ft) below the surface. The hydraulic head in the upper and lower layers is identical, indicating predominantly horizontal flow. Average linear groundwater flow velocity was estimated at 15 cm/day (0.5 ft/day).

Groundwater remediation was done by pumping the contaminated water and treating it in an air-stripper tower. The number of remedial wells, their spacing along the length of the plume, and their flow rate were established using the MODFLOW computer model. Different model scenarios were tested to design a remedial well field that would encompass the plume, form a continuous trough in the water table by overlapping the individual cones of depressions (Figure 13–3, Plate I), and expedite the remediation process in a cost-effective manner. A total of nine remedial wells were installed to pump approximately 3456 liters (913 gallons) of contaminated groundwater per minute. After air stripping, the water was discharged into a major surface-water body. TCE concentrations in the stripper influent were in the range of 400 µg/L; those of the effluent were below the detection limit (0.5 µg/L).

Approximately 80 percent of the TCE contamination was removed from the aquifer in the first six years after the remediation program was begun in 1987. Nearly 11.4 billion liters (3 billion gallons) of contaminated groundwater have been pumped and treated. Sampling of groundwater over time revealed a progressive decline in the TCE concentration. Figure 13–4 (Plate II) shows the drop in TCE levels at three time intervals. The TCE concentrations in several of the monitoring wells located at the perimeter of the plume have been reduced to nondetectable levels. It is planned to vary the pumping rate of individual wells to shift the cone of depressions between the wells to expedite recovery from the zones between the wells. The entire remediation program is expected to extend over 10 to 12 years.

A special aspect of this remediation project relates to the absence of TCE in the deeper parts of the aquifers. As a DNAPL (density = 1.46 g/mL), TCE would be expected to sink to the bottom of the aquifer and collect at the top of the bedrock surface. However, nested wells in the upper and lower aquifer layers, as well as wells screened at the base of the aquifer, showed clearly that the highest concentrations were in the uppermost part of the aquifer; the concentrations in the groundwater at the bedrock surface were all <10 µg/L, even directly beneath the high-concentration zone of the upper aquifer plume. The reason for this departure from the classical DNAPL scenario is the high solubility (>1,000,000 µg/L) and the low loading rate (contamination occurred over a 20- to 30-yr period) of TCE. High solubility and small quantities caused the TCE to dissolve as soon as it entered the groundwater, and advective flow moved it downgradient, preventing its vertical movement (sinking).

This case history illustrates the important principle that dissolved DNAPLs with high water solubility may not sink, but may flow with the groundwater, provided the loading rate is slow enough to allow their advective flow.

(Materials for the case history were provided by Burns & McDonnell Waste Consultants, Inc., Overland Park, KS.)

Reverse Osmosis

Normal osmosis involves flow of a solvent (e.g., water) from a weaker (dilute) solution to a stronger (concentrated) solution through a semipermeable membrane. This results in the reduction of the concentration of the stronger solution.

By applying pressure to the stronger solution greater than the osmotic pressure of the solution (osmotic pressure is the pressure that, when applied to the semipermeable membrane, will prevent the passage of the solvent through the membrane), the solvent can be made to flow from the stronger (concentrated) to the weaker (dilute) solution—hence the term *reverse osmosis*. The process can be used to produce brine concentrate and high-purity water from aqueous salt wastes.

CHEMICAL METHODS

Chemical treatment of hazardous waste utilizes chemical reaction(s) to convert the waste into a chemically neutral or less hazardous material. Treatment technologies are based on common chemical principles, and are discussed briefly in the following sections.

Neutralization

Acids can be used to neutralize caustic (base) waste, and bases can be used to neutralize acidic waste. Neutralization reaction produces a salt and water. The treated material is considered neutral if it has pH range of 6–8. The resulting product(s) should be carefully evaluated for their characteristics before considering them safe for disposal.

Acidic wastewater can be neutralized with slaked lime [$Ca(OH)_2$], caustic soda ($NaOH$), or soda ash (Na_2CO_3). Slaked lime, because of its low cost, is used more frequently than other bases. *Alkaline wastewater* can be neutralized with a strong acid, such as H_2SO_4 or HCl, or with CO_2, which forms carbonic acid with water. CO_2 is injected into the neutralizing tanks, where it combines with water to form carbonic acid (H_2CO_3) that neutralizes alkaline substances. Flue gas contains CO_2 and should be used whenever possible to reduce the treatment cost of alkaline hazardous waste.

Neutralization reactions are exothermic. Therefore, adequate safety measures must be taken to control the buildup of high temperatures.

Precipitation

The addition of certain chemicals to liquid hazardous waste may cause undesirable heavy metals to form a precipitate, which can be easily removed. In addition, the hydroxides of heavy metals are generally insoluble in water; therefore, lime (CaO) can be used to precipitate them. Metal carbonates and sulfides are less soluble than the hydroxides, but may also be precipitated.

Coagulation and Flocculation

When a hazardous waste solution contains very fine particles that cannot be separated by dewatering or clarification processes, coagulating or flocculating agents may be used to separate them.

Inorganic chemicals such as alum [$Al_2(SO_4)_3$], ferric chloride ($FeCl_3$), ferric sulfate [$Fe_2(SO_4)_3$], or polymers are mixed with the hazardous waste solution to allow fine colloidal solids to agglomerate and settle out of the solution. Water-soluble organic polymers are commonly used for coagulation because they are more efficient than alum or iron salts.

Flocculation is agglomeration of colloidal particles that have been coagulated. Gentle mixing helps the colloidal particles to form large flocs. Essentially, flocculation results in formation of large particles that makes separation relatively easy.

Oxidation and Reduction

Oxidation involves an increase in the valence from the loss of electrons; reduction occurs due to an increase in valence from the gain of electrons. Chemical reactions that involve both oxidation and reduction are called *redox reactions*. The following reaction equation shows how oxidation can transform toxic sodium cyanide into nontoxic constituents by treating it with sodium hydroxide and chlorine. This is a two-step reaction: in the first step, highly toxic sodium cyanide is oxidized to the less toxic sodium cyanate, in the second it is oxidized to non-toxic carbon dioxide and nitrogen gas.

$$NaCN + Cl_2 + 2NaOH \rightarrow NaCNO + 2NaCl + H_2O$$
(Na-cyanide)   (an alkali)   (Na-cyanate)

$$2NaCNO + 3Cl_2 + 4NaOH \rightarrow 2CO_2 + N_2 + 6NaCl + 2H_2O$$

Oxidation reactions can also be produced by treating the waste with ozone. Contaminated water containing organic compounds such as alcohols, alkanes, aldehydes, benzene, ketones, and phenols can be effectively treated by ozone and made biodegradable. Introducing ultraviolet (UV) radiation in the reactor chamber enhances ozone-oxidation reactions. Ozone, in combination with UV radiation, has been effectively used to treat groundwater contaminated with low concentrations of chlorinated hydrocarbons and pesticides. Most organic compounds, however, require a second treatment, such as a biological treatment, to completely detoxify the organic compounds.

Reduction reaction can be used to change the highly toxic hexavalent chromium into less toxic trivalent chromium. This can be done by treating a hazardous waste containing hexavalent chromium with SO_2 and slaked lime, $Ca(OH)_2$. The first reaction generates sulfurous acid, which, in the next step, converts the highly toxic hexavalent chromium into the trivalent variety. Finally, slaked lime precipitates trivalent chromium as chromium hydroxide, which is far less toxic.

$$SO_2 + H_2O \rightarrow H_2SO_3$$
(sulfurous acid)

$$2CrO_3 + 3H_2SO_3 \rightarrow Cr_2(SO_4)_3 + 3H_2O$$
(hexavalent Cr)   (trivalent Cr)

$$Cr_2(SO_4)_3 + 3Ca(OH)_2 \rightarrow 2Cr(OH)_3 + 3CaSO_4$$
(Cr-hydroxide, trivalent)

Ion Exchange

Heavy metals and other toxic substances in wastewaters can be removed by an ion-exchange process. The process results in the replacement of the toxic ion with a nontoxic one. In one frequently used process, wastewater is passed through a bed, usually of an organic polymer resin. Both cation exchange and anion exchange resins are used; the former to exchange positive ions, such as Na^+ or H^+, for other positive ions in the wastewater. Anion exchange resins exchange negative ions, such as $(OH)^-$ or Cl^- ions, with other negative ions in the wastewater. Zeolite, a common mineral of the silicate family, is an inorganic ion exchanger that is used in many ion-exchange processes.

BIOLOGICAL METHODS

Biological treatment is the use of microorganisms to degrade or convert hazardous wastes into nonhazardous or less hazardous materials. Biological treatment can be used to remove hazardous constituents from contaminated waste-water, contaminated groundwater, landfill leachates, and contaminated soil. *Bioremediation* is a term used for cleanup of a contaminated site through the use of microbes.

Microorganisms (bacteria and fungi) utilize organic matter as nutrients for their growth and survival. Existing biological treatment methods work most effectively for organic contaminants present in wastewaters, in groundwater, or when they exist as a separate solid phase. The method has not been widely used for treatment of inorganic contaminants, although current research results indicate that, in the future, bioremediation may be used for treating inorganic constituents in hazardous wastes.

Biological treatment was successfully used to clean up beach contamination following the March 23, 1989, release of 11 million gallons (41.6 million liters) of crude oil, covering 2331 km^2 (900 mi^2) of water at Bligh Island in Prince William Sound in south Alaska. Using the hydrocarbon-oxidizing microbes in combination with inorganic nutrients such as nitrogen and phosphorus, large tracts of oil-contaminated beaches and waters were cleaned up within a few months of the spill.

Aerobic bacteria use the organic constituents in the hazardous waste as nutrients, converting them into carbon dioxide and water. This process is called *mineralization.* Anaerobic bacteria decompose the organic contaminants into methane and carbon dioxide through the process known as *methanogenesis.*

Ideally, an organic waste could be degraded to the point that it either does not leave any residue or leaves only recyclable residues. The term mineralization is often used for this type of biodegradation, which results in the formation of CO_2, H_2O, and other ions. Mineralization can be defined as the process that causes conversion of complex molecules present in organic compounds into simple chemical forms.

Mineralization usually involves several steps of chemical reactions where the complex organic molecules progressively change into simpler compounds. For example, bioremediation of alkane (an aliphatic compound, such as methane or ethane) undergoes four intermediate reactions before it is completely mineralized:

$$CH_3-(CH_2)_n-CH_3 + O_2, 2H \rightarrow CH_3-(CH_2)_n-CH_2OH + H_2O$$
(alkane)   (water)   (alcohol)

$$CH_3-(CH_2)_n-CH_2OH-2H \rightarrow CH_3-(CH_2)_n-CHO$$
(alcohol)   (reduction by microbe)   (aldehyde)

$$CH_3-(CH_2)_n-CHO-2H \rightarrow CH_3(CH_2)_n-COOH$$
(aldehyde)   (reduction by microbe)   (fatty acid)

$$CH_3(CH_2)_n-COOH + \text{ß-oxidation} \rightarrow CO_2 + H_2O$$
(fatty acid)   (by microbe)   (mineralized)

Microbial activities leading to mineralization may take place under aerobic or anaerobic conditions, or both. Table 13–1 lists some organic compounds and their oxygen requirements.

TABLE 13–1   Oxygen Requirements for Microbial Degradation (adopted from Schneider and Billingsley, 1990)

Organic Compound/Microbial Metabolic Processes	Oxygen Requirement
Biodegradation of organic pollutants	
Petroleum hydrocarbons	Aerobic
Alkylpyridines	Aerobic/anaerobic
Creosote chemicals	Aerobic/anaerobic
Coal gasification products	Aerobic
Sewage effluent	Aerobic
Halogenated organic compounds	Aerobic/anaerobic
Nitriloacetate (NTA)	Aerobic/anaerobic
Pesticides	Aerobic/anaerobic
Nitrification	Aerobic
Denitrification	Anaerobic
Sulfur oxidation	Aerobic
Sulfur reduction	Anaerobic
Iron oxidation	Aerobic
Iron reduction	Anaerobic
Manganese oxidation	Aerobic
Methanogenesis	Anaerobic

Microorganisms

Microorganisms are made up of common elements that are also found in other living forms. Table 13–2 gives the elemental composition of microorganisms. Microorganisms can be classified in several ways, depending on their occurrence, utilization of oxygen, and nutrient source:

- Based on occurrence:

    1. *Indigenous.* Occur at the site *in situ*

    2. *Exogenous.* Imported to the (remediation) site

TABLE 13–2   Elemental Composition of Microorganisms (adopted from Schneider and Billingsley, 1990)

Element	Percent of Total Dry Weight
Oxygen	65.0
Carbon	18.0
Hydrogen	10.0
Nitrogen	3.0
Phosphorus	1.0
Potassium	0.4
Sulfur	0.3

Indigenous microorganisms are preferable for biological treatment because of lower project cost and acceptability (indigenous microorganisms alleviate the concern arising from the introduction of foreign bacteria to the site).

- Based on their utilization of oxygen:

    1. *Strict aerobes.* Must have oxygen to survive

    2. *Obligate anaerobes.* Cannot survive in the presence of free oxygen (use the oxygen present in combination with other elements, such as nitrates, NO_3, phosphates, PO_4 and sulfates, SO_4)

    3. *Microaerophilic.* Can survive only in a limited supply of oxygen

    4. *Facultative anaerobes.* Can survive both in the presence or absence of oxygen (use aerobic process when O molecules are available; anaerobic process when O is lacking)

- Based on their source of nutrients:

    1. *Heterotrophic.* Use organic compounds for nutrient source

    2. *Autotrophic.* Use inorganic compounds, PO_4, NO_3, and SO_4

Table 13–3 lists microorganisms that may be used for biodegradation of aliphatic and aromatic compounds.

Phases of Bacterial Growth

In a batch biological treatment system, four distinct phases of microorganism growth have been recorded (Jackman and Powell, 1991) (Figure 13–5). In the beginning, when the hazardous waste and the microorganisms have been combined, the reproduction rate is very slow because the microorganisms take time to acclimate to the changed chemical environment presented by the hazardous waste. After the acclimation period is over, the microorganisms grow extremely rapidly because of the plentiful supply of nutrients (food). This is the logarithmic growth phase, B. The source of food is called *organic substrate.* After some time,

TABLE 13–3   Organisms Capable of Degrading Hydrocarbons (adopted from Schneider & Billingsley, 1990)

Aliphatic		Aromatic	
Acinetobacter calcoaceticus	*Mycobacterium parafficum*	*Acineobacter*	*Phanerochaete chrysoporium*
Arthrobacter paraffineus	*Mycobacterium smegmatis*	*Agrobacterium*	*Pseudomonas acidovorans*
Arthrobacater simplex	*Nocardia petrooleophila*	*Alcaligenes*	*Pseudomonas putida*
Canada lypolyticum		*Aspergillus*	*Pseudomonas testosteroni*
Cephalosporium roseum	*Pseudomonas aeruginosa*	*Azotobacter*	*Pseudomonas testosteroni*
Corynebacterium glutamicum	*Pseudomonas fluorescens*	*Bacillus*	*Rhizobium*
Flavobacterium	*Torulopsis colliculosa*	*Bradyrhizobium*	*Rhodotorula*
		Candida	*Trichosporon*
		Moraxella	
		Neurospora	
		Nocardia	

FIGURE 13–5 Microbial growth and nutrient consumption (adopted from Jackman and Powell, 1991)

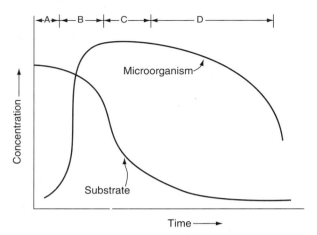

the substrate becomes limited, causing a decline in the growth rate and leading to death of a part of the microbe population. Activity does not completely cease, because the dead microbes serve as the substrate for the remaining population; this is the third phase, C. In the fourth phase, the *endogenous phase*, because of reduced substrate the microbes are forced to metabolize the polysaccharide slime layer that coats their cell walls, creating denser and heavier cells. Collision of such cells causes them to agglomerate and settle. On the other hand, if the substrate is minimal, the slime layer will decrease to the extent that no adhesion of cells is possible, and agglomeration will cease.

These four phases are of great importance in designing a biological treatment system. For instance, the activated sludge system, which is an example of the *suspended growth system*, performs optimally when the cell density is maximal and a thin slime layer is maintained; cells' adhesion upon collision will result in agglomeration. Other treatment systems, called *supported growth systems*, exemplified by trickling filters and rotating biological contactors, are designed for optimal performance in the logarithmic growth phase, when the system is capable of supporting a dense, sticky slime. Biological treatment utilizing aerated lagoons and anaerobic digestion takes advantage of the endogenous phase, which enables the system to have a long retention time.

Factors Affecting Bacterial Growth

Bacteria reproduce by fission, i.e., the parent cell splits into new cells. When a good supply of nutrients is available, they reproduce rapidly. The most important nutrients are nitrogen, phosphorus, sulfur, potassium, calcium, and magnesium; these are called the *macronutrients*, and are utilized in large quantities. There are other nutrients, which, despite being utilized in very minute amounts, are nonetheless essential to support bacterial growth. These elements are designated *micronutrients*, and include iron, boron, copper, manganese, zinc, chromium, and cobalt (Jackman and Powell, 1991). In addition, pH, temperature, and moisture availability are important factors that control bacterial population.

pH

A pH range of 6 to 8 is most suitable for maintaining the required bacterial population. Only a few species can grow at pH values of less than 2 or greater than 10. Ideally, pH of 7.0—neutral—is the best condition. According to Dibble and Bartha (1979), a soil pH of 7.8 is optimal for bioremediation of petroleum-contaminated soils. If, during biological treatment, pH changes beyond the 6–8 range occur, it must be controlled and conditions must be adjusted.

Temperature

Microorganisms are known to survive in a temperature range of less than 0°C to over 100°C. In general, though, microbial growth is drastically reduced when the temperature falls below 0°C. Freezing temperatures prevent microbial growth, but do not always result in death of the microorganisms. Figure 13–6 shows the relationship between microbial population and temperature. Common bacteria have optimal activity in the temperature range of 15°C to 50°C. Some bacteria thrive in warm to hot temperatures, such as the *thermophiles* (40°–60°C) and *hyperthermophiles* (80°C). Below 15°C and above 50°C, significant decline in microbial activities of common bacteria occcurs. Aerobic treatment systems, routinely used for municipal and industrial wastewater treatment, operate at a temperature range of 20°C to 50°C; anaerobic treatment systems operate at temperature range of 35°C to 50°C.

Freezing temperatures present the greatest threat to bioremediation. For small sites, additional heating (electric heating cables, infrared lamps) may be used to prevent freezing. For larger sites, covering the treatment site with polyethylene sheets may be advantageous.

Moisture Availability

Moisture (or water) availability is one of the most imporant factors controlling microbial growth. In a given situation, the availability of water to the microbes is more important than the total amount of water in the medium. For example, a considerable amount of water may be adsorbed to clayey soil, but it may not be

FIGURE 13–6 Relationship between temperature and microbial population

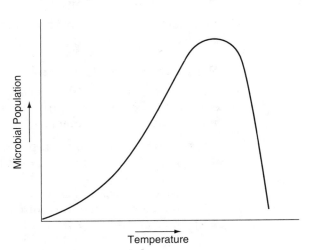

readily available for microbial growth. The term *water activity (a_w)*, or *water potential* is used to express the physiological water availability. Generally, microbes thrive in the a_w range of 0.9 to 1 (water activity in agricultural soils ranges between 0.9 and 1). Most bacteria are unable to survive in environments of low water activity, and either die or remain dormant. Only a few specialized bacteria can live under conditions of low water activity; e.g., the halophile *Halobacter*, which is known to grow in the Dead Sea, where a_w is 0.7 (Baker and Herson, 1994).

Concentrations of Toxic Substances

High concentrations of inorganic constituents, such as arsenic, cyanide, copper, lead, and zinc, inhibit microbial enzyme formation resulting in death of the microorganisms. Therefore, only the wastes containing low concentrations of these inorganic contaminants are suitable for biological treatment.

Concentrations of Organic Substances

Two parameters—the BOD (Biochemical Oxygen Demand) and COD (Chemical Oxygen Demand)—are commonly used to characterize organic substances in wastewater. BOD can be defined as the amount of oxygen used by microorganisms to remove organic substances from a unit volume of wastewater; it is measured as milligrams per liter of oxygen consumed by microorganisms over a five-day period. As the amount of organics in the wastewater increases, more oxygen is used, resulting in a higher BOD, but a greater depletion of oxygen in the water. The reduced oxygen level proves detrimental to other aquatic life forms, such as fish, which become threatened or may actually die. Dissolved oxygen concentration of >9 mg/L promotes a good fish population; below 5 mg/L the water is considered polluted (Keller, 1992). Chemical Oxygen Demand is based on utilization of certain chemicals, instead of microorganisms, to oxidize organic pollutants in the wastewater. Unlike the BOD test, which takes five days to complete, the COD test only takes three hours.

Oxygen Concentrations

Most aerobic treatment processes operate efficiently when the dissolved oxygen in the wastewater is >1 mg/L. No free oxygen is required for anaerobic processes.

Nature of Organic Constituents

Most organic constituents in wastewater are amenable to bioremediation; however, certain compounds are easier to biodegrade than others. The following generalizations can be made:

- Aliphatic organic compounds (carbon atoms arranged as straight or branched chains, e.g., *alkenes:* at least one carbon is doubly bonded [-C=C- bonds], such as ethylene; *alkynes:* at least one C has triple bond, [-C≡C-], such as acetylene, alcohols, and aldehydes, which are easier to biodegrade than the aromatic compounds containing single or multiple benzene rings, i.e., phenols [$C_6H_5O_6$], toluene [$C_6H_5CH_3$], etc. ; *alkanes:* all C is singly bonded [-C-H- and -C-C-], amines, and nitro-group compounds [nitrous oxide bonded to C, e.g., -CNO_2]).

- Halogenation makes organic compounds more resistant to biodegradation. Such compounds have the lowest biodegradability. In general, the greater the degree of halogenation, the more difficult the biodegradation of the compound is. Table 13–4 lists chemical compounds that are generally biodegradable.
- Introduction of oxygen into the organic compound makes it easy to biodegrade.

Shock Loading

Shock loading refers to a sudden increase in concentrations of contaminants in the media being bioremediated. Concentrations of suspended solids, BOD, and

TABLE 13–4 Organic Compounds Generally Recognized as Biodegradable by Oxidation, Co-Metabolism, or Anaerobic Degradation (adopted from Schneider and Billingsley, 1990)

Atrazine	Ethyl benzene
Acenaphthalene	Etyl glycol
Acenanaphthane	Fluroanthenefluorene
Acetone	Indeno(1,2,3)pyrene
Acrylonitrile	Lindane
Anthracene	2-methyl naphthalene
Benzene	Methylene chloride
Benzo anthracene	Methylene ketone
Benzoic acid	Methylmethacrylate
Benzo fluoranthene	Naphthalene
Benzo pyrene	Nitroglycerine
Benzo perylene	Nonane
Butanol	Octane
Butylcellosolve	Pentachlorophenol
Carbon tetrachloride	Phenantherene
Chlordane	Phenol
Chloroform	Phytane
Chrysene	Pristane
P-cresol	Pyrene
Dibenzo anthracene	Styrene
DDT	Tetrachloroethylene
Dicholorobenzene	1,1,1-trichloroethane
Dichloroethane	Trichloroethylene
Dichloroethylene	Tridecene
Dioxane	Trinitrotoluene
Dioxin	Vinyl chloride
Dodecane	Xylene

toxic materials must be increased in a controlled manner to allow the microorganisms to tolerate shock loading.

Future Trends in Bioremediation

Research in bioremediation is progressing rapidly. In addition to using known microorganisms, a great deal of attention is being focused on plants and on genetically engineered microorganisms (GEMs) for bioremediation of hazardous wastes. It is well known that some plants are capable of selectively removing certain metals from the soil; such plants are called *accumulator plants.* As far back as 1949, a plant species of the genus *alyssum* was used to remove nickel from soil. Since then, many other plants that are capable of accumulating large quantities of metals (Co, Cu, Ni, and other elements), called *hyperaccumulators*, have been identified. These plants can accumulate 0.1 to 5 percent of metals in their dry leaves. It has even been suggested that such plants could serve as a source of metals (Colwell, et al., 1993).

Development of new strains of GEMs raises the issues of safety and public concern. If the GEMs were to be released into the natural environment, the consequences could be drastic and far reaching. To avoid this possibility researchers have proposed incorporating a "suicide gene" into bacteria, combining the killing function with the biodegradative function so that the bacteria will die after metabolization of the substrate—when the nutrient supply from the hazardous waste is depleted. Research is underway to produce bacteria that will self-annihilate after completion of biodegradation (Colwell, et al., 1993).

The bioremediation market is on the upswing. It generated about $125 million worth of business in 1992, and is expected to reach the $550 million mark by the year 2000. At 159 hazardous waste sites in the United States, bioremediation either is being implemented or is under consideration (Glass, 1994). At the same time, larger sums of money are being allocated for bioremediation research. According to Glass (1994), the EPA's fiscal 1993 budget for bioremediation research was about $10 million; the Department of Defense was to spend $11 million. The Department of Energy had set aside more than $30 million for biotechnology research and development.

With the growing market and intensive research, combined with successful remediation, bioremediation will become more popular in the years ahead. Use of indigenous bacteria will probably still lead the treatment processes, but some GEMs may also be approved, at least on a trial basis. However, batch treatment of a hazardous waste site—using a mixture of bacteria, each with a preference for biodegrading a particular compound, applied over a period of time—may gain more popularity. For example, a hazardous waste stream containing volatile, semivolatile, and aliphatic organic compounds has been successfully treated by sequentially introducing one microbe species that performed biodegradation of the volatiles (Colwell, et al., 1993), followed by the introduction of another set of microbes to biodegrade the nonvolatiles, and so on. It was also demonstrated that biodegradation was greatly inhibited when *all* microbial species were used simultaneously. Furthermore, it was shown that selecting the proper species of bacteria that thrive over certain temperature ranges (cold, winter temperatures or warmer spring and fall temperatures) enabled the bioremediation process to operate throughout the year.

MISCELLANEOUS METHODS

Waste Immobilization/Stabilization

Stabilization involves reducing or limiting the ultimate release of hazardous constituents from a waste stream. A hazardous waste in liquid form must first be solidified by removing water. The solid matrix can then be mixed with lime, Portland cement, fly ash, bottom ash, or molten thermoplastic materials (asphalt, bitumen, polyethylene, polypropylene), glass, and the like, to bind the solids together to form a stabilized mass.

Stabilization generally increases the volume of waste, but the stabilized mass can be more easily and safely transported to a disposal facility, and can be disposed of in a conventional landfill. Prior to disposal, however, the stabilized waste must be tested to ensure that hazardous constituents will not leach out of the mass.

Ground Freezing

Ground freezing involves use of a refrigerant to freeze the moisture in the pores of soils to produce a hard, rock-like mass that can stand steep to vertical cuts for a short period of time. This technology has been used in geotechnical and mining applications for shaft sinking and excavation through loose, water-saturated soil materials for several decades. As recently as 1992, the ground-freezing technique was used for containment of hazardous waste and for reducing the volatilization of harmful contaminants in a hazardous waste. Normally, removal of soil contaminated with toxic chemicals with high vapor pressures will be difficult, but solidification of contaminated soil by freezing eliminated toxic vapors and allowed easy removal of the soil (Sopko and Aluce, 1993).

Soil Washing

This treatment technology has been in use in Europe since the 1980s; it is now being introduced in the United States. Soil washing dissolves, or forms a suspension of, the contaminants in a solution. The technology, initially used for treatment of coarse-grained soils, which have less surface area (sand and gravel), contaminated with organic and inorganic contaminants, has been modified to treat fine-grained soils. Depending on the nature of the contaminants, either water or leachants can be used to dissolve or make a suspension of the contaminants for subsequent removal. These operations greatly reduce the volume of the contaminated soils that must be disposed of. At a Superfund site (No. NJD 980505341) in Camedon County, New Jersey, fine-grained soil contaminated with copper, chromium, and nickel, was successfully remediated with soil washing. The treatment reduced the volume of contaminated soil by 80 percent (Environmental Protection, 1993). As of late 1993, there were more than 14 vendors offering soil-washing services, and with increased EPA recommendations for this technology, soil washing appears to hold great promise.

STUDY QUESTIONS

1. What is hazardous waste treatment? Why is it important to treat the hazardous waste before its disposal? What kind of hazardous waste can be disposed of without treatment?
2. What are the main categories of the methods of hazardous waste treatment? Give examples of each.
3. What is a multimedia filter system? Describe one such system. How can the system be regenerated?
4. How does the activated charcoal remove contaminants from a liquid hazardous waste? Is it possible to reclaim any contaminants from the waste stream using this method? Explain.
5. What is batch distillation? Explain how VOCs having different vapor pressures can be reclaimed from the waste stream using batch distillation.
6. What is air stripping? What other gases can be used to remove hazardous substances from a waste stream? Is this method suitable for removing VOCs from fine-grained soils? What is the lower limit of hydraulic conductivity where air stripping can be used?
7. Explain the difference between air stripping and air sparging. What is biological air sparging?
8. Referring to the case history on the remediation of a TCE-contaminated aquifer, why was TCE, despite being a DNAPL, found to be present at shallow depth in the aquifer and not at the bottom?
9. What is coagulation? Which chemicals are commonly used in coagulation? How is flocculation different from coagulation?
10. What is bioremediation? Discuss the various phases of microbial growth in a biological treatment system designed for cleanup of a hydrocarbon-contaminated soil.
11. What factors control microbial growth? What is the general range of pH and temperature over which biological treatment works optimally? How does the concentration of toxic substances in the waste affect microbial activities?
12. What is the future of bioremediation? Do you think the use of GEMs will become common? Explain your answer.
13. What is waste immobilization? What materials can be used to immobilize harmful constituents in a hazardous waste?

REFERENCES CITED

Baker, K.H., and D.S. Herson, eds., 1994, Bioremediation: New York, McGraw-Hill, Inc., 375 p.

Brock, T.D., M.T. Madigan, J.M. Martinko, and J. Parker, 1994, Biology of Microorganisms (7th ed.): Englewood Cliffs, NJ, Prentice-Hall, Inc., 667 p.

Colwell, R., M.A. Levin, and M.A. Gealt, 1993, Future direction in bioremediation, *in* R. Levin and M.A. Gealt (eds.): Biotreatment of Industrial and Hazardous Waste, New York, McGraw-Hill, Inc., p. 309–321.

Dibble, J.T., and R. Bartha, 1979, Effect of environmental parameters on the biodegradation of oil sludge: Applied Environmental Microbiology, vol. 37, p. 729–739.

Environmental Protection, 1993, Cleaning up with borrowed technology: Environmental Protection, v. 4, n. 10, p. 8.

Glass, D.J., 1994, Exploring the U.S. Market for Hazardous Waste Bioremediation: TNEJ (The National Environmental Journal), v. 4, n. 1, p. 56.

Jackman, A.P., and R.L. Powell, 1991, Hazardous Waste Treatment Technologies: Park Ridge, NJ, Noyes Publications, 276 p.

Keller, E.A., 1992, Environmental Geology (6th ed.): New York, Macmillan Publishing Co., 521 p.

Sopko, J.A., and G.F. Aluce, 1993, Ground Freezing for Containment and Remediation: Peawaukee, WI, Layne-Northwest Company, brochure, unpaginated.

Schneider, D., and R. Billingsley, 1990, Bioremediation: A Desk Manual for the Environmental Professional: Northbrook, IL, Pudvan Publishing Co., 97 p.

U.S. Environmental Protection Agency, 1987, Compendium of Technologies Used in the Treatment of Hazardous Waste: U.S. EPA Center for Environmental Research Information, Cincinnati, OH, Report No. EPA/625/8-87/014, 55 p.

U.S. Environmental Protection Agency, 1990, The Superfund Innovative Technology Evaluation Program: U.S. EPA Office of Solid Waste and Emergency Response, Report No. EPA/540/5-90/001, 83 p.

SUPPLEMENTAL READING

American Chemical Society Symposium Series No. 422 (1989), 468 (1990), 518 (1991) and 554 (1992): Emerging Technologies in Hazardous Waste Management; Washington, D.C., American Chemical Society.

Freeman, H.M., ed., 1989, Standard Handbook of Hazardous Waste Treatment and Disposal: New York, McGraw-Hill, Inc., 14 chapters plus index.

Wentz, C.A., 1989, Hazardous Waste Management: New York, McGraw-Hill Publishing Co., 461 p.

C
H
A
P
T
E
R

14

Methods of Hazardous Waste Disposal

INTRODUCTION

Disposal of hazardous waste constitutes a vital link in the cradle-to-grave concept of hazardous waste management. The current emphasis on pollution prevention (Chapter 6) and development of innovative treatment technologies has resulted in the reduction of the volume of hazardous waste that must be disposed of, but hazardous wastes cannot be completely eliminated; there will always be some quantity of such waste that will need disposal.

Hazardous wastes can be disposed of in a variety of ways. Most of the commonly used methods fall into three categories:

1. Thermal Methods:
   - Incineration
   - Pyrolysis
   - Other

2. Land Disposal:
   - Landfill
   - Land farm (land treatment)
   - Waste pile
   - Surface impoundment
   - Deep well injection
   - Underground disposal (mines, caverns)
   - Concrete vaults and bunkers

3. Ocean disposal (dumping):

THERMAL METHODS

Incineration

Incineration utilizes heat and oxygen (from air) to destroy the organic fraction of a waste stream. Incineration requires high temperatures, generally 900°C or more. From a chemical point of view, incineration represents an *exothermic oxidation process* that converts organic compounds into carbon dioxide and water (steam), with an accompanying release of heat.

The waste streams that are fed into incinerators may not be entirely comprised of organic compounds; in fact, inorganic materials, including metals and glass, are frequently found in the waste streams. Hazardous metals and nonmetals in the waste stream undergo oxidation as a result of incineration; for example, metallic copper changes into copper oxide, sodium into sodium hydroxide, potassium into potassium hydroxide, fluoride into hydrogen fluoride or fluorine, chloride into hydrogen chloride, and carbon into carbon dioxide. Other waste constituents either go into the residual ash or into the flue gas after incineration, and require special handling, treatment, and disposal.

Performance standards of incinerators for hazardous waste disposal are regulated under RCRA. The key requirements are

- An incinerator must achieve a destruction and removal efficiency (DRE) of 99.99 percent for each principal organic hazardous constituent (POHC) for each waste feed. DRE has been defined as

$$[(W_{in} - W_{out}) \div W_{in}] \times 100 \tag{14-1}$$

  where W_{in} = the mass feed rate of one principal POHC in the waste stream, and W_{out} = the mass emission rate of the same POHC in exhaust emissions prior to release into the atmosphere.

- An incinerator burning hazardous waste and producing stack emission of hydrogen chloride (HCl) must control its emission to less than 1.8 kg/hr (4.0 lb/hr), or 1 percent of the HCl in the stack gas prior to its entry into any pollution-control equipment.

- An incinerator must not emit particulate matter exceeding 180 mg/dscm (day standard cubic meter) corrected for the amount of oxygen in the stack gas, according to the formula

$$P_c = P_m \frac{14}{21-Y} \tag{14-2}$$

  where P_c = the corrected concentration of particulate matter, P_m = the measured concentration of particulate matter, and Y = the measured concentration of oxygen in the stack gas, using the Orsat method for oxygen analysis of dry flue gas.

RCRA covers all hazardous wastes except PCBs, which are considered a toxic material and are regulated under the Toxic Substances Control Act (TSCA). Rules for destruction of PCBs using incineration, established under TSCA, require that PCBs be disposed of by incineration unless its concentration is less than 50 ppm. If the concentration is between 50 and 500 ppm, the PCB waste can be used as fuel in high-efficiency boilers. The general requirements of TSCA for incineration of liquid PCBs are

- Liquids fed into the incinerator must be maintained for a residence time of 2 seconds at a temperature of 1200°C (±100°C) and 3 percent excess oxygen in the stack gas. Or the liquids must be maintained for a residence time of 1.5 seconds at 1600°C (±100°C) and 2 percent excess oxygen in the stack gas. Using these conditions, a DRE of ≥ 99.9999 percent must be achieved for liquid PCBs.
- Combustion efficiency shall be at least 99.99 percent, computed by the equation

$$C_{co_2} \div (C_{co_2} + C_{co}) \times 100 \qquad (14\text{-}3)$$

where C_{co_2} = the concentration of carbon dioxide, and C_{co} = the concentration of carbon monoxide.

For nonliquid PCBs, the requirement is that mass air emission from the incinerator must not be >0.001 g of PCB per kg of the PCB introduced into the incinerator; this corresponds to a DRE of ≥99.9999 percent.

In addition, other requirements concerning monitoring of gases in the emissions, incinerator temperature, PCB feed rate, and installation of scrubbers have been set forth for both liquid and nonliquid PCBs.

Both RCRA and TSCA require a trial burn, or test data equivalent to a trial burn, to demonstrate the capability of a hazardous waste incinerator to comply with the performance standards.

Incineration Technologies

Incineration technology is expensive; an incinerator may cost up to $70 million. Incinerators are gaining popularity, despite their high capital cost, because at present they offer the only method of detoxifying certain wastes; e.g., PCBs and dioxins, and all combustible carcinogens, mutagens, teratogens, and pathological wastes that otherwise may result in transmission of serious diseases. As of August 1993, there were 184 incinerators in the United States burning 5 million tons of hazardous waste per year. In addition, there were 171 industrial furnaces burning hazardous materials. Incineration of hazardous waste also eliminates the vexing problem of leachate formation in landfills, and their subsequent migration that contaminates ground and surface waters. Incineration significantly reduces the volume of waste destined for disposal.

Some of the disadvantages of incineration include downtime of about 10–15 percent, the large amount of ash produced (25 to 35 percent by weight and 10–15 percent by volume), and high cost. Figure 14–1 shows the 1987 cost of disposal of hazardous waste using various methods. Incineration cost is 200 percent greater than the most expensive of the other disposal options, namely landfills. However, stringent design and compliance monitoring have led to a sharp increase in landfilling cost, thus making incineration a competitive alternative for hazardous waste disposal. Various types of incinerators are commercially available. Each is designed to meet specific requirements, to deal with a

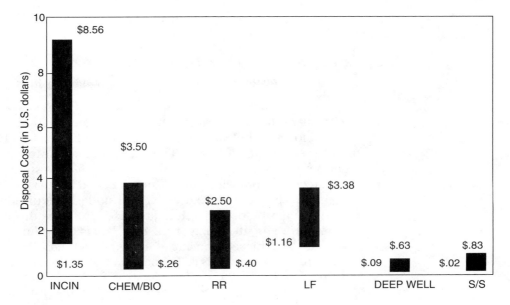

FIGURE 14–1 Comparison of disposal costs, 1987 (adapted from Dempsey and Oberacker, 1988)

particular problem, or to handle a certain physical form of the hazardous waste. Incinerator types include

- Auger combustor
- Catalytic
- Cement kiln
- Cyclonic
- Fluidized bed
- Fume
- Infrared (moving bed)
- Large industrial boiler
- Liquid injection
- Molten salt
- Multiple chamber
- Multiple hearth
- Oxygen-enriched
- Recirculating fluidized bed
- Rotary kiln
- Two-stage (starved air)

Of these, liquid injection, rotary kiln, fixed hearth, and fluidized bed incinerators are most common. (For details, see Freeman, 1989.)

Components of a Hazardous Waste Incinerator

The major components incorporated in the design of hazardous waste incinerators are

- *Waste preparation and feeding.* Design of the feed method is based on the physical form of the hazardous waste. Bulk solids usually require shredding to reduce particle size. These are then fed into the combustion chamber via rams, air-lock feeders, gravity feed systems, vibratory or screen feeders, and belt feeders. Containerized waste is usually gravity or ram fed. Cavity pumps are used to feed sludges. Liquid wastes are converted to a gas before injection, atomized into fine droplets, and then incinerated. The ideal size of the droplets is between 40 and 100 µm, which can be achieved by using atomizers or nozzles.
- *Combustion chamber.* This is the main part of the incinerator, where temperatures on the order of 824°C to 1650°C are maintained for effective destruction of the hazardous waste. Design of combustion chambers is based on the physical form of the waste and its ash content.
- *Air-pollution control.* This is an integral part of an incinerator. Following the Clean Air Amendments Act of 1990, strict requirements have been established for air emission of gases and particulate matter (PM). Air-pollution control systems are designed to capture combustion gases and other products for further treatment before their release into the atmosphere. Common air-pollution control devices, in order of frequency of use, are: quenching systems for cooling and conditioning of gases, high-energy venturi scrubbers for removal of PM, wet scrubbers, wet electrostatic precipitators, packed tower absorbers, and spray tower absorbers.
- *Residue/ash handling system.* The inorganic fractions of hazardous waste produce ash and PM. The PM goes into the flue gas, and the ash usually settles in the bottom of the combustion chamber, in scrubber waters or in other air-pollution control devices. The ash is periodically removed and suitably treated before disposal. Figure 14–2 is a generalized flow diagram of an incinerator system.

Pyrolysis

The process of pyrolysis also utilizes heat for destruction of hazardous waste constituents. However, it differs from incineration in that the decomposition process involves an *endothermic* reaction that occurs at a lower temperature in the absence of oxygen. The general temperature range for pyrolysis is between 425°C and 750°C.

Pyrolysis is a two-step process. In the first step, heating the mixed waste at lower temperatures (425°–750°C) results in the separation of volatile fractions from the nonvolatiles. In the second step, the volatiles are burned in a fume incinerator to achieve DRE of 99.9999 percent, leaving behind ash (solid residue). The two-step process allows for precise control of temperature and requires smaller equipment.

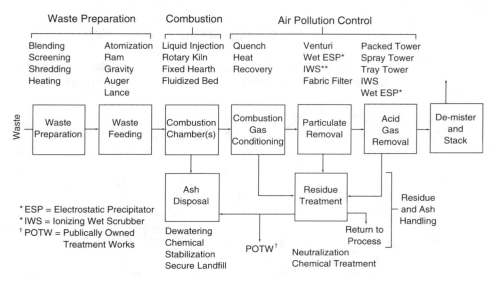

FIGURE 14–2   Generalized flow diagram of an incinerator system

Like incineration, several variations of pyrolysis technology are commercially available. Various modifications are designed to allow recovery of useful compounds from the volatile fraction and/or from the solid residue.

The types of hazardous wastes that are especially amenable to pyrolysis are

- Containerized wastes; e.g., drums that cannot be easily drained
- Sludges or liquids with (a) high ash contents, (b) volatile inorganic compounds (NaCl, FeCl$_2$, Zn and Pb), or (c) high concentrations of Cl, S, and/or N

Other Thermal Destruction Methods

Other methods of destruction of hazardous waste using high temperature include: molten salt combustion, calcination, wet air oxidation, industrial boilers, and furnaces. Facilities using these methods are regulated under RCRA (40 CFR, Part 270) and are required to perform trial burn or similar tests.

Molten salt combustion utilizes simultaneous combustion and sorption to burn organic constituents and sorb the objectionable combustion products from the gas stream, respectively. The hazardous waste, in the presence of air, is mixed together in the combustion chamber, where molten sodium carbonate is maintained at a temperature of 815°C to 1100°C. Organic constituents, such as the hydrocarbons, oxidize to carbon dioxide and water; the inorganic constituents, such as arsenic, halogens, phosphorus, and sulfur, react with sodium carbonate to form salts that are retained in the melt. Sodium carbonate has to be periodically replaced to maintain its efficiency and its ability to absorb the acidic products.

LAND DISPOSAL

Land disposal refers to disposal of hazardous waste, after proper treatment, on the land—both at the surface and in the subsurface. Land disposal requires prior conversion of hazardous waste into nonhazardous products through the processes of degradation, transformation, or immobilization. Landfills, surface impoundments, waste piles, land farms, and concrete vaults and bunkers are examples of disposal on the land; deep well injection and disposal in salt domes, bedded salt formations, and mines and caverns are examples of subsurface (underground) disposal.

Regulatory Aspects of Land Disposal

In view of the threat to the environment, particularly to the quality of groundwater resulting from improper disposal of hazardous wastes on the land, severe restrictions were imposed for land disposal of hazardous waste under the 1984 Hazardous and Solid Waste Amendments (HSWA) to the 1976 Resource Conservation and Recovery Act (RCRA). The law required the EPA to develop appropriate regulations to enforce these restrictions, which came to be known as the Land Disposal Restrictions (LDR). Details of the LDR are given in 40 CFR, Parts 124 and 260–271 (40 CFR, 1993). The EPA publication *Land Disposal Restrictions: Summary of Requirements* (EPA, 1991) is a good and concise source of information on LDR.

All land disposal is regulated under RCRA except deep well injection, which is regulated under the 1986 Amendments to the Safe Drinking Water Act (SDWA). The LDR were implemented over a period of four years to allow the EPA sufficient time to evaluate hazardous wastes that will be covered under the new regulations and to develop appropriate rules and methodologies, such as design of secure landfills, groundwater monitoring requirements, and leachate management systems. Several regulations were developed during this period to address land disposal of wastes containing certain hazardous constituents. These are explained in the following sections:

- *Solvents and Dioxins Rule.* This rule restricts land disposal of hazardous wastes containing solvents (codes F001 to F005) and dioxins (codes F020 to F023, and F026 to F028). It requires that spent solvent wastes be treated prior to disposal on land. Spent solvent was defined as a discarded solvent that is no longer usable without being regenerated, reclaimed, or otherwise reprocessed. Degreasers, cleaners, fabric scourers, diluents, extractants, and reaction and synthetic media are examples of spent solvents. Details are included in 40 CFR, §§ 268.30 and 268.31. The rule groups solvent wastes into two categories, defined as

  1. *Wastewaters.* Solvent-water mixtures containing less than one percent (by weight) of total organic carbon (TOC), or less than one percent (by weight) of total solvent constituents

  2. *Nonwastewaters.* All other spent solvent wastes that contain more than one percent TOC

  Solvent- and dioxin-containing wastes cannot be land disposed unless the waste extract meets the Maximum Concentration Limits (MCL) or Treatment Technology-Based Standards. Waste extracts are obtained using the criteria set under the Toxicity Characteristic Leaching Procedure (TCLP). MCL for wastewater and nonwastewaters for F001–F005 spent solvent wastes are given in Table 14–1. Any waste

TABLE 14–1 Concentration Limits for Restricted Chemicals Found in Wastewaters and Nonwastewaters* (adapted from EPA, 1991)

Constituents of F001–F005 Spent Solvent Wastes	Extract Concentrations (mg/L)	
	Wastewaters	Nonwastewaters
Acetone	0.05	0.59
n-Butyl alcohol	5.00	5.00
Carbon disulfide	1.05	4.81
Carbon tetrachloride	0.05	0.96
Chlorobenzene	0.15	0.05
Cresols (cresylic acid)	2.82	0.75
Cyclohexanone	0.125	0.75
1,2-Dichlorobenzene	0.65	0.125
Ethyl acetate	0.05	0.75
Ethylbenzene	0.05	0.053
Ethyl ether	0.05	0.75
Isobutanol	5.00	5.00
Methanol	0.25	0.75
Methylene chloride	0.20	0.96
Methyl ethyl ketone	0.05	0.75
Methyl isobutyl ketone	0.05	0.33
Nitrobenzene	0.66	0.125
Pyridine	1.12	0.33
Tetrachloroethylene	0.079	0.05
Toluene	1.12	0.33
1,1,1-Trichloroethane	1.05	0.41
1,1,2-Trichloro-1,2,2-trifluorethane	1.05	0.96
Trichloroethylene	0.062	0.091
Trichlorofluoromethane	0.05	0.96
Xylene	0.05	0.15

*Solvent-water mixtures containing <1 weight percent of total organic carbon (TOC) or <1 weight percent of total solvent constituents are called wastewaters. All other spent solvent wastes containing >1 weight percent of TOC are called nonwastewaters.

exceeding the MCL cannot be land disposed unless treated to bring the concentration down to the levels listed in Table 14–1. Dioxin-containing wastes that exceed the maximum concentration levels (Table 14–2) are also prohibited from land disposal. Dilution of a waste stream as a substitute for treatment is prohibited by law.

The EPA has also established Treatment Technology-Based Standards for certain wastes. For example, standards for dioxins are based on incineration technology that achieves a DRE of 99.9999 percent (commonly referred to as six nines—6-9s). Any dioxin-containing waste that, after

TABLE 14–2   Concentration Limits of Constituents in Dioxin Waste Extract (adopted from 40 CFR, 268.41)

Constituent	Concentration
HxCDD - All Hexachlorodibenzo-*p*-dioxins	< 1 ppb
HxCDF - All Hexachlorodibenzofurans	< 1 ppb
PeCDD - All Pentachlorodibenzo-*p*-dioxins	< 1 ppb
PeCDF - All Pentachlorodibenzofurans	< 1 ppb
TCDD - All Tetrachlorodibenzo-*p*-dioxins	< 1 ppb
TCDF - All Tetrachlorodibenzofurans	< 1 ppb
2,4,5-Trichlorophenol	< 0.05 ppm
2,4,6-Trichlorophenol	< 0.05 ppm
2,3,4,6-Tetrachlorophenol	< 0.10 ppm
Pentachlorophenol	< 0.01 ppm

incineration, meets the 6-9s criteria will be acceptable for land disposal. For a complete list of technology-based treatment standards for RCRA Hazardous Wastes, refer to 40 CFR, § 268.42

- *California List Rule.* The state of California (in 1983) had banned land disposal of certain hazardous wastes comprised mainly of liquids containing cyanides, heavy metals, PCBs, HOCs, and acids of pH ≤ 2.0. The EPA included the California List Rule in their rules promulgated on July 8, 1987. The following categories of hazardous wastes have been banned from land:

1. Any liquid hazardous waste containing free cyanides at a concentration ≥1000 mg/L

2. Liquid hazardous wastes containing any of the following metals and their compounds at concentration ≥

Arsenic (as As)	500 mg/L
Cadmium (as Cd)	100 mg/L
Chromium (as Cr VI)	500 mg/L
Lead (as Pb)	500 mg/L
Mercury (as Hg)	20 mg/L
Nickel (as Ni)	134 mg/L
Selenium (as Se)	100 mg/L
Thallium (as Tl)	130 mg/L

3. Liquid hazardous waste having a pH ≤ 2.0

4. Liquid hazardous wastes containing PCBs at concentrations ≥ 50 ppm.

5. Hazardous waste containing halogenated organic compounds (HOCs) in total concentrations ≥ 1000 mg/L or 1000 mg/kg

The California List Rule requires using the Paint Filter Liquids Test to determine whether a waste may be classified as liquid or nonliquid (EPA, 1986).

- *The First-Third Rule.* This regulation includes provisions for one-third of the listed RCRA wastes (numbering over 700), excluding those covered under the Solvents and Dioxins and the California List rules. Accordingly, it came to be known as the First-Third Rule, and became effective on August 17, 1988.
- *The Second-Third and Third-Third Rules.* These were published on June 23, 1989, and June 1, 1990, respectively. 40 CFR §§ 268.41, 268.42, and 268.43 gives concentration limits and treatment technology-based standards for these wastes.

In addition to the standards discussed in the above sections, details of certification, recordkeeping, exemptions and variances, responsibilities of TSD owners and operators, and testing and permitting requirements were also established.

With the restrictions imposed by the LDR rules, over 90 percent of all hazardous waste listed in RCRA Codes of Hazardous Wastes (40 CFR) may require pretreatment before land disposal. The intent of these rules was to prohibit land disposal of hazardous wastes that are known to cause environmental problems and pose a threat to human health. Waste generators are encouraged to pretreat the waste, using available technologies to bring the concentrations down to acceptable levels, before considering land disposal.

Landfills

The first sanitary landfill was established in the United Kingdom in 1912. In the United States, landfills as a preferred method of waste disposal became common in the 1930s. Since then, thousands of sanitary landfills have been built. It is estimated that at one time there were 70,000 landfills in the United States. The majority of these accepted mixed wastes in landfills that were neither properly designed nor managed to control leachate migration. All kinds of unwanted trash—including acidic and alkaline substances, organic chemicals, pesticides, and paints—were put in landfills all across the country. Hazardous wastes were commonly mixed with domestic waste and placed in landfills. The main concern at the time was to put these materials out of sight, and thereby avoid the nuisance that results from blowing refuse, odors, flow of liquid wastes, and attracting birds and small animals that scavenge food and related wastes from the landfill. The more serious problems of formation of leachate and its subsequent migration were not considered, nor was any thought given to potential impact on the environment from hazardous constituents. Since the 1980s, however, the number of landfills has been on the decline because of stringent design, operation, and maintenance requirements enforced by the EPA and the states. However, in some developed countries, this practice of mixing MSW with hazardous waste is still common. The following case history relates to aquifer contamination resulting from improper disposal of MSW and industrial wastes, near Vienna, in Austria, and illustrates how past practices of uncontrolled disposal of hazardous wastes in geologically unfavorable locations have resulted in contamination of groundwater. It also highlights the practice of waste disposal prevalent until the early 1980s in some European countries. Use of a computer model, combined with a fast-track investigation, resulted in a quick assessment of the extent of aquifer contamination and recommendations for remediation alternatives.

Case History: *Aquifer Contamination from an Uncontrolled Landfill near Vienna*

Location and History

The disposal site known as the Theresienfeld Landfill is located in a rural area about 30 km (19 mi) south of Vienna. Theresienfeld is a small community with a population of a few hundred (1982) with agriculture as its dominant economy. Extraction of gravel by surface mining commenced at Theresienfeld in the early 1960s. By 1966, mining operations had resulted in the formation of a large elongated pit, 750 m (2460 ft) long, 100 m (328 ft) wide, and 20 m (66 ft) deep. Mining operations ceased around 1970, because of a declining market demand for gravel. In 1972, the gravel pit was leased by a paint and solvent manufacturer and recycling plant owner, who obtained a permit from local authorities to dispose of drummed waste materials in the pit. The permit required the drummed waste to be deposited in layers with 20 cm (8 in.) of fill material placed between each layer. There was no provision to prevent contaminant migration; no impermeable liners, leachate or gas collection systems were required. The site even lacked basic security. The owner was later allowed to receive and dispose of wastes from other industries, including paint and solvent residues, sludges, shredded rubber, metal, and other manufacturing discards. Shredded materials were covered with used oils.

Disposal of hazardous waste was carried out in an uncontrolled manner, and there were instances of illegal disposal. For example, 200 drums were illegally buried in trenches dug in the bottom of the unfilled half of the pit (Figure 14–3). Problems from uncontrolled disposal of hazardous wastes began to surface within a year from the beginning of landfilling operations. Pools of colored liquid formed as a result of chemicals leaking from drums and other containers, and several chemical fires erupted at the landfill. By 1980, when one-half of the pit was full of mixed solids and hazardous wastes, the landfill started accepting household wastes (MSW) as well.

During the mid-1980s, drinking water wells, located downgradient from the landfill, were installed to meet the increased water demand of the city of Vienna. All wells were found to be contaminated with chlorinated solvents in a concentration range of 20 to 30 µg/L. In subsequent investigations, monitoring wells installed 400 m (1312 ft) downstream from the landfill showed concentrations in the range of 500 to 1000 µg/L. A systematic investigation of water quality in the Vienna basin led to the conclusion that the Theresienfeld Landfill, along with industrial plants located upgradient from the landfill, was the source of groundwater contamination.

Geologic Setting

Theresienfeld is located in the middle of the Vienna basin, a structural depression in the basin-and-range topography of the region. The basin is about 60 km (37 mi) long, and nearly 10 km (6 mi) wide in the Theresienfeld area. The Vienna basin represents the graben part of the classical horst-and-graben structure. Tertiary rock formations comprising beds of clays, clay marls, and conglomerate, overlain by gravel deposits of Quaternary age, were involved in

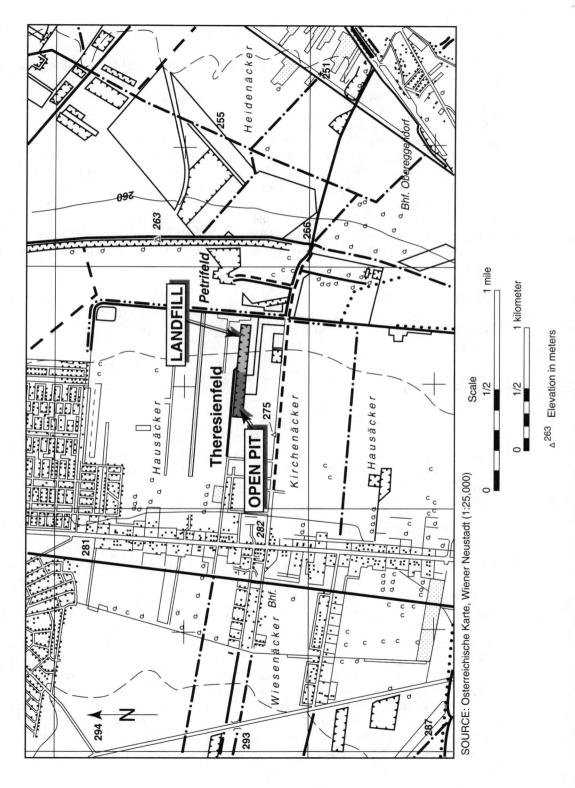

SOURCE: Osterreichische Karte, Wiener Neustadt (1:25,000)

Scale

0   1/2   1 mile

0   1/2   1 kilometer

△263   Elevation in meters

FIGURE 14-3   Details of the gravel pit showing landfill on the eastern half and open pit on the western side (illustration provided by Richard Rudy, Ecology & Environment, Inc., Tallahassee, FL)

downthrown movement along the faults, and occur in the structural basin. Older Jurassic and early Tertiary rocks were upthrown along the fault and comprise the horst. Tectonic movements that took place after the formation of horst-and-graben structure have resulted in the formation of four subbasins. The Theresienfeld area lies in the Mittendorfer Senke—one of the four subbasins within the Vienna basin.

Quaternary gravel deposits overlie the Tertiary rock formations; gravel beds are fairly homogeneous, with lenses of clays and sands occurring sporadically. The gravel beds range in thickness from 3 to 150 m (9.8 to 492.2 ft) in the Vienna basin. In the Mittendorfer Senke, the gravel is about 100 m (328 ft) thick.

The Tertiary formation consists of blue-gray clay beds, locally intermixed with limestone fragments and layers of sand. The average thickness of clay in the Vienna basin is approximately 300 m (984 ft). The contact between the clay and the gravel is very distinct, but the depth is highly variable. It is estimated to occur at a depth of 100 m (328 ft) at Theresienfeld and at 25 m (82 ft) in areas to the north and south of Theresienfeld.

Hydrogeology

The Quaternary gravels constitute one of the best freshwater unconfined aquifers in the region. The groundwater table in the Theresienfeld area is approximately 20 m (65.6 ft) below the surface. The gravels are highly permeable, with average hydraulic conductivity of 1.34×10^{-3} cm/s (28.41 gpd/ft^2). The groundwater flow rate is rather high. Dye tracer tests gave flow velocity ranging between 6 and 10 m/day (20–33 ft/day) near Theresienfeld, and up to 20 m/day (66 ft/day) at the southwestern end of the basin. The general groundwater flow direction is in a north-northeast direction, parallel to the longitudinal axis of the Vienna basin. Aquifer recharge is mainly from the Scharzau and the Piesting Rivers, which drain the area to the southeast and east of the basin. These and other streams that empty into the north–south-flowing Danube River, northwest of the basin, are the primary surface water sources in the area.

Aquifer Contamination

Several groundwater wells in the Vienna basin, including those in the Theresienfeld area, were sampled in 1982. Groundwater was analyzed for chlorinated hydrocarbons and metals. Test results showed the presence of trichloroethylene, perchloroethane, toluene, and 1,2-dichloroethane, along with minor concentrations of other compounds.

In order to fully evaluate the extent of aquifer contamination, a number of monitoring wells were installed by the Austrian government around the perimeter of the landfill between 1982 and 1985. These wells and the nearby private wells were sampled on several occasions. Chemical analytical results were used to delineate a lateral contamination plume in the aquifer (Figure 14–4, Plate III). Soil gas sampling was also conducted in the landfill area, and the results were similar to the groundwater results. It was concluded that the leachate was migrating from the landfill into the aquifer, resulting in significant contamination of the groundwater. Based on test results, it was obvious that the contamination plume was migrating downgradient.

PLATE III

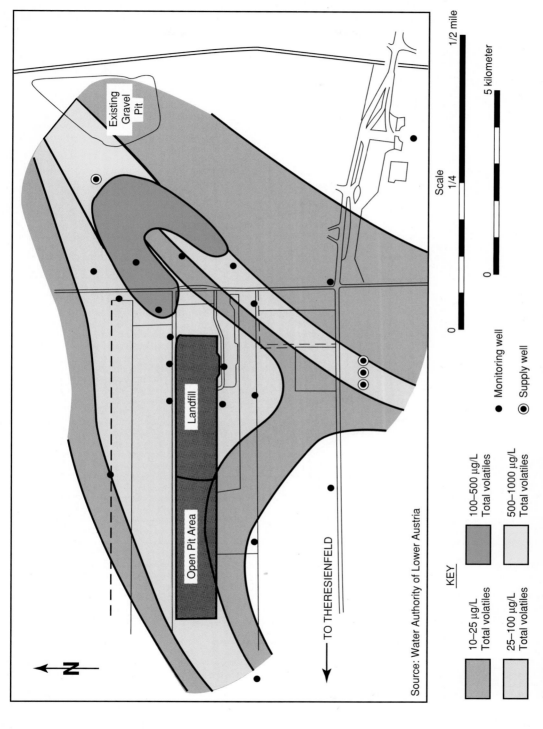

Source: Water Authority of Lower Austria

KEY

	10–25 µg/L Total volatiles
	25–100 µg/L Total volatiles
	100–500 µg/L Total volatiles
	500–1000 µg/L Total volatiles

● Monitoring well

◉ Supply well

Scale

0   1/4   1/2 mile

0   5 kilometer

FIGURE 14–4 Extent of contamination plume at the Theresienfeld Landfill (illustration provided by Richard Rudy, Ecology & Environment, Inc; Tallahassee, FL.)

PLATE IV

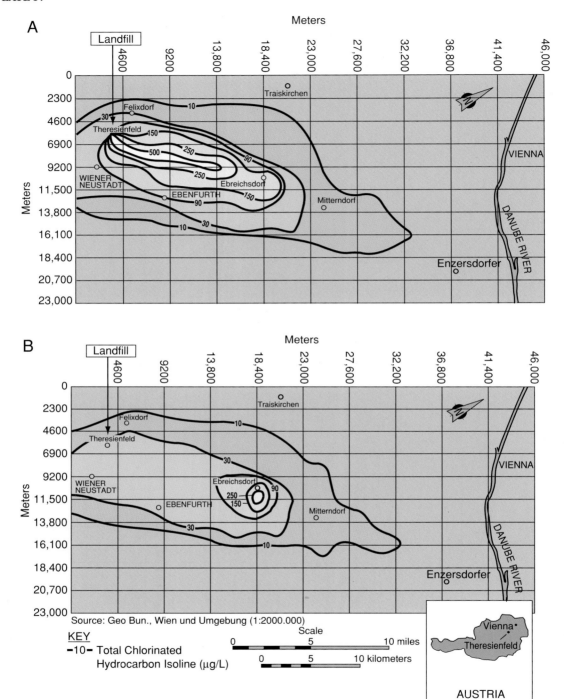

FIGURE 14–5 (A) Contamination plume in the year 2040, assuming no remediation; (B) contamination plume in the year 2040 after remediation (illustrations provided by Richard Rudy, Ecology & Environment, Inc., Tallahassee, FL.)

Modeling Studies

Computer modeling to predict the solute transport through the aquifer was carried out using the RANDOM-WALK model. The model was used to predict contaminant transport without any remediation, and contaminant transport after remediation. The first case assumed that, without any remediation, contaminants will continue to leach into the aquifer from the landfill and from upgradient sources. This will cause the plume to advance and grow over time, contaminating larger volumes of groundwater. Figure 14–5(A), Plate IV, shows the areal extent of the plume in the year 2040 (50 years after study date). The contamination will extend about 19.5 km (12 mi) downgradient from the landfill, attaining a maximum width of approximately 10 km (6 mi). This would result in contamination of about 195 km^2 (75 mi^2) area of the aquifer at a concentration level of 40 µg/L. It was also found that the plume will reach the Danube River, a distance of 40 km (25 mi), in about 150 years.

The second simulation was done assuming that the site will be remediated so that further leaching from landfill will not occur, although contamination will continue to occur from other regional sources located upgradient from Theresienfeld. The model showed that the contaminant plume will attenuate significantly over time. Figure 14–5(B), Plate IV, shows extent of the plume in the year 2040. It can be noted that although the plume will extend over the same distance, about 19.5 km, its width and area will be much less. The aquifer area with contaminant concentrations at the 40 µg/L level will be about 95 km^2 (36.7 mi^2)—nearly 100 km^2 (38.6 mi^2) less than in the first case.

Remediation Program

The results of the modeling study provided a sound basis for drawing up alternative remediation plans. The feasibility study proved to be less complex than similar work in the United States conducted under EPA guidelines. The Austrian government did not require public input on selection of the remedial measures. In addition, only proven and effective remedial techniques were considered, and innovative or untested technologies were excluded from consideration. The final remediation program comprised two phases: (a) immediate measures, and (b) long-term measures.

Immediate (short-term) measures. Short-term measures taken included simple things, such as fencing the site, capping the landfill, and treating the well water with activated carbon before piping it for residents' use. Other measures, including installation of a leachate management system and subsurface barrier walls to control groundwater flow, were not adopted in view of the high cost and uncertainty about their effectiveness in the absence of complete site-characterization data.

Long-term measures. A broad range of long-term cleanup strategies were evaluated. These included: construction of a 100-m (328-ft) deep slurry wall, taken down to the aquifer confining layer, for waste containment; a perimeter slurry wall with bottom seal; excavation of all waste and placing it in a secure landfill; construction of an onsite secure landfill for hazardous waste; construction of an onsite secure landfill for nonhazardous waste and incineration of hazardous waste at an onsite temporary incinerator; onsite incineration of the waste; and offsite inciner-ation at Vienna—and pump-and-treat remediation. The last remediation option was chosen by the Austrian government.

Remediation was accomplished by installing 9 extraction wells, 7 injection wells, and 21 monitoring wells. Contaminated groundwater, after being pumped to the surface, is first led through a pretreatment plant comprising sedimenta-tion, flocculation, biofiltration, and ozonation units. In the second phase, the water is treated in a bioreactor, using burnt clay, to remove VOCs. Granulated activated charcoal (GAC) is used in conjunction with a two-bed mineral filter to remove all other contaminants. Final disinfection is achieved by using UV light. Pretreatment is done only when the level and nature of contaminants warrant it; otherwise, the contaminated groundwater is directly led to the second-stage treatment plant. After treatment, clean water is injected back into the aquifer through the injection wells. The remediation program began in late 1989 and was expected to be completed in 1995. It is claimed that the remediation method adopted will result in a complete cleanup of the groundwater to drinking water standards set by the Austrian government (Urban and Frischherz, 1992).

(Materials for this case history were provided by Richard Rudy, Principal Hydrogeologist, Ecology & Environment Inc., Tallahassee, FL.)

The Secure Landfill

The 1984 Hazardous and Solid Waste Amendments Act requires all new land dis-posal units—landfills, surface impoundments, and waste piles—to include a minimum of two liners for containment of leachate, a leachate collection and removal system, a leak detection system, and a landfill gas management system. A landfill that includes all these components is called a *secure landfill*. The law requires existing landfills to be retrofitted with these features, or else closed down. Obviously it is not possible for many of the existing small landfill owners to meet these requirements. As a result, such facilities are on their way out. Figure 14–6(A) shows the main features of a secure landfill.

The geologic and hydrogeologic criteria for selection of a hazardous waste landfill site are discussed in Chapter 11; additional details of liners, leachate and gas management systems, and closure requirements, are presented in the fol-lowing sections.

Landfill Liners

Pursuant to the passage of the Hazardous and Solid Waste Amendments (HSWA) of 1984, the EPA developed technological standards for hazardous waste landfills. At least two liners—a bottom liner and a top liner—are required for all hazardous waste landfills, surface impoundments, and waste piles. The bottom liner usually comprises a layer of compacted soil, at least 0.91 m (3 ft) thick and having hydraulic conductivity of $\leq 1 \times 10^{-7}$ cm/s. This is covered with an upper flexible membrane liner (FML) at least 0.76 mm (30 mil), and preferably between 1.5–2.54 mm (60 and 100 mil), thick (EPA, 1991). This barrier is called a *composite liner.* FMLs are synthetic sheets made of polymers such as thermoplastics, elastomers, and thermoplastic elastomers. *Geosynthetic* is a generic term for manufactured polymer sheets that are designed to serve various functions (separation, drainage, filter, reinforcement, and barrier) in engineered structures; FML is a

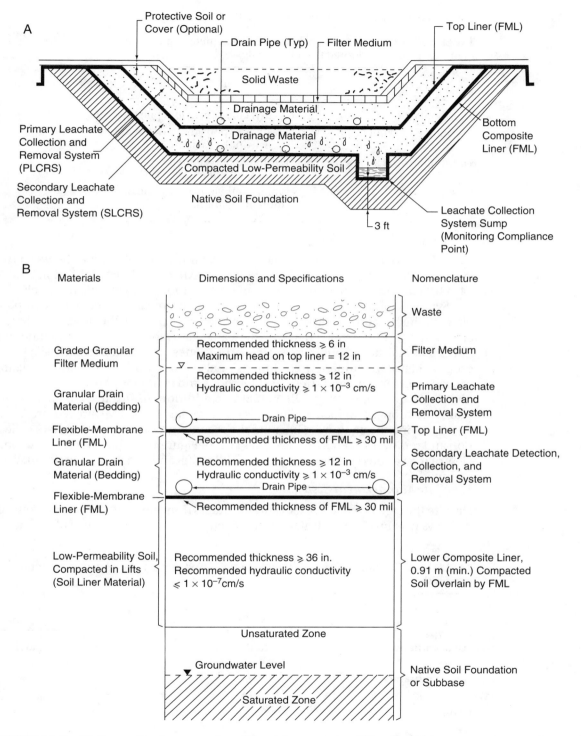

FIGURE 14–6 (A) Generalized cross sectional view of a secure landfill (adopted from EPA, 1989), and (B) details of various layers (adopted from EPA, 1991)

TABLE 14–3   Common Polymers Used for Manufacturing Flexible Membrane Liners (adopted from EPA, 1989)

Polymer	Liner Material
Thermoplastic	Polyvinyl chloride, nylon, polyester
Crystalline thermoplastic	High-density polyethylene, linear low-density polyethylene
Thermoplastic elastomers	Chlorinated polyethylene, chlorosulfonated polyethylene
Elastomers	Neoprene, ethylene propylene diene monomer, urethane

special type of geosynthetic that is used as an impermeable barrier between two layers. Table 14–3 lists some of the common FMLs and their applications. Table 14–4 lists the main functions of some of the common geosynthetic materials.

Synthetic liners selected for waste containment must be chemically compatible with the hazardous waste and the leachate, and possess the desired hydraulic conductivity, stress-strain characteristics, and survivability. The latter includes the FML's tear resistance, puncture strength, brittleness, seam strength, tensile properties at elevated temperatures, dimensional stability, and resistance to cracking caused by environmental stress. Test methods and guidelines for evaluating FMLs have been developed by the EPA (1986). In addition, *in situ* tests to ensure that seams are properly welded and there is no other flaw in the FML are required.

Overlying the composite liner are the secondary and primary leachate collection and removal systems (SLCRS and PLCRS). Figure 14–7 shows a cross sectional view of the various components of a secure landfill and details of its various layers.

Leachate Management System

Despite the care taken to prevent surface waters from entering the landfill, some of it always gets inside. This water either directly reacts with the variety of chemicals

TABLE 14–4   Common Geosynthetic Materials Used in Waste Management (adopted from EPA, 1991)

Type of Geosynthetic	Primary Function				
	Separate	Reinforce	Filter	Drain	Barrier
Geomembrane					●
Geotextile	●	●	●	●	
Geonet				●	
(Geo) pipe				●	
Geocomposite				●	
Geogrid		●			

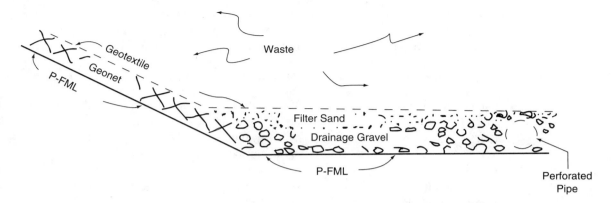

FIGURE 14–7   Details of various filter materials used in the primary leachate collection system (adopted from EPA, 1989)

that are present in the landfill or promotes reactions that result in a large number of chemicals getting dissolved in the water. *Leachate* is a term used to describe this highly mineralized liquid, usually of a dark brown or rust color and containing a high concentration of contaminants. This "soup" of contaminants has been responsible for affecting the quality of many of the aquifers all over the world, besides impacting surface waters. The chemical nature of leachate varies with the nature of hazardous materials in the landfill—from strongly acidic to strongly basic. Table 14–5 lists chemical compounds found in MSW landfills in Wisconsin. It should be noted that the chemicals found in leachate from hazardous waste landfills are far more complex and toxic than those found in MSW landfills. Proper design and efficient functioning of the leachate collection and removal systems is therefore critical. Details of design requirements to ensure that the system is functioning as intended are discussed in the following paragraphs.

EPA specifications require that the primary (or upper) leachate collection and removal system be designed to ensure that the depth of leachate that would collect above the liner will not exceed 0.30 m (1 ft). The secondary (or lower) leachate collection and removal system is designed to detect, collect, and remove any leachate that might percolate through the upper liner.

The design of PLCRS incorporates installation of filters. Conventional filter is made of selected soil materials that are installed underneath or adjacent to structures where seepage of water or other liquids is anticipated. Proper drainage of such soils is essential for preventing buildup of excessive pore-water pressure that may result in the instability of the structure. A filter has to satisfy two requirements: (a) allow adequate drainage of water or other liquid, and (b) prevent movement of fine particles to clog the voids. Filters are common in dams and other hydraulic structures. Geosynthetic fabrics are finding greater use as filters, because of the ease of installation and economic considerations. Unlike soil materials, a very small thickness, about 2 cm (<1 in.) of the synthetic drainage material, fulfills the same hydraulic design requirements as a large thickness (0.3 m or 1 ft) of soil drainage material. Their use in landfills increases the capacity to hold additional volumes of waste. This increased capacity is preferred by landfill owners because of the increased revenue resulting from additional space.

TABLE 14–5 Leachate Composition from Sanitary Landfills in Wisconsin (adapted from Fetter, 1994)

Parameter	Overall Range* (mg/L)	Number of Analyses
Total Dissolved Solids (TDS)	584 – 50,430	172
Specific conductance	480 – 72,500**	1167
Total suspended solids	2 – 140,900	2700
BOD	ND – 195,000	2905
COD	6.6 – 97,900	467
TOC	ND – 30,500	52
pH	5 – 8.9	1900
Total alkalinity ($CaCO_3$)	ND – 15,050	328
Hardness ($CaCO_3$)	52 – 225,000	404
Chloride	2 – 11,375	303
Calcium	200 – 2500	9
Sodium	12 – 6010	192
Total Kjeldahl nitrogen	2 – 3320	156
Iron	ND – 1500	416
Potassium	ND – 2800	19
Magnesium	120 – 780	9
Ammonia-nitrogen	ND – 1200	263
Sulfate	ND – 1850	154
Aluminum	ND – 85	9
Zinc	ND – 731	158
Manganese	ND – 31.1	67
Total phosphorus	ND – 234	454
Boron	0.87 – 13	15
Barium	ND – 12.5	73
Nickel	ND – 7.5	133
Nitrate-nitrogen	ND – 250	88
Lead	ND – 14.2	142
Chromium	ND – 5.6	138
Antimony	ND – 3.19	76
Copper	ND – 5.6	138
Thallium	ND – 0.78	70
Cyanide	ND – 6	86
Arsenic	ND – 70.2	112
Molybdenum	0.01 – 1.43	7
Tin	ND – 0.16	3
Nitrite-nitrogen	ND – 1.46	20
Selenium	ND – 1.85	121
Cadmium	ND – 0.4	158
Silver	ND – 1.96	106
Beryllium	ND – 0.36	76
Mercury	ND – 0.01	111

*ND indicates no data.

**In μ ohms/cm.

Whether the drainage material is a synthetic geonet or natural soil, it must have a high hydraulic conductivity of 1 cm/s. A 0.30-m (1-ft) thick gravel layer, without fines and with individual particles in the size range of 6–12 mm (.25–.5 in.), serves as the required drainage material to allow for unimpeded flow of liquid into perforated drain pipes. However, in order to prevent clogging of the void spaces, the upper surface of this drainage medium has to be covered with suitable filter material. Figure 14–7 is a cross sectional view of a primary leachate collection system using synthetic and natural drainage and filter materials. A series of perforated pipes and a sump complete the design of the PLCRS.

The top of the SLCRS is covered with an FML at least 0.76 mm (30 mil) thick. Underlying the PLCRS, and above the composite bottom liner, is the secondary leachate detection, collection, and removal system. It has the same type of drainage material as the PLCRS, except that an FML covers its upper surface (Figure 14–6). In addition, one or more sumps with submersible pumps for leachate removal are installed in the SLCRS.

Landfill Gas Control

Landfill gases are produced as a result of anaerobic decomposition of organic constituents. Once produced, landfill gases can migrate either laterally or vertically through soil under a pressure gradient or a concentration gradient (i.e., by diffusion). Landfills exclusively used for hazardous wastes are relatively new, and gas has not been reported in such landfills. According to the EPA (1989), it may take 40 years or more for gas to develop in a closed secure hazardous waste landfill. Landfill gas (LFG) is primarily a mixture of methane and carbon dioxide with a small concentration of VOCs such as vinyl chloride. Methane is odorless and nontoxic, but becomes explosive in the presence of air at concentrations of between 5 and 15 percent. At higher concentrations in air, methane is flammable.

A secure landfill should have the provision to collect and vent these gases. Landfill gases from hazardous waste sites may be vented into the air, collected and flared, or incinerated (if it is too odiferous or toxic). A typical LFG vent system is shown in Figure 14–8. A 30-cm (12-in.) thick layer of gravel overlies the compacted waste; upward-sloping perforated PVC pipes laid in the gravel layer are connected to vertical riser pipes with a vent at the surface. The layer of gravel is covered with a geonet. Overlying this filter is a layer of compacted soil, with low hydraulic conductivity (1×10^{-7} cm/s) and at least 61 cm (2 ft) thick, which is in intimate contact with a minimum 0.5-mm (20-mil) thick FML on its top (Figure 14–8). The FML is covered with a layer of coarse soil 30 cm (12 in.) thick with hydraulic conductivity of 1×10^{-2} cm/s or greater. The coarse soil material may be replaced by a geosynthetic material with the same hydraulic conductivity. The purpose of the drainage layer is to intercept the water and move it away from the waste. This is achieved by providing a minimum of 3 percent slope, which causes the percolating water to move rapidly by gravity to a toe drain, away from the perimeter of the waste cells. A geosynthetic filter layer is placed above the drainage layer.

Instead of being vented into the air, the LFG can also be extracted by installing a gas extraction well. Such systems use exhaust blowers. Figure 14–9 shows a gas extraction well.

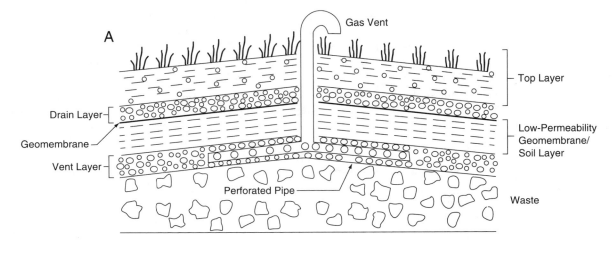

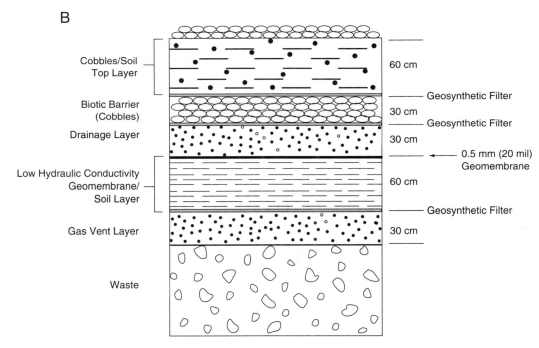

FIGURE 14–8   (A) A typical landfill gas venting system, and (B) details of the cover layers (adopted from EPA, 1991)

Full evaluation of settlement and subsidence potentials, and the maximum depth of frost penetration, are other factors that must be considered in the design of the gas control system.

Surface Water Control

To minimize run-on of water over the landfill and its subsequent infiltration into the landfill through the cover layers, adequate diversion structures have to be

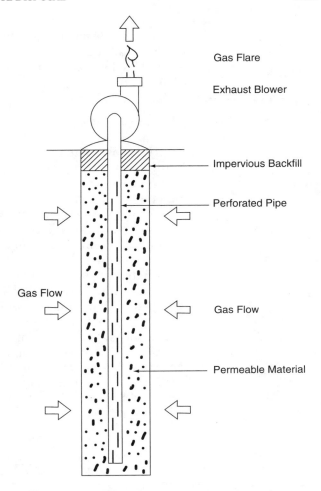

FIGURE 14–9 Gas extraction well for landfill gas control (adopted from EPA, 1991)

Gas Flare

Exhaust Blower

Impervious Backfill

Perforated Pipe

Gas Flow

Gas Flow

Permeable Material

constructed around the landfill. Runoff can be controlled in one of several ways, depending on the area it is coming from:

- Runoff coming from areas that are not landfilled can be diverted directly from the facility to an offsite location.
- Runoff from exposed excavation areas should be diverted to a siltation basin constructed at the site, and then discharged offsite.
- Runoff from an active landfill area should be diverted to nearby holding sumps for sampling to detect any contamination. If the water is found to be uncontaminated, it can be directed offsite. If it is found contaminated, it has to be properly treated before being discharged.
- Runoff from a closed landfill is managed in the same way as in the active landfill area, except that the sumps are of much larger capacity to accommodate larger volume of runoff.

Closure and Post-Closure

The EPA has established closure and post-closure standards to ensure that hazardous waste landfills are closed in a manner that will (a) protect human health and the environment, and (b) control, minimize, or eliminate escape of hazardous materials or leachate into the ground or the atmosphere after closure.

After the hazardous waste landfill has become full, and has stopped receiving waste, it must be closed in accordance with the design standards established by the EPA for the final cover. Figure 14–10 shows a cross sectional view of the cover design, which consists of (from top of the waste to surface):

- The gas venting layer, 30 cm (12 in.) thick, covered with geonet.
- A 61 cm (2 ft) layer of compacted soil with a hydraulic conductivity of 1×10^{-7} cm/s. The compacted soil may be replaced by suitable geosynthetics with the same hydraulic conductivity as the compacted soil. The upper surface of the compacted soil should have a slope of 3 percent and should be below the frost line. The compacted soil should be covered with an FML at least 0.5 mm (20 mil) thick.
- A 30 cm (12 in.) drainage layer above the compacted soil–FML. This should have hydraulic conductivity of 1×10^{-2} cm/s or greater. If a geosynthetic material is substituted for soil, it should have the same hydraulic conductivity and meet all other requirements for FML. A layer of filter material should be installed above the drainage layer; the filter may be designed with natural soils or it may be a geosynthetic material.
- The top vegetation and soil layer constitute the uppermost component of the closure system. A medium-textured soil, such as loam, is the preferred type. The soil layer should be at least 61 cm (2 ft) thick to support vegetation and should comprise at least 15 cm (6 in.) of topsoil. To allow the plant roots to develop, the topsoil should not be compacted. It should also have a minimum of 3 percent slope. When selecting the type of vegetation care must

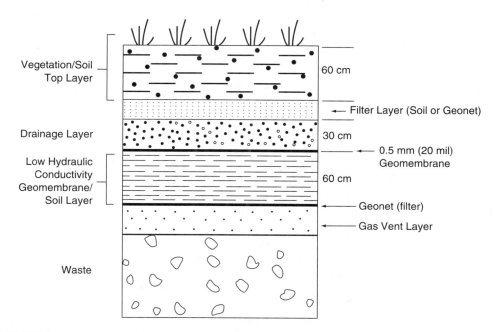

FIGURE 14–10   Cover layers for hazardous waste landfill (adopted from EPA, 1991)

be taken to see that: (a) the roots of the mature plants will not penetrate into the drainage layer, (b) the plants need low or no maintenance, (c) the plants can survive in soils with low nutrients, and (d) the soil has adequate density to limit the rate of soil erosion to under 2 ton/ac/yr (5.5 MT/ha/yr).

At locations where vegetation cannot be maintained, a 30 cm (1 ft) layer of 5–10 cm (2–4 in.) cobbles should be placed above the drainage layer. This should be covered with a 61-cm (2-ft) thick layer of soil capped by a layer of cobbles or stones (Figure 14–8(B)).

Post-Closure Monitoring

The law requires a hazardous waste landfill to be monitored for 30–50 years after its closure (40 CFR, Part 264, Subpart G). This includes groundwater monitoring and inspection, and care to ensure that the leachate management and the run-on and runoff management systems are functioning properly. A written post-closure plan has to be kept at the facility; it should describe the frequency of monitoring and maintenance activities to be conducted after closure of the landfill. The post-closure plan must include, at a minimum, the following information:

- A description of the planned groundwater monitoring requirements, including the frequency of sampling
- The integrity of the cap and final cover and other containment systems
- A description of the planned maintenance activities for the cap, containment, and monitoring equipment
- The name and address of a facility contact person overseeing post-closure care

The owners must demonstrate that there is sufficient money available for proper post-closure care of the units.

Land Treatment

Terms like land treatment, land farming, land cultivation, land application, and sludge spreading refer to the practice of using the soil (the land) as a medium for simultaneous treatment and disposal of hazardous waste. Land treatment aims at the degradation, transformation, or immobilization of hazardous waste. The natural biodegradation process brought about by the organisms present in the soil, or by chemical reaction, is taken advantage of in rendering the waste nonhazardous. The process is optimized by tilling the soil for aeration, controlling the soil pH and water content, and fertilization (to provide nutrient supply for organisms). Land treatment is best suited for listed petroleum wastes (codes K048 to K052), although sewage sludge is also amenable to land treatment. Halogenated organic compounds and hazardous wastes containing heavy metals are not amenable to land treatment; for this reason, sewage sludge containing heavy metals is also not suitable for land treatment and disposal.

The law requires the owner or operator of a land treatment facility to

- Carry out analyses of the waste prior to its placement on the land to determine the concentrations of hazardous and toxic constituents.
- Conduct a pretreatment demonstration to verify that the hazardous constituents of the waste will be properly treated.

- Monitor the soil and the vadose zone to ensure that no migration of hazardous constituents is occurring. If migration is indicated from monitoring, a permit application has to be filed with the EPA indicating changes in the operating practice to rectify the problem. Migration is considered to have occurred when the monitoring results show a higher concentration of constituents compared to the background concentrations in untreated soils.
- Carry out proper closure procedures at the facility, which include: maintaining the optimum conditions to maximize degradation, transformation, and immobilization of the wastes; minimization of precipitation and runoff from the treatment zone; and control of wind dispersion. A vegetative cover should be placed over the treated soil to promote plant growth to protect the site from wind or water erosion.

Regulations prohibit growing food-chain crops in land-treated areas where the hazardous waste is known to contain arsenic, cadmium, lead, and mercury, unless it is demonstrated that they will not be transferred to the food portion of the crop or occur in concentrations greater than in identical crops grown on untreated land in the same region.

Waste Piles

A waste pile is a noncontainerized accumulation of nonflowing, solid hazardous waste. Depending on whether the waste pile is used for storage, treatment, or disposal, different requirements have to be satisfied. The owner or operator of a *treatment* or *storage waste pile* must meet the following requirements:

- Perform waste analyses for overall characterization of the hazardous waste. In addition, representative samples from each incoming shipment have to be analyzed to prevent inadvertent mixing of incompatible wastes in the pile.
- Protect the pile from wind dispersion.
- Locate the waste pile on an impermeable base.
- No liquid should be placed on the pile.
- Separate the waste pile from other potentially incompatible materials.
- Pile must not be exposed to precipitation (must have a cover).
- A run-on control and collection system to prevent water flow onto any active portion of the waste pile from a peak discharge from a 25-year storm.
- No ignitible or reactive waste should be placed on the pile, unless it has been treated and rendered nonignitible or nonreactive.
- All contaminated soils, liners, equipment, and structures must be removed or decontaminated in accordance with the closure plan. If this is not possible, the waste pile must be closed as a landfill.

A *disposal waste pile*, in addition to the above, must comply with the following requirements:

- Have a liner on a supporting base or foundation to prevent migration of any waste in vertical and horizontal directions.
- Have a leachate management system installed above the liner to remove all leachate.
- Have a system to control water run-on resulting from a 25-year storm for at least 24 hours.

- Weekly inspection, as well as after every storm.
- Removal or decontamination of all contaminated equipment, structures, liners, and soils, in accordance with the closure plan, to *clean close*. If the facility cannot be clean-closed, it must comply with closure requirements for a landfill.

Waste piles containing nonmagnetic materials (called *fluff*) generated from automobile shredding operations are a good example of hazardous waste piles because they contain concentrations of lead and cadmium above the levels prescribed in the TCLP toxicity test. They also contain synthetic fabrics, rubber, plastic, and insulation material. These are usually saturated with oil and have a tendency to auto-ignite.

Surface Impoundments

Surface impoundment is a facility designed to hold an accumulation of liquid wastes or wastes containing free liquids. The impoundment may be in a natural topographic depression, manmade excavation, or diked area (basin). Holding, storage, and aeration pits, ponds, and lagoons are examples of surface impoundments. Surface impoundments are different from tanks in that impoundments cannot maintain their structural integrity without support from the surrounding earth materials, unlike tanks, which maintain their structural integrity upon removal of the surrounding earth materials.

A surface impoundment may be used for storage or disposal of hazardous wastes. The following requirements apply to the owners or operators of surface impoundments:

- Carry out analyses of the waste to be treated or stored.
- Maintain at least 0.61 m (2 ft) of free board, which must be inspected daily to ensure compliance. In addition, weekly inspection must be conducted of the dike, surrounding vegetation, and the impoundment area to detect any leaks, deterioration, or failures in the impoundment.
- Must have two or more liners, along with a leachate collection and removal system.
- Groundwater monitoring system.
- No ignitible or reactive waste can be placed in a surface impoundment.

The impoundment may be clean-closed by removing or decontaminating all equipment, liners, structures, underlying and surrounding soils and groundwater, if any. If clean closure cannot be achieved, the impoundment must be closed as a landfill.

Except for hazardous waste landfills, other land disposal facilities such as waste piles and surface impoundments may be either clean-closed or closed. For clean-closing, the owners and operators of such units have to ensure that all hazardous wastes and waste residues have been removed from the site, that the facility and equipment have been effectively decontaminated, and that the air, soil, surface water, and groundwater samples have not indicated a level of concentration exceeding the EPA's recommended limits. Upon approval by the EPA, the owners or operators of clean-closed units do not have to carry out post-closure care for the unit.

Deep Well Injection

In the United States, disposal of industrial wastes into underground wells by the oil and chemical industry has been carried out since the 1950s. With the passage of the Safe Drinking Water Act of 1974 and the Hazardous and Solid Waste Amendments of 1984, strict requirements have been placed on waste disposal using deep wells. The purpose is to protect the nation's underground sources of drinking water (USDW) from possible contamination by improper disposal of hazardous waste into subsurface geologic formations. Philosophically, deep well injection, though permitted by law, should be considered only after source reduction, waste minimization, and other disposal methods have been evaluated. Hazardous waste restricted from land disposal cannot be disposed of in underground injection wells unless it has been treated or exemption has been granted by the EPA or the delegated state authority.

Several types of wells have been used for underground disposal of waste. The EPA has classified these into the following five categories, based on the type of waste and the relationship of the injection zone to a USDW (Shannon, 1993):

1. *Class I wells.* These are wells that are used to inject hazardous or nonhazardous industrial or municipal waste into a geologic formation located below the lowermost formation that contains a USDW within 0.4 km (0.25 mi) of the well bore. The majority of industrial and municipal wells belong to this class.

2. *Class II wells.* These wells are used in petroleum exploitation for injection of formation fluids that are brought up during petroleum production, and for enhanced recovery of petroleum. Class II wells are also used for storage of hydrocarbons that are liquid at standard temperatures and pressures.

3. *Class III wells.* These wells are used for mineral extraction, including those used in solution mining of certain minerals, such as sulfur, salt, and potash.

4. *Class IV wells.* Injection wells, used for disposal of radioactive or hazardous wastes into or above a USDW within 0.4 km (0.25 mi) of the well bore, comprise Class IV wells. This class of wells has been prohibited by the EPA under HSWA.

5. *Class V wells.* All other wells that do not fit into the above four classes are grouped as Class V wells. These include, but are not restricted to, drainage, recharge, or geothermal re-injection wells that inject nonhazardous wastes into or above USDWs.

In 1993, there were 438 Class I wells in the United States, of which 39 percent (172) were used for disposal of hazardous waste and 61 percent for disposal of nonhazardous waste. Table 14–6 gives the distribution of Class I wells used for hazardous waste disposal in the United States.

Stringent requirements have been imposed on the disposal of hazardous waste in deep injection wells (DIWs) by the EPA. Owners must demonstrate that the waste will not migrate from the injection zone before it is rendered nonhazardous by hydrolysis, chemical interaction, or other means. Another standard requires that the injected fluids will remain in the injection zone for at least 10,000 years.

TABLE 14–6   Distribution of Deep Injection Wells
for Hazardous Waste Disposal in the United States
(adapted from Shannon, 1993)

State	Number of Wells	State	Number of Wells
Arkansas	4	Florida	1
Illinois	4	Indiana	4
Kansas	5	Kentucky	1
Louisiana	29	Michigan	8
Mississippi	7	Ohio	2
Oklahoma	4	Texas	89
Wyoming	14	TOTAL	172

Criteria for selection of well injection sites, design and construction of injection wells, and operating and closure requirements have been set forth by the EPA. Deep well injection sites should not be located in areas that are geologically unstable. No faults, shear zones, joint systems, or unplugged or improperly plugged abandoned wells that might provide an upward migration path for the injected waste should be located in the vicinity of an injection well.

A clear understanding of geologic and hydrogeologic factors is critical in the location, construction, operation, and maintenance of deep well injection units. The general guidelines for site selection (Chapter 11) should also serve as a guideline for selection of sites for deep injection wells. A review of the local and regional stratigraphy is very helpful in the initial stage of project planning. In the U.S. Midwest and other oil-producing areas, a great deal of subsurface geologic information is available from rock cores and geophysical logs of oil wells; this information should be used to supplement generalized geologic information for preliminary site selection. Important geologic factors to be considered in selecting sites for deep injection wells are

Regional Geologic Setting:
- Geomorphology
- Structure and tectonics
- Stratigraphy
- Mineral resources
- Seismicity, including active faults

Local Geologic and Hydrogeologic Settings:
- Geologic structures
- Lithology of subsurface rock units
- Geotechnical characteristics of subsurface rock units—thickness and lateral distribution, RQD, porosity and permeability, chemical characteristics of rock units and the fluids, along with reservoir pressure and temperature and hydrodynamics

- Geohydrology—occurrence of freshwater aquifers, their depth, thickness, water quality, whether a USDW, and potential use if not a USDW
- Mineral resources—past and present exploitation of coal, petroleum, and other minerals, and the prospects for any future development

Protection of water quality is the main objective of the laws controlling deep well injection. Accordingly, elaborate standards for well design and monitoring have been established. Figure 14–11 shows the details of a deep injection well. Injection holes are cased with two steel casings. The outer one extends well below the lowermost aquifer; the inner one goes down to the

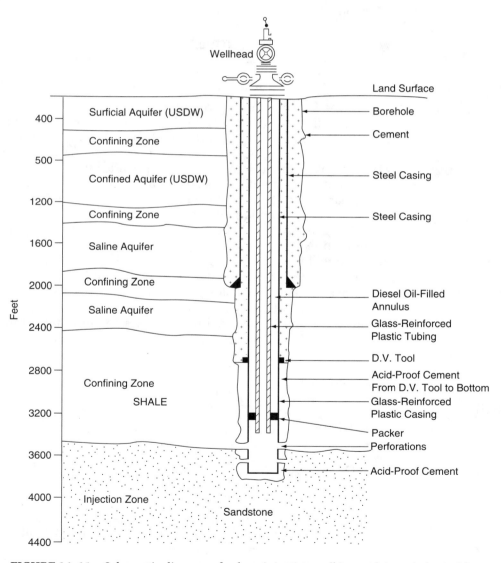

FIGURE 14–11   Schematic diagram of a deep injection well in sandstone (adopted from Warner and Lehr, 1977)

injection zone. Tubings, installed inside the inner casing, serve as conduits for the injected waste. Packers are placed in the annular space between the tubing and the inner casing near the bottom of the hole in the injection zone. The rest of the annular space between the top of the packers and the surface is filled with fluid that is continuously monitored under pressure. Cement is used to backfill the space between the casings and the side walls of the hole.

Known cases of contamination of near-surface aquifers from failure of deep injection wells are rare. The greatest hazard comes from unplugged abandoned wells in the vicinity of deep injection wells.

Detailed monitoring of the injection well is required for controlling the injection pressure, flow rate, and volume of injection fluid. Fluids are generally injected at very low pressure, because high-pressure injection is known to trigger earthquake activities. Periodic demonstration of structural integrity and monitoring of groundwater quality in the overlying aquifers are also required. Because of their depths—0.3 km to 2 km (984–6560 ft) and more—and elaborate construction and monitoring requirements, deep injection wells are very expensive and cost well over one million dollars to construct.

Mine Disposal

Extraction of mineral resources creates large volumes of mined-out space in the subsurface. Such spaces offer an attractive alternative for storage and/or disposal of hazardous waste (Hasan, 1990). Existing mined-out spaces are generally more economical for hazardous waste disposal/storage than space specially created by mining. However, if the mined-out earth material can find a ready market then the cost of mining will be offset by the sale of the material, justifying creation of the space for hazardous waste disposal/storage.

Disposal/Storage in Conventional Mines

Many rock formations, such as limestone, sandstone, bedded salt and potash, and granite, are suitable for storage/disposal of hazardous wastes. Sandstone, limestone, and granites may be mined by drilling and blasting, which results in creation of a patterned space in which the void space, called *room*, is created from extraction of the geologic material. The excavated areas are supported by columns of geologic material, called *pillars*; this mining technique is known as the *room-and-pillar* method. The ratio of the volume of room to the volume of pillars determines the available space for disposal/storage. Generally, an extraction ratio of 75 percent or more would mean that a sufficiently large volume of void space is available for storage/disposal. Figure 14–12 shows a limestone mine with rooms being used for storage of merchandise and for roadways, and pillars supporting the overburden above the mine roof.

Important factors to be considered in determining suitability of mined-out space for hazardous waste storage/disposal include

- *Hydrogeologic conditions.* The mine should be dry and should not be hydrologically connected to the surface. Small amounts of water discharge can be handled by installing proper water collection and removal structures. However, if the amount of water discharging into the mine is large, it may make the mine unsuitable for storage/disposal of hazardous wastes.

FIGURE 14–12   Room-and-pillar mining of limestone in Kansas City, Missouri. Space created by mining can be used for roadways (A) or developed into office or storage space (B) (photographs provided by the author)

Hazardous wastes disposed/stored in shallow mines will be above the groundwater table; in deep mines they will be below the GWT. In either case, all requirements for the protection of the potable aquifer must be satisfied.

- *Depth.* Ideally, the mine should be located at or close to the surface to allow easy access of trucks and haulage vehicles. At many places, such as in Kansas City, Missouri, the natural geologic and topographic settings are such that the limestone beds form hills and bluffs, while the valley floor, comprising shale, provides level grounds for extending roads and railways into the mines (Figure 14–13). When the mined-out space occurs at some depth below the surface, shafts can be used for entry and exit. Deeper mines may be desirable if the purpose is to isolate the waste; shallow mines, because of their ease of access, are better suited for storage and disposal.

- *Mine stability.* The inherent geomechanical nature of rock formations, effect of blasting, hydrogeologic conditions, nature of ambient earth stresses, and size (span) of the opening control stability of the mine. The main concern in using mined-out spaces for hazardous waste disposal/storage is the stability of the space. Where the natural conditions do not result in the desired stability of the mines, the mine can be stabilized by a variety of reinforcement methods, such as rock bolts, shotcrete, concrete, and steel-concrete supports. Periodic monitoring of the structural conditions to ensure overall integrity of the mined-out space is very important if it is to be used for hazardous waste storage/disposal.

- *Safety.* Unlike surface structures, underground structures have limited entry and exit points. This poses special problems in terms of emergency exits and cleanup operations following a spill. Simple measures, like filling the area farthest from the exit point first and retreating toward the exit, will facilitate emergency evacuation. Spill contingency and evacuation plans should be drawn up and maintained at the site.

Ocean Disposal

Prior to the passage of the Marine Protection Research and Sanctuaries Act (MPRSA) in 1972, ocean dumping of hazardous waste was quite common. MPRSA, also known as the Ocean Dumping Act, regulates waste disposal in oceans. Certain categories of waste materials have been prohibited from ocean dumping; these include

- High-level nuclear wastes
- Materials produced for biological, chemical, or radiological warfare
- Materials that are incompletely characterized
- Persistent inert synthetic or natural materials that float or remain suspended in the ocean and are capable of causing interference with fishing, navigation, or other uses of the ocean

The EPA issues permits for dumping hazardous wastes and other materials in the ocean. Depending on the intended use, it may issue a general, special, or research permit. In cases where materials pose unacceptable risks to human health and where no other feasible method of disposal is available, an emergency

FIGURE 14–13   Favorable geologic setting for easy rail (A) and road (B) access into underground limestone mine, Kansas City, Missouri (photographs provided by the author)

dumping permit may be issued by the EPA. Even then, wastes containing the following constituents are prohibited from emergency disposal in the ocean:

- Organohalogen compounds
- Mercury and mercury compounds
- Cadmium and cadmium compounds
- All kinds of oil and oil mixtures
- Materials either suspected or known to be carcinogens, mutagens, or teratogens

These restrictions led to the development of ocean incineration technology, which was first put into operation by Germany in 1967. Since then, other European nations, such as Belgium and the Netherlands, have used sea incineration on a large scale for disposal of halogenated organic wastes.

In the United States, Chemical Waste Management (CWM) has conducted several at-sea test burns in specially designed ships (*Vulcanus I* and *Vulcanus II*). The first test burn was carried out at the Gulf Coast Ocean Incineration Site

(GCOIS), aboard the *Vulcanus I,* from October 1974 through January 1975. This site is located 209 km (130 mi) south of Sabina Pass, Texas, in the Gulf of Mexico. Four shiploads, each with 4000 metric tons of chlorinated organic wastes (trichloropropane, trichloroethane, and dichloroethane), were disposed of by incineration. The first two loads were incinerated under a research permit, and the last two under a special permit granted by the EPA. The second at-sea incineration took place during March and April, 1977, on the *Vulcanus I* at the GCOIS. The third incineration, also of chlorinated organic wastes—Herbicide Orange— occurred in the Pacific Ocean, about 193 km (120 mi) west of Johnston Atoll, from July 14 to September 2, 1977. DRE of >99.9999 percent were achieved for the main constituents of the Herbicide Orange.

During December 1981 and January 1982, PCBs were incinerated at the GCOIS aboard the *Vulcanus I* under a research permit issued by the EPA. A second burn took place, in August 1982, in which an extensive testing and monitoring program was undertaken by the EPA. DRE of >99.9966 percent was achieved. Samples of stack gas were found to have no trace of PCB, CB, TCCD, or TCDF. Testing of the marine environment and the ambient air showed no trace of PCBs. Test animals, placed in cages, did not show elevated levels of PCBs.

All test burns were monitored by the EPA and other federal agencies. In all cases, both CWM and the EPA claimed that test burns aboard the *Vulcanus I* met and even exceeded the DRE. However, certain uncertainties relating to the minimum-residence-time requirement and removal of heavy metals and acids from the emissions, coupled with public opposition, have restrained the EPA from issuing commercial permits for ocean incineration of hazardous waste. Sea incineration is commonly practiced in European countries.

STUDY QUESTIONS

1. What are the methods for disposal of hazardous waste? What law controls land disposal of hazardous waste?

2. What are the two common thermal methods of hazardous waste disposal? Discuss the advantages and the disadvantages of incineration.

3. What requirements have been established, under the TSCA, for disposal of PCBs using incinerators?

4. What is the land disposal restriction (LDR)? What types of hazardous wastes are regulated by the LDR? What is the aim of the LDR?

5. What is the California List? When and why did the EPA include it in its rules? Do the rules allow dilution of a hazardous waste (to lower the concentration) as a substitute for treatment? Explain.

6. What are the new standards for landfill that were developed under the HSWA in 1984? Draw a sketch to show the main components of a secure landfill.

7. What are geosynthetics, and what is their application in the geotechnical/environmental engineering projects? Why are they preferred over natural materials to serve as a filter or drainage medium? How may their use result in an increase in the capacity of a landfill?

8. What requirements must be complied with during a landfill closure? What post-closure monitoring has to be done and for how long?

9. What is "clean" closing? How can an operator of a land treatment facility ensure that he has met the requirements for clean closure of a waste pile?

10. What is deep well injection? What are the various classes of wells used for underground disposal of wastes? What is the main purpose of regulating waste disposal using deep wells?

11. What geologic and related factors should be considered in selecting sites for deep well disposal of hazardous wastes?

12. What factors control the storage/disposal of hazardous wastes in mined-out areas? Is mine disposal a viable alternative? Explain.

13. Why is ocean disposal of hazardous waste not being done on a commercial basis in the United States? Is it permissible to dispose of high-level nuclear waste in the ocean anywhere in the world?

REFERENCES CITED

40 CFR, 1993, Parts 124 and 260–271; Washington, D.C., U.S. Government Printing Office.

Dempsey, C. R., and D.A. Oberacker, 1988, Overview of Incineration Performance, Cincinnati, OH, U.S. Environmental Protection Agency, Report No. EPA/600/D-88/230, Risk Reduction Engineering Laboratory, 24 p.

Freeman, H.M., ed., 1989, Standard Handbook of Hazardous Waste Treatment and Disposal, New York, McGraw-Hill, Inc., 14 chapters plus index.

Hasan, S.E., 1990, Possibility of using abandoned mines for waste disposal in urban area (Abst.); Association of Engineering Geologists, 33d Annual Meeting, Pittsburgh, PA, October 1–5, p. 97.

Rudy, R., G. Strobel, and W. Widmann, 1989, RANDOM-WALK modeling of organic contaminant migration from the Theresienfeld Landfill located in the Vienna basin aquifer: Silver Spring, MD, Proceedings 1989 Superfund Conference, Hazardous Materials Control Research Institute, p. 163–180.

Rudy, R. J., R.A. Marszalkowski, and H. Bayer, 1989, Remedial planning at an uncontrolled hazardous waste site in Austria: Hazardous Materials Control, November/December, 1989, Hazardous Materials Control Research Institute, p. 66–72.

Shannon, L.B., 1993, Injection: an option for industrial waste: Environmental Protection, v. 4, n. 9, p. 42–48.

Urban, W., and H. Frischherz, 1992, Remediation of groundwater resources contaminated by seepage from hazardous waste dump by extraction from wells, treatment and recharge of clean water, Proceedings, International Symposium on Environmental Contamination in Central and Eastern Europe, October 12–16, Budapest, Hungary, p. 646–650.

U.S. Environmental Protection Agency, 1991, Land Disposal Restrictions: Summary of Requirements, Office of Solid Waste and Emergency Response, Report No. OSWER 9934.0-1A, 26 p. plus appendices.

U.S. Environmental Protection Agency, 1986, Method 9095 paint filter liquid; Test Methods for Evaluating Solid Waste, Office of Solid Waste and Emergency Response, Report No. SW-846, 3d ed., p. 9095-1–9095-4.

U.S. Environmental Protection Agency, 1986, Method 9090A compatibility test for wastes and membrane liners; Test Methods for Evaluating Solid Waste, Office of Solid Waste and Emergency Response, Report No. SW-846, 3d ed., p. 9090A-1–9090A-16.

U.S. Environmental Protection Agency, 1991, Design and construction of RCRA/CERCLA final covers; Office of Research and Development, Report No. EPA/625/4-91/025, 145 p. plus 2 appendices.

U.S. Environmental Protection Agency, 1989, Requirements for hazardous waste landfill design, construction, and closure; Report No. EPA/625-4-89/022, 127 p.

Warner, D.L., and J.H. Lehr, 1977, An introduction to the technology of subsurface waste water injection: U.S. EPA, Report No. 600/2-77-240, 345 p.

SUPPLEMENTAL READING

Freeman, H.M., ed., 1989, Standard handbook of hazardous waste treatment and disposal, New York, McGraw-Hill, Inc., 14 chapters plus index.

Wentz, C.A., 1989, Hazardous Waste Management: New York, McGraw-Hill Publishing Co., 461 p.

APPENDIX A

Listing of Hazardous Wastes

TABLE A–1 Hazardous Wastes from Nonspecific Sources (adapted from 40 CFR, Part 261, 1990)

EPA Waste Number	Hazardous Waste	Hazard Code*
F001	The following spent halogenated solvents used in degreasing: tetrachloroethylene, trichloroethylene, methylene chloride, 1,1,1-trichloroethane, carbon tetrachloride, chlorinated fluorocarbons, all spent solvent mixtures/blends used in degreasing containing, before use, a total of 10 percent or more (by volume) of one or more of the above halogenated solvents or those listed in F002, F004, and F005; and still bottoms from the recovery of these spent solvents and spent solvent mixtures	(T)
F002	The following spent halogenated solvents: tetrachloroethylene, methylene chloride, trichloroethylene, 1,1,1-trichloroethane, chlorobenzene, 1,1,2-trichloro-1,2,2,-trifluoroethane, *o*-dichlorobenzene, and trichlorofluoromethane; all spent solvent mixtures/blends containing, before use, a total of 10 percent or more of the above halogenated solvents or those listed in F001, F004, or F005; and still bottoms from the recovery of these spent solvents and spent solvent mixtures	(T)
F003	The following spent nonhalogenated solvents: xylene, acetone, ethyl acetate, ethyl benzene, ethyl ether, methylisobutyl ketone, *n*-butyl alchohol, cyclohexanone, methanol; all spent solvent mixtures/blends containing, before use, one or more of the above nonhalogenated solvents, and a total of 10 percent or more (by volume) of one or more of those solvents listed in F001, F002, F004, and F005; and still bottoms from the recovery of these spent solvents and spent solvent mixtures	(I)
F004	The following spent nonhalogenated solvents: cresols and cresylic acid, nitrobenzene; all spent solvent mixtures/blends containing, before use, a total of 10 percent or more (by volume) of one or more of the above nonhalogenated solvents or those solvents listed in F001, F002, and F005; and the still bottoms from the recovery of these spent solvents and spent solvent mixtures	(T)
F005	The following spent nonhalogenated solvents: toluene, methyl ethyl ketone, carbon disulfide, isobutanol, pyridine; all spent solvent mixtures/blends containing, before use, a total of 10 percent or more (by volume) of one or more of the above nonhalogenated solvents or those listed in F001, F002, and F004; and still bottoms from the recovery of these spent solvents and spent solvent mixtures	(I,T)

*C = corrosive, H = acutely hazardous, I = ignitible, R = reactive, and T = toxic.

EPA Waste Number	Hazardous Waste	Hazard Code*
F006	Wastewater treatment sludges from electroplating operations except from the following processes: (1) sulfuric acid anodizing of aluminum; (2) tin plating on carbon steel; (3) zinc plating (segregated basis) on carbon steel; (4) aluminum or zinc-aluminum plating on carbon steel; (5) cleaning/stripping associated with tin, zinc, and aluminum plating on carbon steel; and (6) chemical etching and milling of aluminum	(T)
F019	Wastewater treatment sludges from the chemical conversion coating of aluminum	(T)
F007	Spent cyanide plating bath solutions from electroplating operations (except for precious metals electroplating spent cyanide plating bath solutions)	(R,T)
F008	Plating bath sludges from the bottom of plating baths from electroplating operations for which cyanides are used in the process (except for precious metals electroplating plating bath sludges)	(R,T)
F009	Spent stripping and cleaning bath solutions from electroplating operations for which cyanides are used in the process (except for precious metals electroplating spent stripping and cleaning bath solutions)	(R,T)
F010	Quenching bath sludges from oil baths from metal heat treating, operations for which cyanides are used in the process (except for precious metals heat-treating quenching bath sludges)	(R,T)
F011	Spent cyanide solutions from salt bath pot cleaning from metal heat-treating operations (except for precious metals heat-treating spent cyanide solutions from salt bath pot cleaning)	(R,T)
F012	Quenching wastewater treatment sludges from metal heat-treating operations for which cyanides are used in the process (except for precious metals heat treating quenching wastewater treatment sludges)	(T)
F024	Wastes including but not limited to distillation residues, heavy ends, tars, and reactor cleanout wastes from the production of chlorinated aliphatic hydrocarbons, having carbon content from one to five, utilizing free radical catalyzed processes (does not include light ends, spent filters and filter aids, spent desiccants, wastewater, wastewater treatment sludges, spent catalysts, and wastes listed in 261.32)	(T)
F020	Wastes (except wastewater and spent carbon from hydrogen chloride-purification) from the production or manufacturing use (as a reactant, chemical intermediate, or component in a formulating process) of tri- or tetrachlorophenol or of intermediates used to produce their pesticide derivatives (does not include wastes from the production of hexachlorophene from highly purified 2,4,5-trichlorophenol)	(H)
F021	Wastes (except wastewater and spent carbon from hydrogen chloride-purification) from the production or manufacturing use (as a reactant, chemical intermediate, or component in a formulating process) of pentachlorophenol or of intermediates used to produce its derivatives	(H)
F022	Wastes (except wastewater and spent carbon from hydrogen chloride purification) from the manufacturing use (as a reactant, chemical intermediate, or component in a formulating process) of tetra-, penta-, or hexachlorobenzenes under alkaline conditions	(H)
F023	Wastes (except wastewater and spent carbon from hydrogen chloride purification) from the production of materials on equipment previously used for the production or manufacturing use (as a reactant, chemical intermediate, or component in a formulating process) of tri- and tetrachlorophenols (does not include wastes from equip-	(H)

*C = corrosive, H = acutely hazardous, I = ignitible, R = reactive, and T = toxic.

EPA Waste Number	Hazardous Waste	Hazard Code*
	ment used only for the production or use of hexachlorophene from highly purified 2,4,5-trichlorophenol)	
F026	Wastes (except wastewater and spent carbon from hydrogen chloride purification) from the production of materials on equipment previously used for the manufacturing use (as a reactant, chemical intermediate, or component in a formulating process) of tetra-, penta-, or hexachlorobenzene under alkaline conditions	(H)
F027	Discarded unused formulations containing tri-, tetra-, or penta-chlorophenol or discarded unused formulations containing compounds derived from these chlorophenols (does not include formulations containing hexachlorophene synthesized from pre-purified 2,4,5-trichlorophenol as the sole component	(H)
F028	Residues resulting from the incineration or thermal treatment of soil contaminated with EPA hazardous wastes numbered F020, F021, F022, F023, F026, and F027	(T)

TABLE A–2 Hazardous Wastes from Specific Sources (adapted from 40 CFR, Part 261, 1990)

EPA Waste Number	Hazardous Waste	Hazard Code*
	Wood Preservatives	
K001	Bottom sediment sludge from the treatment of wastewaters from wood preserving processes that use creosote and/or pentachlorophenol	(T)
	Inorganic Pigments	
K002	Wastewater treatment sludge from the production of chrome yellow and orange pigments	(T)
K003	Wastewater treatment sludge from the production of molybdate orange pigments	(T)
K004	Wastewater treatment sludge from the production of zinc yellow pigments	(T)
K005	Wastewater treatment sludge from the production of chrome green pigments	(T)
K006	Wastewater treatment sludge from the production of chrome oxide green pigments (anhydrous and hydrated)	(T)
K007	Wastewater treatment sludge from the production of iron blue pigments	(T)
K008	Oven residue from the production of chrome oxide green pigments	(T)
	Organic Chemicals	
K009	Distillation bottoms from the production of acetaldehyde from ethylene	(T)
K010	Distillation side cuts from the production of acetaldehyde from ethylene	(T)
K011	Bottom stream from the wastewater stripper in the production of acrylonitrile	(R,T)
K013	Bottom stream from the acetonitrile column in the production of acrylonitrile	(R,T)

*C = corrosive, H = acutely hazardous, I = ignitible, R = reactive, and T = toxic.

EPA Waste Number	Hazardous Waste	Hazard Code*
K014	Bottoms from the acetonitrile purification column in the production of acrylonitrile	(T)
K015	Still bottoms from the distillation of benzyl chloride	(T)
K016	Heavy ends or distillation residues from the production of carbon tetrachloride	(T)
K017	Heavy ends (still bottoms) from the purification column in the production of epichlorohydrin	(T)
K018	Heavy ends from the fractionation column in ethyl chloride production	(T)
K019	Heavy ends from the distillation of ethylene dichloride in tethylene dichloride production	(T)
K020	Heavy ends from the distillation of vinyl chloride in vinyl chloride monomer production	(T)
K021	Aqueous spent antimony catalyst waste from fluoromethanes production	(T)
K022	Distillation bottom tars from the production of phenol/acetone from cumene	(T)
K023	Distillation light ends from the production of phthalic anhydride from naphthalene	(T)
K024	Distillation bottoms from the production of phthalic anhydride from naphthalene	(T)
K093	Distillation light ends from the production of phthalic anhydride from o-xylene	(T)
K094	Distillation bottoms from the production of phthalic anhydride from o-xylene	(T)
K025	Distillation bottoms from the production of nitrobenzene by the nitration of benzene	(T)
K026	Stripping still tails from the production of methyl ethyl pyridines	(T)
K027	Centrifuge and distillation residues from toluene diisocyanate production	(R,T)
K028	Spent catalyst from the hydrochlorinator reactor in the production of 1,1,1-trichloroethane	(T)
K029	Waste from the product steam stripper in the production of 1,1,1-trichloroethane	(T)
K095	Distillation bottoms from the production of 1,1,1-trichloroethane	(T)
K096	Heavy ends from the heavy ends column from the production of 1,1,1-trichloroethane	(T)
K030	Column bottoms or heavy ends from the combined production of trichloroethylene and perchloroethylene	(T)
K083	Distillation bottoms from aniline production	(T)
K103	Process residues from aniline extraction from the production of aniline	(T)
K104	Combined wastewater streams generated from nitrobenzene/aniline production	(T)
K085	Distillation or fractionation column bottoms from the production of chlorobenzenes	(T)
K105	Separated aqueous stream from the reactor product washing step in the production of chlorobenzenes	(T)
K111	Product washwaters from the production of dinitrotoluene via nitration of toluene	(C,T)
K112	Reaction by-product water from the drying column in the production of toluenediamine via hydrogenation of dinitrotoluene	(T)
K113	Condensed liquid light ends from the purification of toluenediamine in the production of toluenediamine via hydrogenation of dinitrotoluene	(T)

*C = corrosive, H = acutely hazardous, I = ignitible, R = reactive, and T = toxic.

EPA Waste Number	Hazardous Waste	Hazard Code*
K114	Vicinals from the purification of toluenediamine in the production of toluenediamine via hydrogenation of dinitrotoluene	(T)
K115	Heavy ends from the purification of toluenediamine in the production of toluenediamine via hydrogenation of dinitrotoluene	(T)
K116	Organic condensate from the solvent recovery column in the production of toluene diisocyanate via phosgenation of toluenediamine	(T)
K117	Wastewater from the reactor vent gas scrubber in the production of ethylene dibromide via bromination of ethene	(T)
K118	Spent adsorbent solids from purification of ethylenedibromide via bromination of ethene	(T)
K136	Still bottoms from the purification of ethylene dibromide in the production of ethylene dibromide via bromination of ethene	(T)
	Inorganic Chemicals	
K071	Brine purification muds from the mercury cell process in chlorine production for which separately pre-purified brine is not used	(T)
K073	Chlorinated hydrocarbon waste from the purification step of the diaphragm cell process using graphite anodes in chlorine production	(T)
K106	Wastewater treatment sludge from the mercury cell process in chlorine production	(T)
	Pesticides	
K031	By-product salts generated in the production of MSMA and cacodylic acid	(T)
K032	Wastewater treatment sludge from the production of chlordane	(T)
K033	Wastewater and scrub water from the chlorination of cyclopentadiene in the production of chlordane	
K034	Filter solids from the filtration of hexachlorocyclopentadiene in the production of chlordane	(T)
K097	Vacuum stripper discharge from the chlordane chlorinator in the production of chlordane	(T)
K035	Wastewater treatment sludges generated in the production of creosote	(T)
K036	Still bottoms from toluene reclamation distillation in the production of disulfoton	(T)
K037	Wastewater treatment sludges from the production of disulfoton	(T)
K038	Wastewater from the washing and stripping of phorate production	(T)
K039	Filter cake from the distillation of diethylphosphorodithioic acid in the production of phorate	(T)
K040	Wastewater treatment sludge from the production of phorate	(T)
K041	Wastewater treatment sludge from the production of toxaphene	(T)
K098	Untreated process wastewater from the production of toxaphene	(T)
K042	Heavy ends or distillation residues from the distillation of tetrachlorobenzene in the production of 2,4,5-T	(T)
K043	2,6-Dichlorophenol waste from the production of 2,4-D	(T)
K099	Untreated wastewater from the production of 2,4-D	(T)
	Explosives	
K044	Wastewater treatment sludges from the manufacturing and processing of explosives	(R)
K045	Spent carbon from the treatment of wastewater containing explosives	(R)
K046	Wastewater treatment sludges from the manufacturing, formulation, and loading of lead-based initiating compounds	(R)
K047	Pink/red water from TNT operations	(R)

*C = corrosive, H = acutely hazardous, I = ignitible, R = reactive, and T = toxic.

EPA Waste Number	Hazardous Waste	Hazard Code*
	Petroleum Refining	
K048	Dissolved air floatation (DAF) float from the petroleum refining industry	(T)
K049	Slop oil emulsion solids from the petroleum refining industry	(T)
K050	Heat exchanger bundle cleaning sludge from the petroleum refining industry	(T)
K051	API separator sludge from the petroleum refining industry	(T)
K052	Tank bottoms (leaded) from the petroleum refining industry	(T)
	Iron and Steel	
K061	Emission control dust/sludge from the primary production of steel in electric furnaces	(T)
K062	Spent pickle liquor generated by steel finishing operations of facilities within iron and steel industry SIC codes 331 and 332	(C,T)
K064	Acid plant blowdown slurry/sludge resulting from the thickening of blowdown slurry from primary copper production	(T)
K065	Surface impoundment solids contained in and dredged from surface impoundments at primary lead smelting facilities	(T)
K066	Sludge from treatment of process wastewater and/or acid blowdown from primary zinc production	(T)
K088	Spent potliners from primary aluminum reduction	(T)
K090	Emission control dust or sludge from ferrochromium silicon production	(T)
K091	Emission control dust or sludge from ferrochromium production	(T)
	Secondary Lead	
K069	Emission control dust/sludge from secondary lead smelting	(T)
K100	Waste leaching solution from acid leaching of emission control dust/sludge from secondary lead smelting	(T)
	Veterinary Pharmaceuticals	
K084	Wastewater treatment sludges generated during the production of veterinary pharmaceuticals from arsenic or organoarsenic compounds	(T)
K101	Distillation tar residues from the distillation of aniline-based compounds in the production of veterinary pharmaceuticals from arsenic or organoarsenic compounds	(T)
K102	Residue from the use of activated carbon for decolorization in the production of veterinary pharmaceuticals from asrenic or organoarsenic compounds	(T)
	Ink Formulation	
K086	Solvent washes and sludges, caustic washes and sludges, or water washes and sludges from cleaning tubs and equipment used in the formulation of ink from pigments, driers, soaps, and stabilizers containing chromium and lead	(T)
	Coking	
K060	Ammonia still lime sludge from coking operations	(T)
K087	Decanter tank tar sludge from coking operations	(T)

*C = corrosive, H = acutely hazardous, I = ignitible, R = reactive, and T = toxic.

TABLE A–3   Commercial Chemical Products (adapted from 40 CFR, Part 261, 1990)

EPA Waste Number	Hazardous Waste
The following P code wastes are considered acutely hazardous (H):	
P023	Acetaldehyde, chloro-
P002	Acetamide, N-(aminothioxomethyl)-
P057	Acetamide, 2-fluoro-
P058	Acetic acid, fluoro-, sodium salt
P066	Acetimidic acid, N-[(methylcarbamoyl)oxy]thio-, methyl ester
P001	3-(alpha-acetonylbenzyl)-4-hydroxy-coumarin and salts, when present at concentration greater than 0.3 percent
P002	1-Acetyl-2-thiourea
P003	Acrolein
P070	Aldicarb
P004	Aldrin
P005	Allyl alcohol
P006	Aluminum phosphide
P007	5-(Aminomethyl)-3-isoxazolol
P008	4-aAminopyridine
P009	Ammonium picrate (R)
P119	Ammonium vanadate
P010	Arsenic acid
P012	Arsenic (III) oxide
P011	Arsenic (V) oxide
P011	Arsenic pentoxide
P012	Arsenic trioxide
P038	Arsine, diethyl
P054	Aziridine
P013	Barium cyanide
P024	Benzenamine, 4-chloro-
P077	Benzenamine, 4-nitro-
P028	Benzene, (chloromethyl)-
P042	1,2-Benzenediol, 4-[(1-hydroxy-2-(methyl-amino) ethyl)]-
P014	Benzenethiol
P028	Benzyl chloride
P015	Beryllium dust
P016	Bis(chloromethyl) ether
P017	Bromoacetone
P018	Brucine
P021	Calcium cyanide
P123	Camphene, octachloro-
P103	Carbamimidoselenoic acid
P022	Carbon bisulfide
P022	Carbon disulfide
P095	Carbonyl chloride
P033	Chlorine cyanide
P023	Chloroacetaldehyde

EPA Waste Number	Hazardous Waste
P024	*p*-Chloroaniline
P026	1-(*O*-Chlorophenyl)thiourea
P027	3-Chloropropionitrile
P029	Coppercyanides
P030	Cyanides (soluble cyanide salts), not elsewhere specified
P031	Cyanogen
P033	Cyanogen chloride
P036	Dichlorophenylarsine
P037	Dieldrin
P038	Diethylarsine
P039	O,O-Diethyl S-[2-(ethylthio)ethyl] phosphorodithioate
P041	Diethyl-*p*-nitrophenyl phosphate
P040	O,O-Diethyl O-pyrazinyl phosphorothioate
P043	Diisopropyl fluorophosphate
P044	Dimethoate
P045	3,3-Dimethyl-1-(methylthio)-2-butanone, O-[(methylamino) carbonyl] oxime
P071	O,O-Dimethyl O-*p*-nitrophenyl phosphorothioate
P082	Dimethylnitrosamine
P046	alpha, alpha-Dimethylphenethylamine
P047	4,6-Dinitro-*o*-cresol and salts
P034	4,6-Dinitro-*o*-cyclohexylphenol
P048	2,4-Dinitrophenol
P020	Dinoseb
P085	Disphosphoramide, octamethyl
P039	Disulfoton
P049	2,4-Dithiobiuret
P109	Dithiopyrophosphoric acid, tetraethyl ester
P050	Endosulfan
P088	Endothall
P051	Endrin
P042	Epinephrine
P046	Ethanamine, 1,1-dimethyl-2-phenyl-
P084	Ethenamine, N-methyl-N-nitroso-
P101	Ethyl cyanide
P054	Ethylenimine
P097	Famphur
P056	Fluorine
P057	Fluoroacetamide
P058	Fluoroacetic acid, sodium salt

EPA Waste Number	Hazardous Waste
P065	Fulminic acid, mercury(II) salt (R,T)
P059	Heptachlor
P051	1,2,3,4,10,10-Hexachloro-6,7-epoxy-1,4,41,5,6,7,8,8a-octahydroendo, endo-1,4:5,8-dimethanonaphthalene
P060	1,2,3,4,10,10-Hexachloro-1,4,41,5,8,8a-hexahydro-1,4:5,8-endo, endo-dimethanonaphthalene
P004	1,2,3,4,10,10-Hexachloro-1,4,4a,5,8,8a-hexahydro-1,4:5,8-endo, exodimethanonaphthalene
P060	Hexachlorohexahydro-exo, exodimethanonaphthalene
P062	Hexaethyl tetraphosphate
P116	Hydrazinecarbothioamide
P068	Hydrazine, methyl-
P063	Hydrocyanic acid
P063	Hydrogen cyanide
P096	Hydrogen phosphide
P064	Isocyanic acid, methyl ester
P007	3(2H)-isoxazolone, 5-(aminomethyl)-
P092	Mercury, (acetato-0)phenyl-
P065	Mercury fulminate (R,T)
P016	Methane, ixybix(chloro)-
P112	Methane, tetranitro- (R)
P118	Methanethiol, trichloro-
P059	4,7-Methano-1H-indene, 1,4,5,6,7,8,8-heptachloro-3a,4,7, 7a-tetrahydro-
P066	Methomyl
P067	2-Methylaziridine
P068	Methyl hydrazine
P064	Methyl isocyanate
P069	2-Methyllactonitrile
P071	Methyl parathion
P072	alpha-Naphthylthiourea
P073	Nickel carbonyl
P074	Nickle(II) cyanide
P073	Nickle tetracarbonyl
P075	Nicotine and salts
P076	Nitric oxide
P077	p-Nitroaniline
P078	Nitrogen dioxide
P076	Nitrogen(II) oxide
P078	Nitrogen(IV) oxide
P081	Nitroglycerine (R)
P082	N-Nitrosodimethylamine
P084	N-Nitrosomethylvinylamine
P050	5-Norbornene-2,3-dimethanol,1,4,5,6,7,7-hexachloro, cyclic sulfite

EPA Waste Number	Hazardous Waste
P085	Octamethylpyrophosphoramide
P087	Osmium oxide
P098	Osmium tetroxide
P088	7-Oxabicyclo-[2.2.1] heptane-2,3-dicarboxylic acid
P089	Parathion
P034	Phenol, 2-cyclohexyl-4,6-dinitro-
P048	Phenol, 2,4-dinitro-
P047	Phenol, 2,4-dinitro-6-methyl-
P020	Phenol, 2,4-dinitro-6-(1-methyl-propyl)-
P009	Phenol, 2,4,6-trinitro-, ammonium salt (R)
P036	Phenyl dichloroarsine
P092	Phenylmercuric acetate
P093	N-Phenylthiourea
P094	Phorate
P095	Phosgene
P096	Phosphine
P041	Phosphoric acid, diethyl p-nitro-phenyl ester
P044	Phosphorodithioic acid, O,O-dimethyl S-[2-(methylamino)-2-oxyethyl] ester
P043	Phosphorofluoric acid, bis(1-methylethyl)ester
P094	Phosphorothioic acid, O,O-diethyl S-(ethylthio) methyl ester
P089	Phosphorothioic acid, O,O-diethyl O(p-nitrophenyl) ester
P040	Phosphorothioic acid, O,O-diethyl O-pyrazinyl ester
P097	Phosphorothioic acid, O,O-dimethyl)-[p(dimethylamino)-sul-fonyl) phenyl] ester
P110	Plumbane, tetraethyl-
P098	Potassium cyanide
P099	Potassium silver cyanide
P070	Propanal, 2-methyl-2-(methylthio)-O-[(methylamino) carbonyl]-oxime
P101	Propanenitrile
P027	Propanenitrile, 3-chloro
P069	Propanenitrile, 2-hydroxy-2-methyl-
P081	1,2,3-Propanetriol, trinitrate-(R)
P017	2-Propanone, 1-bromo-
P102	Propargyl alcohol
P003	2-propenal
P005	2-propen-1-ol
P067	1,2-Propylenimine
P102	2-Propyn-1-ol
P008	4-Pyridinamine

EPA Waste Number	Hazardous Waste
P075	Pyridine, (S)-3-(1-methyl-2-pyrrolidinyl)-, and salts
P111	Pyrophosphoric acid, tetraethyl ester
P103	Selenourea
P104	Silver cyanide
P105	Sodium azide
P106	Sodium cyanide
P107	Stronitium sulfide
P108	Strychnidin-10-one, and salts
P018	Strychnidin-10-one, 2,3-dimethoxy-
P108	Strychnine and salts
P115	Sulfuric acid, thallium(1) salts
P109	Tetraethyldithiopyrophosphate
P110	Tetraethyl lead
P111	Tetraethylpyrophosphate
P112	Tetranitromethane (R)
P062	Tetraphosphoric acid, hexaethyl ester
P113	Thallic oxide
P113	Thallium(III) oxide
P114	Thallium(I) selenite
P115	Thallium(I) sulfate
P045	Thiofanax
P049	Thiomidodicarbonic diamide
P014	Thiophenol
P116	Thiosemicarbazide
P026	Thiourea, (2-chlorophenyl)-
P072	Thiourea, 1-naphthalenyl-
P093	Thiourea, phenyl
P123	Toxaphene
P118	Trichloromethanethiol
P119	Vanadic acid, ammonium salt
P120	Vandium pentoxide
P120	Vanadium(V) oxide
P001	Warfarin, when present at concentrations greater than 0.3 percent
P121	Zinc cyanide
P122	Zinc phosphide, when present at concentrations greater than 10 percent

The following U code wastes are nonacutely hazardous:

U001	Acetaldehyde (I)
U034	Acetaldehyde, trichloro-
U187	Acetamide, N-(4-ethoxyphenyl)-
U005	Acetamide, N-9H-fluoren-2-yl-
U112	Acetic acid, ethyl ester (I)
U144	Acetic acid, lead salt
U214	Acetic acid, thallium(I) salt
U002	Acetone (I)
U003	Acetonitrile (I,T)

EPA Waste Number	Hazardous Waste
U248	3-(alpha-Acetonylbenzyl)-4-hydroxycoumarin and salts, when present at concentrations of 0.3 percent or less
U004	Acetophenone
U005	2-Acetylaminofluorene
U006	Acetyl chloride (C,R,T,)
U007	Acrylamide
U008	Acrylic acid (I)
U009	Acrylonitrile
U150	Alanine, 3-[p-bis(2-chloroethyl)amino] pheyl-, L-
U328	2-Amino-I-methylbenzene
U353	4-Amino-I-methylbenzene
U011	Amitrole
U012	Aniline (I,T)
U014	Auramine
U015	Azaserine
U010	Azirino (2',3',3',4)pyrrolo (1,2-a)indole-4,7-dione, 6-amino-8-[(aminocarbonyl)oxy)methyl)]-1,1a,2,8,8a,8b-hexahydro-8a-methoxy-5-methyl-
U157	Benz(j)aceanthrylene, 1,2-dihydro-3-methyl-
U016	Benz(c)acridine
U016	3,4-Benzacridine
U017	Benzal chloride
U018	Benz(a)anthracene
U018	1,2-Benzanthracene
U094	1,2-Benzanthracene, 7,12-dimethyl-
U012	Benzenamine (I,T)
U014	Benzenamine, 4,4'-carbonimidoyl-bis(N,N-dimethyl)-
U049	Benzenamine, 4-chloro-2-methyl-
U093	Benzenamine, N,N'-dimethyl-4-phenylazo-
U158	Benzenamine, 4,4'-methylenebis(2-chloro)-
U222	Benzenamine, 2-methyl-,hydrochloride
U181	Benzenamine, 2-methyl-,5-nitro
U019	Benzene (I,T)
U038	Benzeneacetic acid, 4-chloro-alpha-(4-chloro-phenyl)-alpha-hydroxy,ethyl ester
U030	Benzene, 1-bromo-4-phenoxy-
U037	Benzene, chloro
U190	1,2-Benzenedicarboxylic acid anhydride
U028	1,2-Benzenedicarboxylic acid [bis(2-ethyl-hexyl)] ester
U069	1,2-Benzenedicarboxylic acid, dibutyl ester

EPA Waste Number	Hazardous Waste
U088	1,2-Benzenedicarboxylic acid, diethyl ester
U102	1,2-Benzenedicarboxylic acid, dimethyl ester
U107	1,2-Benzenedicarboxylic acid, di-*n*-octyl ester
U070	Benzene, 1,2-dichloro-
U071	Benzene, 1,3-dichloro-
U072	Benzene, 1,4-dichloro-
U017	Benzene, (dichloromethyl)-
U223	Benzene, 1,3-diisocyanatomethyl- (R,T)
U239	Benzene, dimethyl- (I,T)
U201	1,3-Benzenediol
U127	Benzene, hexachloro-
U056	Benzene, hexahydro-(I)
U188	Benzene, hydroxy-
U220	Benzene, methyl-
U105	Benzene, 1-methyl-1,2,4-dinitro-
U106	Benzene, 1-methyl-2,6-dinitro-
U203	Benzene, 1,2-methylenedioxy-4-allyl-
U141	Benzene, 1,2-methylenedioxy-4-propenyl-
U090	Benzene, 1,2-methylenedioxy-4-propyl-
U055	Benzene, (1-methylethyl) (I)
U169	Benzene, nitro-(I,T)
U183	Benzene, pentachloro-
U185	Benzene, pentachloro-nitro-
U020	Benzenesulfonic acid chloride (C,R)
U020	Benzenesulfonyl chloride (C,R)
U207	Benzene, 1,2,4,5-tetrachloro-
U023	Benzene, (trichloromethyl)-(C,R,T)
U234	Benzene, 1,3,5-trinitro (R,T)
U021	Benzidine
U202	1,2-Benzisothiazolin-3-one,1,1-dioxide
U120	Benzo(j,k)fluorene
U022	Benzo(a)pyrene
U022	3,4-Benzopyrene
U197	*p*-Benzoquinone
U023	Benzotrichloride (C,R,T)
U050	1,2-Benzphenanthrene
U085	2,2'-Bioxirane (I,T)
U021	(1,1'-Biphenyl)-4,4'-diamine
U073	(1,1'-Biphenyl)-4,4'-diamine, 3,3'-dichloro-
U091	(1,1'-Biphenyl)-4,4'-diamine, 3,3'-dimethoxy-
U095	(1,1'-Biphenyl)-4-4'-diamine,3,3'-dimethyl-
U024	Bis(2-chloroethoxy) methane
U027	Bis(2-chloroisopropyl) ether
U244	Bis(dimethylthiocarbamoyl) disulfide
U028	Bis(2-ethylhexyl)phthalate (DEHP)
U246	Bromine cyanide

EPA Waste Number	Hazardous Waste
U225	Bromoform
U030	4-Bromophenyl phenyl ether
U128	1,3-Butadiene, 1,1,2,3,4,4-hexachloro
U172	1-Butanamine, N-butyl-N-nitroso-
U035	Butanoic acid, 4-[bis(2-chloroethyl)amino]benzene-
U031	1-Butanol (I)
U159	2-Butanone (I,T)
U160	2-Butanone peroxide (R,T)
U053	2-Butenal
U074	2-Butene, 1,4-dichloro-(I,T)
U031	*n*-Butyl alcohol (I)
U136	Cacodylic acid
U032	Calcium chromate
U238	Carbamic acid, ethyl ester
U178	Carbamic acid, methylnitroso-, ethyl ester
U176	Carbamide, N-ethyl-N-nitroso-
U177	Carbamide, N-methyl-N-nitroso-
U219	Carbamide, thio-
U097	Carbamoyl chloride, dimethyl-
U215	Carbonic acid, dithallium(I)salt
U156	Carbonochloridic acid, methyl ester (I,T)
U033	Carbon oxyfluoride (R,T)
U211	Carbon tetrachloride
U033	Carbonyl fluoride (R,T)
U034	Chloral
U035	Chlorambucil
U036	Chlordane, technical
U026	Chlornaphazine
U037	Chlorobenzene
U039	4-Chloro-*m*-cresol
U041	1-Chloro-2,3-epoxypropane
U042	2-Chloroethyl vinyl ether
U044	Chloroform
U046	Chloromethyl ether
U047	beta-Chloronaphthalene
U048	*o*-Chlorophenol
U049	4-Chloro-*o*-toluidine, hydrochloride
U032	Chromic acid, calcium salt
U050	Chrysene
U051	Creosote
U052	Cresols
U052	Cresylic acid
U053	Crotonaldehyde
U055	Cumene (I)
U246	Cyanogen bromide
U197	1,4-Cyclohexadienedione
U056	Cyclohexane (I)
U057	Cyclohexanone (I)
U130	1,3-Cyclopentadiene, 1,2,3,4,5,5-hexa-chloro-
U058	Cyclophosphamide

EPA Waste Number	Hazardous Waste
U240	2,4-D, salts and esters
U059	Daunomycin
U060	DDD
U061	DDT
U142	Decachloro octahydro-1,3,4-metheno-2H-cyclobuta(c,d) pentalen-2-one
U062	Diallate
U133	Diamine (R,T)
U221	Diaminotoluene
U063	Dibenz(a,h)anthracene
U063	1,2:5,6-Dibenzanthracene
U064	1,2:7,8-Dibenzopyrene
U064	Dibnez(a,i)pyrene
U066	1,2-Dibromo-3-chloropropane
U069	Dibutyl phthalate
U062	S-(2,3-Dichloroallyl)diisopropylthio-carbamate
U070	*o*-Dichlorobenzene
U071	*m*-Dichlorobenzene
U072	*p*-Dichlorobenzene
U073	3,3'-Dichlorobenzidine
U074	1,4-Dichloro-2-butene (I,T)
U075	Dichlorodifluoromethane
U192	2,5-Dichloro-N-(1,1-dimethyl-2-propynly)benzamide
U060	Dichloro diphenyl dichloroethane
U061	Dichloro diphenyl trichloroethane
U078	1,1-Dichloroethylene
U079	1,2-Dichloroethylene
U025	Dichloroethyl ether
U081	2,4-Dichlorophenol
U082	2,6-Dichlorophenol
U240	2,4-Dichlorophenoxyacetic acid, salts and esters
U083	1,2-Dichloropropane
U084	1,3-Dichloropropene
U085	1,2:3,4-Diepoxybutane (I,T)
U108	1,4-Diethylene dioxide
U086	N,N-Diethylene dioxide
U087	O,O-Diethyl-S-methyl-dithiophosphate
U088	Diethyl phthalate
U089	Diethylstilbestrol
U148	1,2-Dihydro-3,6-pyradizinedione
U090	Dihydrosafrole
U091	3,3'-Dimethoxybenzidine
U092	Dimethylamine
U093	Dimethylaminoazobenzene
U094	7,12-Dimethylbenz(a)anthracene
U095	3,3'-Dimethylbenzidine
U096	alpha,alpha-Dimethylbenzylhydroperoxide (R)
U097	Dimethylcarbamoyl chloride
U098	1,1-Dimethylhydrazine

EPA Waste Number	Hazardous Waste
U099	1,2-Dimethylhydrazine
U101	2,4-Dimethylphenol
U102	Dimethyl phthalate
U103	Dimethyl sulfate
U105	2,4-Dinitrotoluene
U106	2,6-Dinitrotoluene
U107	Di-*n*-octyl phthalate
U108	1,4-Dioxane
U109	1,2-Dipheylhydrazine
U110	Dipropylamine (I)
U111	Di-N-propylnitrosamine
U001	Ethanal (I)
U174	Ethanamine, N-ethyl-N-nitroso-
U067	Ethane, 1,2-dibromo-
U076	Ethane, 1,1-dichloro-
U077	Ethane, 1,2-dichloro-
U114	1,2-Ethanediylbiscarbamodithioic acid
U131	Ethane, 1,1,1,2,2,2-hexachloro-
U024	Ethane, 1,1'-[methylenebis(oxy)]bis(2-chloro)-
U003	Ethanenitrile (I,T)
U117	Ethane, 1,1'-oxybis- (I)
U025	Ethane, 1,1'-oxybis(2-chloro)-
U184	Ethane pentachloro-
U208	Ethane, 1,1,1,2-tetrachloro-
U209	Ethane, 1,1,2,2-tetrachloro-
U218	Ethanethioamide
U247	Ethane, 1,1,1-trechloro-2,2-bis(*p*-methoxyphenyl)
U227	Ethane, 1,2,1-trichloro-
U043	Ethene, chloro-
U042	Ethene, 2-chloroethoxy-
U078	Ethene, 1,1-dichloro-
U079	Ethene, *trans*-1,2-dichloro-
U210	Ethene, 1,1,2,2-tetrachloro-
U173	Ethanol, 2,2'-(nitrosoimino)bis-
U004	Ethanone, 1-phenyl-
U006	Ethanoyl chloride (C,R,T)
U112	Ethyl acetate (I)
U113	Ethyl acrylate (I)
U238	Ethyl carbamate (urethan)
U038	Ethyl 4,4'-dichlorobenzilate
U359	Ethylene glycol monoethyl ether
U114	Ethylenebis(dithiocarbamic acid)
U067	Ethylene dibromide
U077	Ethylene dichloride
U115	Ethylene oxide (I,T)
U116	Ethylene thiourea
U117	Ethyl ether
U076	Ethylidene dichloride
U118	Ethylmethacrylate
U119	Ethyl methanesulfonate

EPA Waste Number	Hazardous Waste
U139	Ferric dextran
U120	Fluoranthene
U122	Formaldehyde
U123	Formic acid (C,T)
U124	Furan (I)
U125	2-Furancarboxaldehyde (I)
U147	2,5-Furandione
U213	Furan, tetrahydro- (I)
U125	Furfural (I)
U124	Furfuran (I)
U206	D-Glucopyranose,2-deoxy-2(3-methyl-3-nitro-soureido)-
U126	Glycidylaldehyde
U163	Guanidine, N-nitroso-N-methyl-N'nitro-
U127	Hexachlorobenzene
U128	Hexachlorobutadiene
U129	Hexachlorocyclohexane(gamma isomer)
U130	Hexa chlorocylopentdiene
U131	Hexachloroethane
U132	Hexachlorophene
U243	Hexachlorpropene
U133	Hydrazine (R,T)
U086	Hydrazine, 1,2-diethyl-
U098	Hydrazine, 1,1-dimethyl-
U099	Hydrazine, 1,2-dimethyl-
U109	Hydrazine, 1,2-diphenyl-
U134	Hydrofluoric acid (C,T)
U134	Hydrogen fluoride (C,T)
U135	Hydrogen sulfide
U096	Hydroperoxide, 1-methyl-1-phenylethyl- (R)
U136	Hydroxydimethylarsine oxide
U116	2-Imidazolidinethione
U137	Indeno(1,2,3-cd)pyrene
U140	Isobutyl alcohol (I,T)
U141	Isosafrole
U142	Kepone
U143	Lasiocarpine
U144	Lead acetate
U145	Lead phosphate
U146	Lead subacetate
U129	Lindane
U147	Maleic anhydride
U148	Maleic hydrazide
U149	Malononitrile
U150	Melphalan

EPA Waste Number	Hazardous Waste
U151	Mercury
U152	Methacrylonitrile (I,T)
U092	Methanamine, N-methyl- (I)
U029	Methane, bromo-
U045	Methane, chloro- (I,T)
U046	Methane, chloromethoxy-
U068	Methane, dibromo-
U080	Methane, dichloro-
U075	Methane, dichlorodifluoro-
U138	Methane, iodo-
U119	Methanesulfonic acid, ethyl ester
U211	Methane, tetrachloro-
U121	Methane, trichlorofluoro-
U153	Methanethiol (I,T)
U225	Methane, tribromo-
U044	Methane, trichloro-
U121	Methane, trichlorofluoro-
U123	Methanoic acid (C,T)
U036	4,7-Methanoindan, 1,2,4,5,6,7,8,8-octachloro-3a, 4,7,7a-tetrahydro-
U154	Methanol (I)
U155	Methapyrilene
U247	Methoxychlor
U154	Methyl alcohol (I)
U029	Methyl bromide
U186	1-Methylbutadiene (I)
U045	Methyl chloride (I,T)
U156	Methyl chlorocarbonate (I,T)
U226	Methyl chloroform
U157	3-Methylchlolanthrene
U158	4,4'-Methylenebis(2-chloroaniline)
U132	2,2'-Methylenebis(3,4,6-trichlorophenol)
U068	Methylene bromide
U080	Methylene chloride
U122	Methylene oxide
U159	Methyl ethyl ketone (I,T)
U160	Methyl ethyl ketone peroxide (R,T)
U138	Methyl iodide
U161	Methyl isobutyl ketone (I)
U162	Methyl methacrylate (I,T)
U163	N-Methyl-N'-nitro-N-nitrosoguanidine
U161	4-Methyl-2-pentanone (I)
U164	Methylthiouracil
U010	Mitomycin C
U059	5,12-Naphthacenedione,(8S-cis)-8-acetyl-10-[(3-amino-2,3,6-tri-deoxy-alpha-L-lyxo-hexopyranosyl)oxyl]-7, 8,9,10-tetrahydro-6,8, 11-trihydroxy-1-methyoxy-
U165	Naphthalene
U047	Naphthalene,2-chloro-

EPA Waste Number	Hazardous Waste
U166	1,4-Naphthalenedione
U236	2,7-Naphthalenedisulfonic acid,3,3'-[3,3'-dimethyl-(1,1'biphenyl)-4,4'diyl)]-bis(azo)bis(5-amino-4-hydroxy)-, tetrasodium salt
U166	1,4,Naphthaquinone
U167	1-Napththylamine
U168	beta-Naphthylamine
U026	2-Naphthylamine, N,N'-bis(2-chloromethyl)-
U169	Nitrobenzene (I,T)
U170	*p*-Nitrophenol
U171	2-Nitropropane (I)
U172	N-Nitrosodi-*n*-butylamine
U173	N-Nitrosodiethanolamine
U174	N-Nitrosodiethylamine
U111	N-Nitroso-N-propylamine
U176	N-Nitroso-Noethylurea
U177	N-Nitroso-N-methylurea
U178	N-Nitroso-N-methylurethane
U179	N-Nitrosopyrrolidine
U180	N-Nitrosopyrrolidine
U181	5-Nitro-*o*-toluidine
U193	1,2-Oxathiolane,2,2-dioxide
U058	2H-1,3,2-Oxazaphosphorine,2-[bis(2-chloroethyl)amino] tetrahydro, oxide 2-
U115	Oxirane (I,T)
U041	Oxirane, 2-(chloromethyl)-
U182	Paraldehyde
U183	Pentachlorobenzene
U184	Pentachloroethane
U185	Pentachloronitrobenzene
U186	1,3-Pentadiene (I)
U187	Phenecetin
U188	Phenol
U048	Phenol, 2-chloro-
U039	Phenol, 4-chloro-3-methyl-
U081	Phenol, 2,4-dichloro-
U082	Phenol, 2,6-dichloro-
U101	Phenol, 2,4-dimethyl-
U170	Phenol, 4-nitro-
U137	1,10-(1,2-phenylene)pyrene
U145	Phosphoric acid, lead salt
U087	Phosphorodithioic acid 0,0-diethyl-, S-methylester
U189	Phosphorous sulfide (R)
U190	Phthalic anhydride
U191	2-Picoline
U192	Pronamide
U194	1-Propanamine (I,T)
U110	1-Propanamine, N-propylp- (I)

EPA Waste Number	Hazardous Waste
U066	Propane, 1,2-dibromo-3-chloro-
U149	Propanedinitrile
U171	Propane, 2-nitro- (I)
U027	Propane, 2,2'-oxybis(2-chloro)-
U193	1,3-Propane sultone
U235	1-Propanol, 2,3-dibromo-,phosphate(3:1)
U126	1-Propanol, 2,3-epoxy-
U140	1-Propanol, 2-methyl- (I,T)
U002	2-Propanol, (I)
U007	2-Propanone (I)
U084	Propene, 1,3-cichloro-
U243	1-Propene, 1,1,2,3,3,3-hexachloro-
U009	2-Propenenitrile
U152	2-Propenenitrile, 2-methyl- (I,T)
U008	2-Propenoic acid (I)
U113	2-Propenoic acid, ethyl ester (I)
U118	2-Propenoic acid, 2-methyl-, ethyl ester
U162	2-Propenoic acid, 2-methyl, methyl ester (I,T)
U194	*n*-Propylamine (I,T)
U083	Propylene dichloride
U196	Pyridine
U155	Pyridine, 2-[(2-dimethylamino)-2-thenylamino)]
U179	Pyridine, hexahydro-N-nitroso-
U191	Pyridine, 2-methyl-
U164	4(1H)-Pyrijidinone, 2,3-dihydro-6-methyl-2-thioxo-
U180	Pyrrole, tetrahydro-N-nitroso-
U200	Reserpine
U201	Resorcinol
U202	Saccharin and salts
U203	Sagrole
U204	Selenious acid
U204	Selenium dioxide
U205	Selenium disulfide (R,T)
U015	L-Sernine, diazoacetate (ester)
U089	4,4'-Stilbenediol, alpha, alpha;-diethyl-
U206	Streptozotocin
U135	Sulfur hydride
U103	Sulfuric acid, dimethyl ester
U189	Sulfur phosphide (R)
U205	Sulfur selenide (R,T)
U207	1,2,4,5-Tetrachlorobenzene
U208	1,1,1,2-Tetrachloroethane
U209	1,1,2,2-Tetrachloroethane
U210	Tetrachloroethylene
U213	Tetrahydrofuran (I)

EPA Waste Number	Hazardous Waste	EPA Waste Number	Hazardous Waste
U214	Thallium(I)acetate	U182	1,3,5-Trioxane, 2,4,5-trimethyl-
U215	Thallium(I)carbonate	U235	Tris(2,3-dibromopropyl)phosphate
U216	Thallium(I)chloride	U236	Trypan blue
U217	Thallium(I)nitrate		
U218	Thioacetamide	U237	Uracil, 5[bis(2-chloromethyl)amino]-
U153	Thiomethanol, (I,T)	U237	Uracil mustard
U219	Thiourea		
U244	Thiram	U043	Vinyl chloride
U220	Toluene		
U221	Toluenediamine	U248	Warfarin, when present at concentrations of 0.3 percent or less
U223	Toluenediisocyanate (R,T)		
U328	o-Toluidine		
U222	O-Toluidine hydrochloride	U239	Xylene (I)
U353	p-Toluidine		
U011	1H-1,2,4-Triazol-3-amine	U200	Yohimban-16-carboxylic acid, 11,17-dimethoxy-18-[(3,4,5 trimethoxy-benzoyl)oxy]-methyl ester
U226	1,1,1-Trichloroethane		
U227	1,1,2-Trichloroethane		
U228	Trichloroethene		
U228	Trichloroethylene		
U121	Trichloromonofluoromethane	U249	Zinc phosphide, when present at concentrations of 10 percent or less
U234	sym-Trinitrobenzen (R,T)		

TABLE A–4   Hazardous Constituents (adapted from 40 CFR, Part 261, 1990, p. 90-98)

Common Name	Chemical Abstracts Name	Chemical Abstracts No.	Hazardous Waste No.
Acetonitrile	Same	75-05-8	U003
Acetophenone	Ethanone, 1-phenyl-	98-86-2	U004
2-Acetylaminefluarone	Acetamide, N-9H-fluoren-2-yl-	53-96-3	U005
Acetyl chloride	Same	75-36-5	U006
1-Acetyl-2-thiourea	Acetamide, N-(aminothioxomethyl)-	591-08-2	P002
Acrolein	2-Propenal	107-02-8	P003
Acrylamide	2-Propenamide	79-06-1	U007
Acrylonitrile	2-Propenenitrile	107-13-1	U009
Aflatoxins	Same	1402-68-2	
Aldicarb	Propanal, 2-methyl-2-(methylthio)-, O-[(methylamino)carbonyl]oxime	116-06-3	P070
Aldrin	1,4,5,8- Dimethanonaphthalene, 1,2,3,4,10-10-hexachloro-1,4,4a,5,8,8a-hexahydro-, (1alpha,4alpha,5abeta,5alpha,8alpha, 8abeta)-	309-00-2	P004
Allyl alcohol	2-Propen-1-ol	107-18-6	P005
Allyl chloride	1-Propane, 3-chloro	107-18-6	
Aluminum phosphide	Same	20859-73-8	P006
4-Aminobiphenyl	[1,1'-Biphenyl]-4amine	92-67-1	
5-(Aminomethyl)-3-isoxazolol	3(2H)-Isoxazolone, 5-(aminomethyl)-	2763-96-4	P007
4-Aminopyridine	4-Pyndioamine	504-24-5	P008
Amitrole	1H-1,2,4-Triazol-3-amine	61-82-5	U011

Common Name	Chemical Abstracts Name	Chemical Abstracts No.	Hazardous Waste No.
Ammonium vanadate	Vanadic acid, ammonium salt	7803-55-6	P119
Aniline	Benzenamine	62-53-3	U012
Antimony	Same	7440-36-0	
Antimony compounds, N.O.S.[1]			
Aramite	Sulfurous acid, 2-chloroethyl 2-[4-(1,1-dimethylethyl)phenoxy]-1-methylethyl ester	140-57-8	
Arsenic	Same	7440-38-2	
Arsenic compounds, N.O.S.[1]			
Arsenic acid	Arsenic acid H_3AsO_4	7778-39-4	P010
Arsenic pentoxide	Arsenic oxide AS_2O_3	1303-28-2	P011
Arsenic trioxide	Arsenic oxide As_2O_3	1327-53-3	P012
Auramine	Benzenamine, 4,4′-carbonimidoylbis[N,N-di-methyl.	492-80-8	U014
Azaserine	L-Serine, diazoacetate (ester)	115-02-6	U015
Barium	Same	7440-39-3	
Barium compounds, N.O.S.[1]			
Barium cyanide	Same	542-62-1	P013
Benz[c]acridine	Same	225-51-4	U016
Benz[a]anthracene	Same	56-55-3	U018
Benzal chloride	Benzene, (dichloromethyl)-	98-87-3	U017
Benzene	Same	71-43-2	U019
Benzenearsonic acid	Arsonic acid, phenyl-	98-05-5	
Benzidine	[1,1′-Biphenyl]-4,4[1]-diamine	92-87-5	U021
Benzo[b]fluoranthene	Benz[e]acephenanthrylene	205-99-2	
Benzo[j]fluoranthene	Same	205-82-3	
Benzol[a]pyrene	Same	50-32-8	U022
p-Benzoquinone	2,5-Cyclohexadiene-1,4-dione	106-51-4	U197
Benzotrichloride	Benzene, (trichloromethyl)-	98-07-7	U023
Benzyl chloride	Benzene, (chloromethyl)-	100-44-7	P028
Beryllium	Same	7440-41-7	P015
Beryllium compounds, N.O.S.[1]			
Bromoacetone	2-Propanone, 1-bromo-	598-31-2	P017
Bromoform	Methane, tribromo-	75-25-2	U225
4-Bromophenyl phenyl ether	Benzene, 1-bromo-4-phenoxy-	101-55-3	U030
Brucine	Strychnidin-10-one, 2,3-dimethoxy-	357-57-3	P018
Butyl benzyl phthalate	1,2-Benzenedicarboxylic acid, butyl phenyl-methyl ester	85-68-7	
Cacodylic acid	Arsinic acid, dimethyl-	75-60-5	U136
Cadmium	Same	7440-43-9	
Cadmium compounds, N.O.S.[1]			
Calcium chromate	Chromic acid H_2CrO_4, calcium salt	13765-19-0	U032
Calcium cyanide	Calcium cyanide $Ca(CN)_2$	592-01-8	P021
Carbon disulfide	Same	75-15-0	P022
Carbon oxyfluoride	Carbonic difluoride	353-50-4	U033
Carbon tetrachloride	Methane, tetrachloro-	56-23-5	U211
Chloral	Acetaldehyde, trichloro-	75-87-6	U034
Chlorambucil	Benzenebutanoic acid, 4-[bis(2-chloroethyl)amino]-	305-03-3	U035
Chlordane	4,7-Methano-1H-indene, 1,2,4,5,6,7,8,8-octachloro-2,3,3a,4,7,7a-hexahydro-.	57-74-9	U036
Chlordane (alpha and gamma isomers)			U036
Chlorinated benzenes, N.O.S.[1]			

Common Name	Chemical Abstracts Name	Chemical Abstracts No.	Hazardous Waste No.
Chlorinated ethane, N.O.S.[1]			
Chlorinated fluorocarbons, N.O.S.[1]			
Chlorinated naphthalene, N.O.S.[1]			
Chlorinated phenol, N.O.S.[1]			
Chlornaphazin	Naphthalenamine, N,N'-bis(2-chloroethyl)-	494-03-1	U026
Chloroacetaldehyde	Actaldehyde, chloro-	107-20-0	P023
Chloroalkyl ethers, N.O.S.[1]			
p-Chloroaniline	Benzenamine,4-chloro	106-47-8	P024
Chlorobenzene	Benzene, chloro-	108-90-7	U037
Chlorobenzilate	Benzeneacetic acid, 4-chloro-alpha-(4-chlorophenyl)-alpha-hydroxy-, ethyl ester	510-15-6	U038
p-Chloro-m-cresol	Phenol, 4-chloro-3-methyl-	59-50-7	U039
2-Chloroethyl vinyl ether	Ethene, (1-chloroethoxy)-	110-75-8	U042
Chloroform	Methane, trichloro-	67-66-3	U044
Chloromethyl methyl ether	Methane, chloromethoxy0	107-30-2	U046
beta-Chloronaphthalene	Naphthalene, 2-chloro-	91-58-7	U047
o-Chlorophenol	Phenol, 2-chloro-	95-57-8	U048
1-(o-Chlorophenyl)thioures	Thiourea, (2-chlorophenyl)-	5344-82-1	P026
Chloroprene	1,3-Butadiene, 2-chloro-	126-99-8	
3-Chloropropionitrite	Propanenitrile, 3-chloro-	542-76-7	P027
Chromium	Same	7440-47-3	
Chromium compounds, N.O.S.[1]			
Chrysene	Same	218-01-9	U050
Citrus red No. 2	2-Naphthalenol, 1-[(2,5-dimethoxyphenyl)azo]	6358-53-8	
Coal tar creosote	Same	8007-45-2	
Copper cyanide	Copper cyanide CuCN	544-92-3	P029
Creosote	Same		U051
Cresol (Cresylic acid)	Phenol, methyl-	1319-77-3	U052
Crotonaldehyde	2-Butenal	4170-30-3	U053
Cyanides (soluble salts and complexes) N.O.S.[1]			P030
Cyanogen	Ethanedinitrile	460-19-5	P031
Cyanogen bromide	Cyanogen bromide (CN)Br	506-68-3	U246
Cyanogen chloride	Cyanogen chloride (CN)Cl	506-77-4	P033
Cycasin	beta-D-Glucopyranoside, (methyl-ONN-azoxy)methyl	14901-08-7	
2-Cyclohexyl-4,6-dinitrophenol	Phenol, 2-cyclohexyl-4,6-dinitro-	131-89-5	P034
Cyclophosphamide	2H-1,3,2-Oxazaphosphorin-2-amine, N,N-bis(2-chloroethyl)tetrahydro-,2-oxide	50-18-0	U058
2,4-D	Acetic acid, (2,4-dichlorophenoxy)-	94-75-7	U240
2,4-D, salts, esters			U240
Daunomycin	5,12-Naphthacenedione, 8-acetyl-10-[(3 amino-2,3,6-trideoxy-alpha-L-lyxo-hexopyranosyl)oxy]-7,8,9,10-tetrahydro-6,8,11-trihydroxy-1-methoxy-, (8S-cis)-	20830-81-3	U059
DDD	Benzene, 1,1-(2,2-dichloroethylidene)bis[4-chloro-	72-54-8	U060
DDE	Benzene, 1,1-(dichloroethenylidene)bis[4-	72-55-9	
DDT	Benzene, 1,1-(2,2,2-trichloroethylidene)bis[4-chloro-	50-29-3	U061

Common Name	Chemical Abstracts Name	Chemical Abstracts No.	Hazardous Waste No.
Diallate	Carbamothioic acid, bis(1-methylethyl)-, S-(2,3-dichloro-2-propenyl) ester	2303-16-4	U062
Dibenz[a,h]acridine	Same	226-36-8	
Dibenz[a,j]acridine	Same	224-42-0	
Dibenz[a,h]anthracene	Same	53-70-3	U063
7H-Dibenzo[c,g]carbazole	Same	194-59-2	
Dibenzo[a,e]pyrene	Naphtho[1,2,3,4-def]chrysene	192-65-4	
Dibenzo[a,h]pyrene	Dibenzo[b,def]chrysene	189-64-0	
Dibenzo[a,j]pyrene	Benzo[rst]pentaphene	189-55-9	U064
1,2-Dibromo-3-chloropropane	Propane, 1,2-dibromo-3-chloro-	96-12-8	U066
Dibutyl phthalate	1,2-Benzenedicarboxylic acid, dibutyl ester	84-74-2	U069
o-Dichlorobenzene	Benzene, 1,2-dichloro-	95-50-1	U070
m-Dichlorobenzene	Benzene, 1,3-dichloro-	541-73-1	U071
p-Dichlorobenzene	Benzene, 1,4-dichloro-	106-46-7	U072
Dichlorobenzene, N.O.S.[1]	Benzene, dichloro-	25321-22-6	
3,3′-Dichlorobenzidine	[1,1′-Biphenyl]-4,4′-diamine, 3,3′-dichloro-	91-94-1	U073
1,4-Dichloro-2-butene	2-Butene, 1,4-dichloro-	764-41-0	U074
Dichlorodifluoromethane	Methane, dichlorodifluoro-	75-71-8	U075
Dichloroethylene, N.O.S.[1]	Dichloroethylene	25323-30-2	
1,1-Dichloroethylene	Ethene, 1,1-dichloro-	75-35-4	U078
1,2-Dichloroethylene	Ethene, 1,2-dichlrol-, (E)-	156-60-5	U079
Dichloroethyl ether	Ethane, 1,1′oxybis[2-chloro-	111-44-4	U025
Dichlorosopropyl ether	Propane, 2,2′-oxybis[2-chloro-	108-60-1	U027
Dichloromethyoxy ethane	Ethane, 1,1′-[methylenebis(oxy)]bis[2-chloro-	111-91-1	U024
Dichloromethyl ether	Methane, oxybis[chloro-	542-88-1	P016
2,4-Dichlorophenol	Phenol, 2,4-dichloro-	120-83-2	U081
2,6-Dichlorophenol	Phenol, 2,6-dichloro-	87-65-0	U082
Dichlorophenylarsine	Arsonous dichloride, phenyl-	696-28-6	P036
Dichloropropane, N.O.S.[1]	Propane, dichloro-	26638-19-7	
Dichloropropanol, N.O.S.[1]	Propanol, dichloro-	26545-73-3	
Dichloropropene, N.O.S.[1]	1-Propene, dichloro-	26952-23-8	
1,3-Dichloropropene	1-Propene, 1,3-dichloro-	542-75-6	U084
Dieldrin	2,7:3,6-Dimethanonaphth[2,3-b]oxirene, 3,4,5,6,9,9-hexachloro-1a,2,2a,3,6,6a,7, 7a-octahydro-, (1aalpha,2beta,2aalpha, 3beta,6beta,6aalpha,7beta,7aalpha)-	60-57-1	P037
1,2:3,4-Diepoxybutane	2,2′-Bioxirane	1464-53-5	U085
Diethylarsine	Arsine, diethyl-	692-42-2	P038
1,4-Diethyleneoxide	1,4-Dioxane	123-91-1	U108
Diethylhexyl phthalate	1,2-Benzenedicarboxylic acid, bis(2-ethyl-hexyl) ester	117-81-7	U028
N,N′Diethylhydrazine	Hydrazine, 1,2-diethyl-	1615-80-1	U086
O,O-Diethyl S-methyl dithiophosphate	Phosphorodithioic acid, O,O-diethyl S-methyl ester.	3288-58-2	U087
Dithyl-p-nitrophenyl phosphate	Phosphoric acid, diethyl 4-nitrophenyl ester	311-45-5	P041
Diethyl phthalate	1,2-Benzenedicarboxylic acid, diethyl ester	84-66-2	U088
O,O-Diethyl O-pyrazinyl phosphorothioate	Phosphorothioic acid, O,O-diethyl O-pyrazinyl ester	297-97-2	P040
Diethylstilbesterol	Phenol, 4,4′-(1,2-diethyl-1,2-ethenediyl)bis-, (E)-.	56-53-1	U089
Dihydrosafrole	1,3-Benzodioxole, 5-propyl-	94-58-6	U090

Common Name	Chemical Abstracts Name	Chemical Abstracts No.	Hazardous Waste No.
Diisopropylfluorophosphate (DFP)	Phosphorofluoridic acid, bis(1-methyl-ethyl) ester	55-91-4	P043
Dimethoate	Phosphorodithioic acid, O,o-dimethyl S-[2-(methylanimo)-2-oxoethyl] ester	60-51-5	P044
3,3'-Dimethoxybenzidine	[1,1'-Biphenyl]-4,4'-diamine, 3,3'-dimethoxy-	119-90-4	U091
p-Dimethylaminoazobenzene	Benzenamine, N,N-dimethyl-4-(phenylazo)-	60-11-7	U093
7,12-Dimethylbenz[a]anthracene	Benz[a]anthracene, 7,12-dimethyl-	57-97-6	U094
3,3'-Dimethylbenzidine	[1,1'-Biphenyl]-4,4'-diamine, 3,3'-dimethyl-	119-93-7	U095
Dimethylcarbamoly chloride	Carbamic chloride, dimethyl-	79-44-7	U097
1,1-Dimethylhydrazine	Hydrazine, 1,1-dimethyl-	57-14-7	U098
1,2-Dimethylhydrazine	Hydrazine, 1,2-dimethyl-	540-73-8	U099
alpha,alpha-Dimethylphenethylamine	Benzeneethanamine, alpha,alpha-dimethyl-	122-09-8	P046
2,4-Dimethylphenol	Phenol, 2,4-dimethyl-	105-67-9	U101
Dimethyl phthalate	1,2-Benzenedicarboxylic acid, dimethyl ester	131-11-3	U102
Dimethyl sulfate	Sulfuric acid, dimethyl ester	77-78-1	U103
Dinitrobenzene, N.O.S.[1]	Benzene, dinitro-	25154-54-5	
4,6-Dinitro-o-cresol	Phenol, 2-methyl-4,6-dinitro-	534-52-1	P047
4,6-Dinitro-o-cresol salts			P047
2,4-Dinitrophenol	Phenol, 2,4-dinitro-	51-28-5	P048
2,4-Dinitrotoluene	Benzene, 1-methyl-2,4-dinitro-	121-14-2	U105
2,6-Dinitrotoluene	Benzene, 2-methyl-1,3-dinitro-	606-20-2	U106
Dinoseb	Phenol, 2-(1-methylpropyl)-4,6-dinitro-	88-85-7	P020
Di-n-octyl phthalate	1,2-Benzenedicarboxylic acid, dioctyl ester	117-84-0	U017
Diphenylamine	Benzenamine, N-phenyl-	122-39-4	
1,2-Diphenylhydrazine	Hydrazine, 1,2-diphenyl-	122-66-7	U109
Di-n-propylnitrosamine	1-Propanamine, N-nitroso-N-propyl-	621-64-7	U111
Disulfoton	Phosphorodithioic acid, O,O-diethyl S-[2-(ethylthio)ethyl] ester	298-04-4	P039
Dithiobiuret	Thioimidodicarbonic diamide [(H$_2$N)C(S)]$_2$NH	541-53-7	P049
Endosulfan	6,9-Methano-2,4,3-benzodioxathiepin, 6,7,8,9,10,10-hexachloro-1,5,5a,6,9,9a-hexahydro-, 3-oxide.	115-29-7	P050
Endothall	7-Oxabicyclo[2.2.1]heptane-2,3-dicarboxylic acid	145-73-3	P088
Endrin	2,7:3,6-Dimethanonoaphth[2,3-b]oxirene, 3,4,5,6,9,9-hexachloro-1a,2,2a,3,6,6a,7,7a-octa-hydro-, (1aalpha,2beta,2abeta,3alpha,6alpha, 6abeta,7beta,7aalpha)-	72-20-8	P051
Endrin metabolites			P051
Epichlorohydrin	Oxrane, (chloromethyl)-	106-89-8	U041
Epinephrine	1,2-Benzenediol, 4-[1-hydroxy-2-(methylamino)ethyl]-, (R)-	51-43-4	P042
Ethyl carbamate (urethane)	Carbamic acid, ethyl ester	51-79-6	U238
Ethyl cyanide	Propanenitrile	107-12-0	P101
Ethylenebisdithiocarbamic acid	Carbamodithioic acid, 1,2-ethanediylbis-	111-54-6	U114
Ethylenebisdithiocarbamic acid, salts and esters			U114
Ethylene dibromide	Ethane, 1,2-dibromo-	106-93-4	U067

Common Name	Chemical Abstracts Name	Chemical Abstracts No.	Hazardous Waste No.
Ethylene dichloride	Ethane, 1,2-dichloro-	107-06-2	U077
Ethylene glycol monoethyl ether	Ethanol, 2-ethoxy-	110-80-5	U359
Ethyleneimine	Azridine	151-56-4	P054
Ethylene oxide	Oxirane	75-21-8	U115
Ethylenethiourea	2-Imidazolidinethione	96-45-7	U116
Ethylidene dichloride	Ethane, 1,1-dichloro-	75-34-3	U076
Ethyl methacrylate	2-Propenoic acid, 2-methyl-, ethyl ester	97-63-2	U118
Ethyl methanesulfonate	Methanesulfonic acid, ethyl ester	62-50-0	U116
Famphur	Phosphorothioic acid, O-[4-[(dimethylamino)sulfonyl]phenyl] O, o-di-methyl ester	52-85-7	P097
Fluoranthene	Same	206-44-0	U120
Fluorine	Same	7782-41-4	P056
Fluoroacetamide	Acetamide, 2-fluoro-	640-19-7	P057
Fluoroacetic acid, sodium salt	Acetic acid, fluoro-, sodium salt	62-74-8	P058
Formaldehyde	Same	50-00-0	U122
Formic acid	Same	64-18-6	U123
Glycidylaldehyde	Oxiranecarboxyaldehyde	765-34-4	U126
Halomethanes, N.O.S.[1]			
Heptachlor	4,7-Methano-1H-indene, 1,4,5,6,7,8,8-heptachloro-3a,4,7,7a-tetrahydro-	76-44-8	P059
Heptachlor epoxide	2,5-Methano-2H-indeno[1,2-b]oxirene 2,3,4,5,6,7,7-heptachloro-1a,1b,5,5a,6,6a-hexa- hydro, (1aalpha,1bbeta,2alpha,5alpha, 5abeta,6beta,6aalpha)-		
Heptachlor epoxide (alpha, beta, and gamma isomers)			
Hexachlorobenzene	Benzene, hexachloro-	118-74-1	U127
Hexachlorobutadiene	1,3-Butadiene, 1,1,2,3,4,4-hexachloro-	87-68-3	U128
Hexachlorocyclopentadiene	1,3-Cyclopentadiene, 1,2,3,4,5,5-hexachloro-	77-47-4	U130
Hexachlorodibenzo-p-dioxins			
Hexachlorodibenzolurans			
Hexachloroethane	Ethane, hexachloro-	67-72-1	U131
Hexachlorophene	Phenol, 2,2'-methylenebis[3,4,6-trichloro-	70-30-4	U132
Hexachloropropene	1-Propene, 1,1,2,3,3,3-hexachloro-	1888-71-7	U243
Hexaethyl tetraphosphate	Tetraphosphoric acid, hexaethyl ester	757-58-4	P062
Hydrazine	Same	302-01-2	U133
Hydrogen cyanide	Hydrocyanic acid	74-90-8	P063
Hydrogen fluoride	Hydrofluoric acid	7664-39-3	U134
Hydrogen sulfide	Hydrogen sulfide H_2S	7783-06-4	U135
Indeno[1,2,3-cd]pyrene	Same	193-39-5	U137
Isobutyl alcohol	1-Propanol, 2-methyl-	78-83-1	U140
Isodrin	1,4,5,8-Dimethanonaphthalene, 1,2,3,4,10,10-hexachloro-1,4,4a,5,8,8a-hexahydro, (1alpha,4alpha,4abeta,5beta,8beta, 8abeta)-	465-73-6	P060
Isosafrole	1,3-Benzodioxole, 5-(1-propenyl)-	120-58-1	U141
Kepone	1,3,4-Metheno-2H-cyclobuta[cd]Pentalen-2-one, 1,1a,3,3a,4,5,5,5a,5b,6-decachlorooclahydro-	143-50-0	U142

Common Name	Chemical Abstracts Name	Chemical Abstracts No.	Hazardous Waste No.
Lasiocarpine	2-Bulenoic acid, 2-methyl-,7-[[2,3-dihydroxy-2-(1-methoxyethyl)-3-methyl-1-oxobutoxy]methyl]-2,3,5 7a-tetrahydro-1 H-pyrrolizin-1-yl ester [1S-[1alpha,7(2S*,3R*),7aalpha]]	303-34-1	4143
Lead Lead compounds, N.O.S.[1]	Same	7439-92-1	
Lead acetate	Acetic acid, lead(2+) salt	301--4-2	U144
Lead phosphate	Phosphoric acid, lead(2+) salt (2:3)	7446-27-7	U145
Lead subacetate	Lead, bis(acetato-O)tetrahydroxytri-	1335-32-6	U146
Lindane	Cyclohexane, 1,2,3,4,5,6-hexachloro, (1alpha,2alpha,3beta,4alpha,5alpha,6beta)-	58-89-9	U129
Maleic anhydride	2,5-Furandione	108-31-6	U147
Maleic hydrazide	3,6-Pyridazinedione, 1,2-dihydro-	123-33-1	U148
Malononitrile	Propanedinitrile	109-77-3	U149
Melphalan	L-Phenylalanine, 4-[bis(2-chloroethyl)aminol]	148-82-3	U150
Mercury Mercury compounds, N.O.S.[1]	Same	7439-97-6	U151
Mercury fulminate	Fulminic acid, mercury(2+) salt	628-86-4	P065
Methacrylonitrile	2-Propenenitrile, 2-methyl-	126-98-7	U152
Methapyrilene	1,2-Ethanediamine, N,N-dimethyl-N'-2-pyre-dinyl-N'-(1-thienylmethyl)	91-80-5	U155
Methomyl	Ethanimidothioic acid, N-[[(methylamino)carbonyl]oxy], methyl ester	16752-77-5	P066
Methoxychlor	Benzene, 1,1-(2,2,2-trichloroethylidene)bis[4-methoxy-	72-43-5	U247
Methyl bromide	Methane, bromo-	74-83-9	U029
Methyl chloride	Methane, chloro-	74-87-3	U045
Methyl chlorocarbonate	Carbonochloridic acid, methyl ester	79-22-1	U156
Methyl chloroform	Ethane, 1,1,1-trichloro-	71-55-6	U226
3-Methylcholanthrene	Benz[j]aceanthrylene, 1,2-dihydro-3-methyl-	56-49-5	U157
4,4-Methylenebis(2-chloro-aniline)	Bnezenamine, 4,4'-methylenebis[2-chloro-	101-14-4	U158
Methylene bromide	Methane, dibromo-	74-95-3	U068
Methylene chloride	Methane, dichloro-	75-09-2	U080
Methyl ethyl ketone (MEK)	2-Butanone	78-93-3	U159
Methyl ethyl ketone peroxide	2-Butanone, perioxide	1338-23-4	U160
Methyl hydrazine	Hydrazine, methyl-	60-34-4	P068
Methyl iodide	Methane, iodo-	75-88-4	U138
Methyl isocyanate	Methane, isocyanalo-	624-83-9	P064
2-Methyllactonitrile	Propanenitrile, 2-hydroxy-2-methyl-	75-86-5	P069
Methyl methyacrylate	2-Propenoic acid, 2-methyl-, methyl ester	80-62-6	U162
Methyl methanesulfonate	Methanesulfonic acid, methyl ester	66-27-3	
Methyl parathion	Phosphorothioic acid, O,O-dimethyl O-(1-mitrophenyl) ester	298-00-0	P071
Methylthiouracil	4(1H)-Pyrimidinone, 2,3-dihydro-6-methyl-2-thioxo-	56-04-2	U164
Milomycin C	Azirion[2',3':3,4]pyrrolo[1,2-a]indole-4,7-dione 6-amino-8-[[(aminocarbonyl)oxy]methyl]-	50-07-7	U010

Common Name	Chemical Abstracts Name	Chemical Abstracts No.	Hazardous Waste No.
	1,1a,2,8,8a,8b-hexahydro-8a-methoxy-5-methyl-, [1aS-(1aalpha,8beta,8aalpha,8balpha)]-		
MNNG	Guanidine, N-methyl-N′-nitro-N-nitroso-	70-25-7	U163
Mustard gas	Ethane, 1,1′-thiobis[2-chloro-	505-60-2	U163
Naphthalene	Same	91-20-3	U165
1,4-Naphthoquinone	1,4-Naphthalenedione	130-15-4	U166
alpha-Naphthylamine	1-Naphthalenamine	134-32-7	U167
beta-Naphthylamine	2-Naphthalenamine	91-59-8	U168
alpha-Naphthylthiourea	Thiourea, 1-naphthalenyl-	86-88-4	P072
Nickel	Same	7440-02-0	
Nickel compounds, N.O.S.[1]			
Niclel carbonyl	Nickel carbonly Ni(CO) , (T-4)-	13463-39-3	P073
Nickel cyanide	Nickel cyanide Ni(CN)	557-19-7	P074
Nicotine	Pyridine, 3-(1-methyl-2-pyrrolidinyl)-, (S)-	54-11-5	P075
Nicotine salts			P075
Nitric oxide	Nitrogen oxide NO	10102-43-9	P076
p-Nitroaniline	Benzenamine, 4-nitro-	100-01-6	P077
Nitrobenzene	Benzene, nitro-	98-95-3	U169
Nitrogen dioxide	Nitrogen oxide NO	10102-44-0	P078
Nitrogen mustard	Ethanamine, 2-chloro-N-(2-chloroethyl)-N-methyl-	51-75-2	
Nitrogen mustard, hydrochloride salt			
Nitrogen mustard N-oxide	Ethanamine, 2-chloro-N-2(2-chloroethyl)-N-methyl-, N-oxide	126-85-2	
Nitrogen mustard, N-oxide, hydro-chloride salt			
Nitroglycerin	Phenol, 4-nitro-	100-02-7	U170
2-Nitropropane	Propane, 2-nitro	79-46-9	U171
Nitrosamines, N.O.S.[1]		35576-91-1D	
N-Nitrosodi-n-butylamine	i-Butanamine, N-butyl-N-nitroso-	924-16-3	U172
N-Nitrosodiethanolamine	Ethanol, 2,2′-(nitrosoimino)bis-	1116-54-7	U173
N-Nitrosodiethylamine	Ethanamine, N-ethyl-N-nitroso-	55-18-5	U174
N-Nitrosodimethylamine	Methanamine, N-methyl-N-nitroso-	62-75-9	P082
N-Nitroso-N-ethylurea	Urea, N-ethyl-N-nitroso-	759-73-9	U176
N-Nitrosomethylethylamine	Ethanamine, N-methyl-N-nitroso-	10595-95-6	
N-Nitroso-N-methylurea	Urea, N-methyl-N-nitroso-	684-93-5	U177
N-Nitroso-N-methylurethane	Carbamic acid, methylnitroso-	615-53-2	U178
N-Nitrosomethylvinylamine	Vinylamine, N-methyl-N-nitroso-	4549-40-0	P084
N-Nitrosomorpholine	Orphonline,4-nitroso-	59-89-2	
N-Nitrosonornicotine	Pyridine,3-(1-nitroso-2-pyrrolidinyl)- (S)-	16543-55-8	
N-Nitrosopiperidine	Piperidine,1-nitroso-	100-75-4	U179
N-Nitrosopyrrolidine	Pyrrolidine,1-nitroso-	930-55-2	U180
N-Nitrososarcosine	Glycine, N methyl-N-nitroso-	13256-22-9	
5-Nitro-o-toluidine	Benzenamine, 2-methyl-5-nitro-	99-55-8	U181
Octamethylpyrophosphoramide	Diphosphoramide, octamethyl-	152-16-9	P085
Osmium tetroxide	Osmium oxide OsO , (T-4)-	20816-12-0	P087
Paraldehyde	1,3,5-Trioxane, 2,4,6-trimethyl-	123-63-7	U182
Parathion	Phosphorothioic acid, O,O-diethyl O-(4-nitro-phenyl) ester	56-38-2	P089
Pentachlorobenzene	Benzene, pentachloro-	608-93-5	U183
Pentachlorodibenzo-p-dioxins			

Common Name	Chemical Abstracts Name	Chemical Abstracts No.	Hazardous Waste No.
Pentachlorodibenzofurans			
Pentachloroethane	Ethane, pentachloro-	76-01-7	U184
Pentachloronitrobenzene (PCNB)	Benzene, pentachloronitro-	82-68-8	U185
Pentachlorophenol	Phenol, pentachloro-	87-86-5	See F027
Phenecetine	Acetaminde, N-(4-ethoxyphenyl)-	62-44-2	U187
Phenol	Same	108-95-2	U188
Phenylenediamine	Benzenediamine	5265-76-3	
Phenylmercury acetate	Mercury, (acetato-O)phenyl-	62-38-4	P092
Phenylthiourea	Thiourea, phenyl-	103-85-5	P093
Phosgene	Carbonic dichloride	75-44-5	P095
Phosphine	Same	7803-51-2	P096
Phorate	Phosphorodithioic acid, O,O-diethyl S-[(ethylthio)methyl] ester	298-02-2	P094
Phthalic acid esters, N.O.S.[1]			
Phthalic anhydride	1,3-Isobenzofurandione	85-44-9	U190
2-Picoline	Pyridine, 2-methyl-	109-06-8	U191
Polychlorinated biphenyls, N.O.S.[1]			
Potassium cyanide	Potassium cyanide K(CN)	151-50-8	P098
Potassium silver cyanide	Argentate(1-), bis(cyano-C)-, potassium	506-61-6	P099
Pronamide	Benzamide, 3,5-dichloro-N-(1,1-dimethyl-2-propynyl)-	23950-58-5	U192
1,3-Propane sultone	1,2-Oxathiolane, 2,2-dioxide	1120-71-4	U193
n-Propylamine	1-Propanamine	107-10-8	U194
Propargyl alcohol	2-Propyn-1-ol	107-19-7	P102
Propylene dichloride	Propane, 1,2-dichloro-	78-87-5	U083
1,2-Propylenimine	Aziridine, 2-methyl-	75-55-8	P067
Propylthiouracil	4(1H)-Pyrimidinone, 2,3-dihydro-6-propyl-2-thioxo-	51-52-5	
Pyridine	Same	110-86-1	U196
Reserpine	Yohimban-16-carboxylic acid, 11,17-dimethoxy-18-[(3,4,5-trimethoxybenzoyl)oxy]-smethyl ester, (3beta,16beta,17alpha,18beta,20alpha)-	50-55-5	U200
Resorcinol	1,3-Benzenediol	108-46-3	U201
Saccharin	1,2-Benzisothiazol-3(2H)-one, 1,1-dioxide	81-07-2	U202
Saccharin salts			
Safrole	1,3-Benzodioxole, 5-(2-propenyl)-	95-59-7	U203
Selenium	Same	7782-49-2	
Selenium compounds, N.O.S.[1]			
Selenium dioxide	Selenious acid	7783-00-8	U204
Selenium sulfide	Selenium sulfide SeS$_2$	7488-56-4	U205
Selenourea	Same	630-10-4	P103
Silver	Same	7440-22-4	
Silver compounds, N.O.S.[1]			
Silver cyanide	Silver cyanide Ag(CN)	506-64-9	P104
Silvex (2,4,5-TP)	Propanoic acid, 2-(2,4,5-trichlorophenoxy)-	93-72-1	See F027
Sodium cyanide	Sodium cyanide Na(CN)	143-33-9	P106
Streptozotocin	D-Glucose, 2-deoxy-2-[[(methylnitrosoamino)carbonly]amino]-	18883-66-4	U206
Strontium sulfide[2]	Strontium sulfide SrS	1314-96-1	P107
Strychnine	Strychnidin-10-one	57-24-9	P108
Strychnine salts			P108

Common Name	Chemical Abstracts Name	Chemical Abstracts No.	Hazardous Waste No.
TCDD	Dibenzo[b,e][1,4]dioxin, 2,3,7,8-tetrachloro	1746-01-6	
1,2,4,5-Tetrachlorobenzene	Benzene, 1,2,4,5-tetrachloro-	95-94-3	U207
Tetrachlorodibenzo-p-dioxins			
Tetrachlorodibenzofurans			
Tetrachloroethane, N.O.S.[1]	Ethane, tetrachloro, N.O.S.	25322-20-7	
1,1,1,2-Tetrachloroethane	Ethane, 1,1,1,2-tetrachloro-	630-20-6	U208
1,1,2,2-Tetrachloroethane	Ethane, 1,1,2,2-tetrachloro-	79-34-5	U209
Tetrachloroethylene	Ethene, tetrachloro-	127-18-4	U210
2,3,4,6-Tetrachlorophenol	Phenol, 2,3,4,6-tetrachloro-	58-90-2	See F027
Tetraethyldithiopyrophosphate	Thiodiphosphoric acid, tetraethyl ester	3689-24-5	P109
Tetraethyl lead	Plumbane, tetraethyl-	78-00-2	P110
Tetraethyl pyrophosphate	Diphosphoric acid, tetraethyl ester	107-49-3	P111
Tetranitromethane	Methane, tetranitro-	509-14-8	P112
Thallium	Same	7440-28-0	
Thallium compounds, N.O.S.[1]			
Thallic oxide	Thallium oxide Tl_2O_3	1314-32-5	P113
Thallium(l) acetate	Acetic acid, thallium (1+)salt	563-68-8	U214
Thallium(l) carbonate	Carbonic acid, dithallium(1+) salt	6533-73-9	U215
Thallium(l) chloride	Thallium chloride TlCl	7791-12-0	U216
Thallium(l) nitrate	Nitric acid, thallium(1+) salt	10102-45-1	U217
Thallium selenite	Selenious acid, dithallium(1+) salt	12039-52-0	P114
Thallium(l) sulfate	Sulfuric acid, dithallium(1+) salt	7446-18-6	P115
Thioacetamide	Ethanethioamide	62-55-5	U218
Thiofanox	2-Butanone, 3,3-dimethyl-1-(methylthio)-, O-[(methylamino)carbonyl]oxime	39196-18-4	P045
Thiomethanol	Methanethiol	74-93-1	U153
Thiophenol	Benzenethiol	108-98-5	P014
Thiosemicarbazide	Hydrazinecarbothioamide	79-19-6	P116
Thiourea	Same	62-56-6-	U219
Thiram	Thioperoxydicarbonic diamide [(H$_2$N)C(S)]$_2$ S$_2$, tetramethyl-	137-26-8	U244
Toluene	Benzene, methyl-	108-88-3	U220
Toluenediamine	Bezenediamine, ar-methyl-	25376-45-8	U221
Toluene-2,4-diamine	1,3-Benzenediamine, 4-methyl-	95-80-7	
Toluene-2,6-diamine	1,3-Benzenediamine, 2-methyl-	823-40-5	
Toluene-3,4-diamine	1,2-Benzenediamine, 4-methyl-	496-72-0	
Toluenediisocyanate	Benzene, 1,3-diisocyanatomethyl-	26471-62-5	U223
o-Toluidine	Benzenamine,2-methyl-	95-53-4	U328
o-Toluidinehydrochloride	Benzenamine,2-methyl-, hydrochloride	636-21-5	U222
p-Toluidine	Benzenamine, 4-methyl-	106-49-0	U353
Toxaphene	Same	8001-35-2	P123
1,2,4-Trichlorobenzene	Benzene, 1,2,4-trichloro-	120-82-1	
1,1,2-Trichloroethane	Ethane, 1,1,2-trichloro-	79-00-5	U227
Trichloroethylene	Ethene, trichloro-	79-01-6	U228
Trichloromethanethiol	Methanethiol, trichloro-	75-70-7	P118
Trichloromoroflouromethane	Methane, trichlorofluoro-	75-69-4	U121
2,4,5-Trichlorophenol	Phenol, 2,4,5-trichloro	95-95-4	See F027
2,4,6-Trichlorophenol	Phenol, 2,4,6-trichloro-	88-06-2	See F027
2,3,5-T	Acetic acid, (2,4,5-trichlorophenoxy)-	93-76-5	See F027
Trichloropropane, N.O.S.[1]		25735-29-9	
1,2,3-Trichloropropane	Propane, 1,2,3-trichloro-	96-18-4	
O,O,O-Triethyl phosphoro-thioate	Phosphorothioic acid, O,O,O-triethyl ester	126-68-1	

Common Name	Chemical Abstracts Name	Chemical Abstracts No.	Hazardous Waste No.
1,3,5-Trinitrobenzene	Benzene, 1,3,5-trinitro-	99-35-1	U234
Tris(1-azindinyl)phosphine sulfide	Azindine, 1,1,1'-phosphinothioylidynetris	52-24-4	
Tris(2,3-dibromopropyl) phosphate	1-Propanol, 2,3-dibromo-, phosphate (3,1)	126-72-7	U235
Trypan blue	2,7-Naphthalenedisulfonic acid, 3,3'-[(3,3'-di-methyl[(1,1'-biphenyl]-4,4'-diylbis(azo)] bis[5-amino-4-hydroxy-, tetrasodium salt	72-57-1	U236
Uracil mustard	2,4-(1H,3H)-Pyrimidinedione, 5-[bis(2-chloroethyl)amino]	66-75-1	U237
Vanadium pentoxide	Vanadium oxide V_2O	1314-62-1	P120
Vinyl chloride	Ethene, chloro-	75-01-4	U043
Warfarin	2H-1-Benzopyran-2-one, 4-hydroxy-3-(3-oxo-1-phenylbutyl)-, when present at concentrations less than 0.3%	81-81-2	U248
Warfarin	2H-1-Benzopyran-2-one, 4-hydroxy-3(3-oxo-1-phenylbutyl)-, when present at concentrations greater than 0.3%	81-81-2	P001
Warfarin salts, when present at concentrations less than 0.3%			U248
Warfarin salts, when present at concentrations greater than 0.3%			P001
Zinc cyanide	Zinc cyanide $Zn(CN)_2$	557-21-1	P121
Zinc phosphide	Zinc phosphide Zn_3P_2, when present at concentrations greater than 10%	1314-84-7	P122
Zinc phosphide	Zinc phosphide Zn_3P_2, when present at concentrations of 10% or less	1314-84-7	U249

[1]The abbreviation N.O.S. (not otherwise specified) signifies those members of the general class not specifically listed by name in this appendix.

[2]At 53 FR 43884, October 31, 1988, Appendix VIII was amended by removing the listing: "Strontium sulfide . . . Same . . . 1314-96-1." The amendatory instruction in that document was incorrect and the Environmental Protection Agency will publish a correction document in the *Federal Register* at a later date.

TABLE A–5* Basis for Listing Hazardous Wastes (adapted from 40 CFR, Part 316, 1990; p. 88-90

F001	Tetrachloroethylene, methylene chloride trichloroethylene, 1,1,1-trichloroethane, carbon tetrachloride, chlorinated fluorocarbons	F006	Cadmium, hexavalent chromium, nickel, cyanide (complexed)
F002	Tetrachloroethylene, methylene chloride, trichloroethylene, 1,1,1-trichloroethane, 1,1,2-trichloroethane, chlorobenzene, 1,1,2-trichloro-1,2,2-trichfluoroethane, ortho-dichlorobenzene, trichlorofluoromethane	F007	Cyanide (salts)
		F008	Cyanide (salts)
		F009	Cyanide (salts)
		F010	Cyanide (salts)
		F011	Cyanide (salts)
		F012	Cyanide (complexed)
F003	N.A.	F019	Hexavalent chromium, cyanide (complexed)
F004	Cresols and cresylic acid, nitrobenzene	F020	Tetra- and pentachlorodibenzo-*p*-dioxins; tetra and pentachlorodi-benzolfurans; tri- and tetrachlorophenols and their chlorophenoxy derivative acids, esters, ethers, amine, and other salts
F005	Toluene, methyl ethyl ketone, carbon disulfide, isobutanil, pyridine, 2-ethoxyethanol, benzene, 2-nitropropane		

* Some of the listed constituents are the same as those in Tables A–1 and A–2.

F021   Penta- and hexachlorodibenzo-*p*-dioxins; penta- and hexachlorodibenzofurans; pentachlorophenol and its derivatives

F022   Tetra-, penta-, and hexachlorodibenzo-*p*-dioxins; tetra-, penta-, and hexachlorodibenzofurans

F023   Tetra-, and pentachlorodibenzo-*p*-dioxins; tetra-and pentachlorodibenzofurans; tri- and tetrachlorphenols and their chlorophenoxy derivative acids, esters, ethers, amini, and other salts

F024   Chloromethane, dichloromethane, trichloromethane, carbon tetrachloride, chloroethylene, 1,1-dichloroethane, 1,2-dichloroethane, trans-1-2-dichloroethylene, 1,1-dichloroethylene, 1,1,1-trichloroethane, 1,1,2-trichloroethane, trichloroethylene, 1,1,1,2-tetrachloroethane, 1,1,2,2-tetrachloroethane, tetrachloroethylene, pentachloroethane, hexachloroethane, allyl chloride (3-chloropropene). dichloropropane, dichloropropene, 2-chloro-1,3-butadiene, hexachloro-1,3-butadiene, hexachlorocyclopentadiene, hexachlorocyclohexane, benzene, chlorbenzene, dichlorobenzenes, 1,2,4-trichlorobenzene, tetrachlorobenzene, pentachlorobenzene, hexachlorobenzene, toluene, naphthalene

F025   Chloromethane; Dichloromethane; Trichloromethane; Carbon tetrachloride; Chloroethylene; 1,1-Dichloroethane; 1,2-Dichloroethane; trans-1,2-Dichloroethylene; 1,1-Dichloroethylene; 1,1,1-Trichloroethane; 1,1,2-Trichloroethane; Trichloroethylene; 1,1,1,2-Tetrachloroethane; 1,1,2,2-Tetrachloroethane; Tetrachloroethylene; Pentachloroethane; Hexachloroethane; Allyl chloride (3-Chloropropene); Dichloropropane; Dichloropropene; 2-Chloro-1,3-butadiene; Hexachloro-1,3-butadiene; Hexachlorocyclopentadiene; Benzene; Chlorobenzene; Dichlorobenzene; 1,2,4-Trichlorobenzene; Tetrachlorobenzene; Pentachlorobenzene; Hexachlorobenzene; Toluene; Naphthalene

F026   Tetra-, penta-, and hexachlorodibenzo-*p*-dioxins; tetra-, penta-, and hexachlorodibenzofurans

F027   Tetra-, penta-, and hexachlorodibenzo-*p*-dioxins; tetra-, penta-, and hexachlorodibenzofurans; tri-, tetra-, and pentachlorophenols and their chlorophenoxy derivative acids, esters, ethers, amine and other salts

F028   Tetra-, penta-, and hexachlorodebenzo-*p*-dioxins; tetra-, penta-, and hexachlorodibenzofurans; tri-, tetra-, and pentachlorophenols and their chlorophenoxy derivative acids, esters, ethers, amine, and other salts

F039   All constituents for which treatment standards are specified for multi-source leachate (wastewaters and nonwastewaters) under 40 CFR 268.43(a). Table CCW.

K001   Pentachlorophenol, phenol, 2-chlorophenol, p-chloro-m-cresol, 2,4-dimethylphenyl, 2,4-dinitrophenol, trichlorophenols, tetrachlorophenols, 2,4-dinitrophenol, cresosote, chrysene, naphthalene, fluoranthene, benzo(b)fluoranthene, benzo(a)pyrene, indeno(1,2,3-cd)pyrene, benz(a)anthracene, dibenz(a)anthracene, acenaphthalene

K002   Hexavalent chromium, lead

K003   Hexavalent chromium, lead

K004   Hexavalent chromium

K005   Hexavalent chromium, lead

K006   Hexavalent chromium

K007   Cyanide (complexed), hexavalent chromium

K008   Hexavalent chromium

K009   Chloroform, formaldehyde, methylene chloride, methyl chloride, paraldehyde, formic acid

K010   Chloroform, formaldehyde, methylene chloride, methyl chloride, paraldehyde, formic acid, chloroacetaldehyde

K011   Acrylonitrile, acetonitrile, hydrocyanic acid,

K013   Hydrocyanic acid, acrylonitrile, acetonitrile

K014   Acetonitrile, acrylamide

K015   Benzyl chloride, chlorobenzene, toluene, benzo-trichloride

K016   Hexachlorobenzene, hexachlorobutadiene, carbon tetrachloride, hexachloroethane, perchloroethylene

K017   Epichlorohydrin, chloroethers [bis(chloromethyl) ether and bis (2-chloroethyl) ethers], trichloro-propane, dichloropropanols

K018   1,2-dichloroethane, trichloroethylene, hexachloro-butadiene, hexachlorobenzene

K019   Ethylene dichloride, 1,1,1-trichloroethane, 1,1,1-trichloroethane, tetrachloroethanes (1,1,2,2-tetrichloroethane and (1,1,1,2-tetrachloroethane), trichloroethylene, tetrachloroethylene, carbon tetrachloride, chloroform, vinyl chloride, vinylidene chloride

K020   Thylene dichloride, 1,1,1-trichloroethane, 1,1,2-trichloroethane, tetrachloroethanes (1,1,2,2-tetrachloroethane and 1,1,1,2-tetrachloroethane), trichloroethylene, tetrachloroethylene, carbon tetrachloride, chloroform, vinyl chloride, vinylidene chloride

K021   Antimony, carbon tetrachloride, chloroform

K022   Phenol, tars (polycyclic aromatic hydrocarbons)

K023   Phthalic anhydride, maleic anhydride

K024   Phthalic anhydride, 1,4-naphthoquinone

K025   Meta-dinitrobenzene, 2,4-dinitrotoluene

K026   Paraldehyde, pyridines, 2-picoline

K027	Toluene diisocyanate, toluene-2, 4-diamine
K028	1,1,1-trichloroethane, vinyl chloride
K029	1,2-dichloroethane, 1,1,1-trichloroethane, vinyl chloride, vinylidene chloride, chloroform
K030	Hexachlorobenzene, hexachlorobutadiene, hexachloroethane, 1,1,1,2-tetrachloroethane, 1,1,2,2-tetrachloroethane, ethylene dichloride
K031	Arsenic
K032	Hexachlorocyclopentadiene
K033	Hexachlorocyclopentadiene
K034	Cresosote, chrysene, naphthalene, fluoranthene benzo(b) fluoranthene, benzo(a) pyrene, indeno(1,2,3-cd) pyrene, benzo(a)anthrace, dibenzo(a)anthracene, acenaphthalene
K035	Hexachlorocyclopentadiene
K036	Toluene, phosphorothioic and phosphorothioic acid esters
K037	Toluene, phosphorodithioic and phosphorothioic acid esters
K038	Phorate, formaldehyde, phosphorodithioic and phosphorothioic acid esters
K039	Phosphorodithioic and phosphorothioic acid esters
K040	Phorate, formaldehyde, phosphorodithioic and phosphorothioic acid esters
K041	Toxaphene
K042	Hexachlorobenzene, ortho-dichlorobenzene
K043	2,4-dichlorophenol, 2,6-dichlorophenol, 2,4,6-trichlorophenol
K044	N.A
K045	N.A.
K046	Lead
K047	N.A.
K048	Hexavalent chromium, lead
K049	Hexavalent chromium, lead
K050	Hexavalent chromium
K051	Hexavalent chromium, lead
K052	Lead
K060	Cyanide, napthalene, phenolic compounds, arsenic
K061	Hexavalent chromium, lead, cadmium
K062	Hexavalent chromium, lead
K064	Lead, cadmium
K065	Do
K066	Do
K069	Hexavalent chromium, lead, cadmium
K071	Mercury
K073	Chloroform, carbon tetrachloride, hexachloroethane, trichloroethane, tetrachloroethylene, dichloroethylene, 1,1,2,2-tetrachloroethane
K083	Aniline, diphenylamine, nitrobenzene, phenylene-diamine
K084	Arsenic
K085	Benzene, dichlorobenzenes, trichlorobenzenes, te-trachlorobenzenes, pentachlorobenzene, hex-achlorobenzene, benzyl chloride
K086	Lead, hexavalent chromium
K087	Phenol, naphthalene
K088	Cyanide (complexes)
K090	Chromium
K091	Do
K093	Phthalic anhydride, maleic anhydride
K094	Phthalic anhydride
K095	1,1,2-trichloroethane, 1,1,1,2-tetra-chloroethane, 1,1,2,2-tetrachloroethane
K096	1,2-dichloroethane, 1,1,1-trichloroethane, 1,1,2-trichloroethane
K097	Chlordane, heptachlor
K098	Toxaphene
K099	2.4-dichlorophenol, 2,4,6-trichlorophenol
K100	Hexavalent chromium, lead, cadmium
K101	Arsenic
K102	Arsenic
K103	Aniline, nitrobenzene, phenylenediamine
K104	Aniline, benzene, diphenylamine, nitrobenzene, phenylenediamine
K105	Benzene, monochlorobenzene, dichloroben-zenes, 2,4,6-trichlorophenol
K106	Mercury
K107	1,1-Dimethylhydrazine (UDMH)
K108	1,1-Dimethylhydrazine (UDMH)
K109	1,1-Dimethylhydrazine (UDMH)
K110	1,1-Dimethylhydrazine (UDMH)
K111	2,4-Dinitrotoluene
K112	2,4-Toluenediamine, *o*-toluidine, *p*-toluidine, ani-line
K113	2,4-Toluenediamine, *o*-toluidine, *p*-toluidine, aniline
K114	2,4-Toluenediamine, *o*-toluidine, *p*-toluidine.
K115	2,4-Toluenediamine
K116	Carbon tetrachloride, tetrachloroethylene, chloroform phosgene
K117	Ethylene dibromide
K118	Ethylene dibromide
K123	Ethylene thiourea
K124	Ethylene thiourea
K125	Ethylene thiourea
K126	Ethylene thiourea
K131	Dimethyl sulfate, Methyl bromide
K132	Methyl bromide
K136	Ethylene dibromide

APPENDIX B

List of Acronyms

Acronyms are commonly used in many technical fields to facilitate communications. These acronyms provide a verbal and written shorthand for otherwise lengthy and cumbersome titles, names, procedures, and descriptions. In the field of hazardous waste management, perhaps more than in any other field, acronyms are not only extensively used but new ones emerge rather frequently. While this list is believed to be as complete as possible, it is certain that many new acronyms will be coined during the elapsed time between the preparation of this manuscript and publication of the book.

Frequently a plural form of the acronym is used by putting a lower case *s* after the acronym, e.g., ACLs for alternate concentration limits or CODs for chemical oxygen demands.

AA Atomic Absorption (*also* Acronyms and Abbreviations)

AAS Atomic Absorption Spectroscopy

ACES Associated Chemical and Environmental Services

ACGIH American Conference of Governmental Industrial Hygienists

ACL Alternate Concentration Limit

ADPC&E Arkansas Department of Pollution Control and Ecology

ADR Alternate Dispute Resolution

AEC Atomic Energy Commission

AIChE American Institute of Chemical Engineers

AIChE–CCPS American Institute of Chemical Engineers–Center for Chemical Process Safety

AIHA American Industrial Hygiene Association

ALARA As Low As Reasonably Achievable

AMEC American Ecology, Inc.

ANPR Advanced Notice of Proposed Rulemaking

ANSI American National Standards Institute

AOPs Advanced Oxidation Processes

APCDs Air Pollution Control Devices

APCE Adequate Air Pollution Control Equipment

APEN Air Pollution Emission Notice

API American Petroleum Institute

APU Acid Purification Unit

AQTX Aquatic Toxicity

ARARs Applicable or Relevant and Appropriate Regulations

ASTM American Society for Testing and Materials

ATA Anaerobic Toxicity Assay

ATSDR Agency for Toxic Substances and Disease Registry

ATTIC Alternate Treatment Technology Information Center

BACT Best Available Control Technology

BDAT Best Demonstrated Available Technology

BEI Biological Exposure Indices

BFI Browning-Ferris Industries, Inc.

BH Baghouse

BMP Biochemical Methane Potential

BOD Biochemical Oxygen Demand

BOD_5 Five-Day Biological Oxygen Demand

BS&W Bottom Sediments and Water

BTEX Benzene, Toluene, Ethylbenzene, and Xylene

BTX Benzene, Toluene, and Xylene

BTU British Thermal Unit

CAA Clean Air Act

CAAA Clean Air Act Amendments

CAMU Corrective Action Management Unit

CAS Chemical Abstracts Service

CB Chlorobenzene

CBC Complete Blood Count

CBO Congressional Budget Office

CCR Chemical Cartridge Respirator

CCW Constituent Concentrations in Waste

CCWE Constituent Concentrations in Waste Extract

CDC Centers for Disease Control

CDI Cave Development Inspection

CE Combustion Efficiency

CEC Cation Exchange Capacity

CECOS CECOS International, Inc.

CEI Compliance Evaluation Inspection

CELS Constant Energy Synchronous Luminescence Spectroscopy

CEMs Continuous Emission Monitors

CEP Council on Economic Priorities

CEPP Chemical Emergency Preparedness Program

CERCLA Comprehensive Environmental Response, Compensation, and Liability Act

CERCLIS CERCLA Information System

CET Certified Environmental Trainer

CFCs Chlorofluorocarbons

CFR Code of Federal Regulations

CFS Continuous Flow System

CGI Combustible Gas Indicator

CHEMTREC Chemical Transportation Emergency Center

CHRIS Chemical Hazard Response Information System

CIAO Citizen Information and Access Office

CKD Cement-Kiln Dust

CLP Contract Laboratory Program

CMA Chemical Manufacturers Association

CMB Completely Mixed Batch

CME Comprehensive (Groundwater) Monitoring Evaluation

CMI Corrective Measures Implementation

CMS Corrective Measures Study

CNS Central Nervous System

COC Chain of Custody

COD Chemical Oxygen Demand

CQA Construction Quality-Assurance (Programs)

CPE Chlorinated Polyethylene

CPSC Consumer Product Safety Commission

CRC Contamination Reduction Corridor

CRZ Contamination Reduction Zone

CSFs Critical Success Factors

CSI Compliance Sampling Inspection

CSPE Chlorosulfonated Polyethylene

CTGs Control Techniques Guidelines

CWA Clean Water Act

CWGs Community Working Groups

CWM Chemical Waste Management (Corp.)

DAF Dilution/Attenuation Factor

DAVE Desorption and Vapor Extraction System

DBP Disinfectants and Disinfection By-Products

DBR Disinfectants and Disinfection By-Products Rule

DDE 1,1-dichloro-2,2-bis(4-chlorophenyl)ethane

DDT Dichloro-diphenyl Trichloroethane

DE Destruction Efficiency

DECON Decontamination

DF Decontamination Factor

DHD Daily Human Dose

DISC Developers International Services Corp.

DNAPL Dense Non-Aqueous Phase Liquid

DO Dissolved Oxygen

DOC Dissolved Organic Carbon

DOD Department of Defense

DOE Department of Energy

DOL Department of Labor

DOT Department of Transportation

DQO Data Quality Objective

DRE Destruction and Removal Efficiency

dscf Dry Standard Cubic Feet

dscfm Dry Standard Cubic Feet Per Minute

dscm Dry Standard Cubic Meter

dscmm Dry Standard Cubic Meter Per Minute

DST Drill-Stem Test

ECC Environmental Control Commission

ECDs Electron Capture Detectors

ECP Environmentally Conscious Product

EDB Ethylene dibromide

EE/CA Engineering Evaluation/Cost Analysis

EEC European Economic Community

EERU Environmental Emergency Response Unit

EIL Environmental Impairment Liability

EIS Environmental Impact Statement

ENSCO Environmental Systems Co., Inc.

EP Extraction Procedure

EP Toxicity Extraction Procedure Toxicity

EPA Environmental Protection Agency

EPC Environmental Protection Corp.

ERRIS Emergency and Remedial Response Information System

ESF Emergency Support Function

ESI Envirosafe Services, Inc.

ESLI End-of-Service-Life-Indicator

ESP Electrostatic Precipitator

ESWTR Enhanced Surface Water Treatment Rule

FDA Food and Drug Administration

FEI Fondessy Enterprises, Inc.

FEMA Federal Emergency Management Agency
FGR Flue Gas Recirculation
FID Flame Ionization Detector
FIFRA Federal Insecticide, Fungicide, and Rodenticide Act
FEPCA Federal Environmental Pesticide Control Act
FML Flexible Membrane Liner
FOIA Freedom of Information Act
FPR Free Product Recovery
FR *Federal Register*
FRP Federal Response Plan
FTIR Fourier Transform Infrared Spectrometer
GAC Granular Activated Carbon
GAO Government Accounting Office
GC Gas Chromatography
GC-MS Gas Chromatograph-Mass Spectrometer
gpd Gallons Per Day
gpm Gallons Per Minute
GW Groundwater
GWT Groundwater Table
HAP(s) Hazardous Air Pollutant(s)
HAZAN Hazard Analysis
HAZMAT Hazardous Material
HAZOP Hazard and Operability Study
HAZWASTE Hazardous Waste
HCOs Halogenated Organic Compounds
HCS Hazard Communication Standard
HDPE High-Density Polyethylene
HEPA High-Efficiency Particle-Filter Air
HHW Household Hazardous Waste
HLW High-Level Waste (Radioactive)
HMIS Hazardous Materials Information System
HMTA Hazardous Materials Transportation Act
HPLC High-Performance Liquid Chromatography
HPVC High-Density Polyvinyl Chloride
HRI Hydrocarbon Recyclers, Inc.
HRS Hazard Ranking System
HS Hazardous Substance
HSDs Halide-Specific Detectors
HSL Hazardous Substance List
HSWA Hazardous and Solid Waste Amendments
HWDMS Hazardous Waste Data Management System
HWERL Hazardous Waste Engineering Research Laboratory
HWF Hazardous Waste Fuel
HxCDDs Hexachlorodibenzodioxins
HxCDFs Hexachlorodibenzofurans
IARC International Agency for Research on Cancer
ICAP Inductively Coupled Argon-Plasma (Emission Spectroscopy)

ICC Interstate Commerce Commission
ICR Information Collection Requirement
ICS Incident Command System
ID Inner Diameter
IDHL Immediately Dangerous to Health or Life
IDNR Iowa Department of Natural Resources
INCIN Incineration
IR Infrared
IPPP Industrial Pollution Prevention Program
ISC Industrial Source Complex
ISO International Standards Organization
ISV In-Situ Vitrification
IT International Technology Corp.
IU IU International Corp.
KDHE Kansas Department of Health and Environment
L Liter
LC Liquid Chromatography
LDF Land Disposal Facility
LDR(s) Land Disposal Restriction(s)
LEA Loss-Excess Air
LEL Lower Explosion Limit
LFL Lower Flammable Limit
LEPC Local Emergency Planning Committee
LF Landfill
LFG Landfill Gas
LFL Lower Flammable Limit
LHV Lower Heating Value
LI Liquid Injection
LL Liquid Limit
LLW Low-Level Waste (Radioactive)
LLNW Low-Level Nuclear Waste
LST Liquid Solids Treatment
LT^3 Low-Temperature Thermal Treatment
LTRA Long-Term Remedial Action
LUSTs Leaking Underground Storage Tanks
MAC Maximum Allowable Concentrations
MACT Maximum Achievable Control Technology
MCL Maximum Contaminant Levels
MDNR Missouri Department of Natural Resources
MEFR Maximal Expiratory Flow Rate
MEO Mediated Electrochemical Oxidation
MESA Mining Enforcement and Safety Administration
MCD Monochlorobenzene
MCL Maximum (permissible) Concentration Limits
MCS Multiple Chemical Sensitivity
MCT Monochlorotoluene
MDQ Minimum Detectable Quantity
MEK Methyl ethyl ketone
MEP Multiple Extraction Procedure

mgd Million Gallons Per Day

mg/L Milligrams Per Liter

MIBK Methyl Isobutyl Ketone

mL Milliliter

MLG Municipal Landfill Gas

MMT Million Metric Tons

MPN Most Probable Number

MPRSA Marine Protection, Research, and Sanctuaries Act

MQL Method Quantitation Limit

MRF Materials Recovery Facility

MS Mass Spectrometry

MSDS Material Safety Data Sheet

MSHA Mine Safety and Health Administration

MSWLF Municipal Solid Waste Landfill

MTG Minimum Technology Guidance

MTR Minimum Technological Requirements

MTU Mobile Treatment Unit

MUC Maximum Use Concentration

MVV Maximal Voluntary Ventilation

NA Not Applicable

NAAQS National Ambient Air Quality Standards

NACE National Association of Corrosion Engineers

NAL National Analytical Laboratories

NAPL Non-Aqueous Phase Liquid

NARM Naturally Occurring and Accelerator-Produced Radioactive Materials

NAS National Academy of Sciences

NBAR Non-Binding Arbitration Responsibility

NCI National Cancer Institute

NCP National Contingency Plan

ND Not Detected

NDEC Nebraska Department of Environmental Control

NE Not Evaluated

NECO Nuclear Engineering Company

NEIC National Enforcement Investigations Center

NEPA National Environmental Policy Act

NESHAPs National Emissions Standards For Hazardous Air Pollutants

NETA National Environmental Training Association

NFPA National Fire Protection Association

NFAS No Further Action Sites

NFRAP No Further Remedial Action Planned

NHV Net Heating Value

NIMBY Not In My Back Yard

NIOSH National Institute for Occupational Safety and Health

NIPDWS National Interim Primary Drinking Water Standards

NO_x Nitrogen Oxides (NO, N_2O, NO_2, etc.)

NOD Notice of Deficiency

NOS Not Otherwise Specified

NPCA National Paint and Coatings Association

NPDES Natural Pollutant Discharge Elimination System

NPL National Priority List

NRC Nuclear Regulatory Commission

NRC National Response Center

NRT National Response Team

NSPS New Source Performance Standards

NSR New Source Review

NSWMA National Solid Waste Management Association

NTIS National Technical Information Service

NTP National Toxicology Program

NVLAP National Voluntary Laboratory Accreditation Program

NWW Non-Wastewater

O&M Operation and Maintenance

OAT Office of Appropriate Technology

OCP Office of Environmental Programs

OD Outer Diameter

ODCs Ozone-Depleting Chemicals

OEL Occupational Exposure Limit

OHMTADS Oil and Hazardous Materials Technical Assistance Database System

ORM-E Other Regulated Material, Class E

OSC On-Scene Coordinator

OSHA Occupational Safety and Health Administration

OSW Office of Solid Waste

OSWER Office of Solid Waste and Emergency Response

OTA Office of Technology Assessment (U.S. Congress)

OVA Organic Vapor Analyzer

OVM Organic Vapor Meter

OWPE Office of Waste Programs Enforcement

PA/SI Preliminary Assessment/Site Investigation

PAC Powdered Activated Carbon

PACT Powdered Activated Carbon Treatment

PAHs Polynuclear Aromatic Hydrocarbons

PAPR Powered Air Purifying Respirator

PAT Purge-and-Trap Analysis

PATs Processes, Activities, and Tasks

PCA Pollution Control Agency

PCB Polychlorinated Biphenyl

PCE Pentachloroethylene

PCDDs Polychlorinated Dibenzodioxins

PCDF Polychlorinated Dibenzofuran

PCLRS Primary Leachate Collection and Removal System

PCP Pentachlorophenol

PeCCDs Pentachlorodibenzidioxins

PeCDFs Pentachlorodibenzofurans
PDS Personnel Decontamination Station
PEGM Polyethylene Glycol Monoalkyl Ether
PEL Permissible Exposure Limit
PF Protection Factor (*also* Plug Flow)
PFLT Paint Filter Liquid Test
pH Hydrogen-Ion Activity
PHA Preliminary Health Assessment
PI Plasticity Index
PICs Products of Incomplete Combustion
PID Photoionization Detector
PIN Product Identification Number
PITs Products of Incomplete Treatment
PL Plastic Limit
PLIA Pollution Liability Insurance Association
PMCC Pensky-Martens Closed Cup
PMN Premanufacture Notice
PNAs Polynuclear Aromatics (hydrocarbons)
PNOC Particulates Not Otherwise Classified
POC Products of Complete Combustion
POHCs Principal Organic Hazardous Constituents
POTW Publicly Owned Treatment Works
PP Priority Pollutants
ppb Parts Per Billion (μg/g; 10^{-9})
PPE Personnel Protective Clothing and Equipment
PPLs Pits, Ponds, and Lagoons
ppm Parts Per Million (10^{-6})
ppq Parts Per Quadarillion (10^{-15})
ppt Parts Per Trillion (10^{-12})
PQL Practical Quantitation Limit
PQM Project Quality Management
PRP Potentially Responsible Party
PRT Probiological Remediation Technology
PSD Prevention of Significant Deterioration
psi Pounds Per Square Inch
psia Pounds Per Square Inch (Absolute)
psig Pounds Per Square Inch (Gage)
PTFE Polytetrafluoroethylene
PUREX Plutonium Uranium Extraction Process
PVC Polyvinyl Chloride
QA/QC Quality Assurance/Quality Control
QC/QA Quality Control/Quality Assurance
QTM Quick Turnaround Method
RA Remedial Action (*also* Risk Analysis)
RACT Reasonably Available Control Technology
RBC Red Blood Count
RBCs Rotating Biological Contactors
RCRA Resource Conservation and Recovery Act

R&D Research and Development
RD Remedial Design
RD&D Research, Development, and Demonstration
RD/RA Remedial Design and Remedial Action
RDF Refuse-Derived Fuel
REL Recommended Exposure Limit
RES Rollins Environmental Services, Inc.
RF Radio Frequency
RFA RCRA Facility Assessment
RFF Remote Fiber Fluorimetry
RFI RCRA Facility Investigation
RFP Request For Proposal
RIA Regulatory Impact Analysis
RI/FS Remedial Investigation/Feasibility Study
RO Reverse Osmosis
ROD Records of Decision
ROPs Radical Oxidation Processes
RPAR Rebuttal Presumption Against Registration
RPM Remedial Project Manager
RQ Reportable Quantity
RSD Risk-Specific Doses
RSE Removal Site Evaluation
RSPA Research and Special Programs Administration
RTECS Registry of Toxic Effects of Chemical Substances
RV Residual Volume
SACM Superfund Accelerated Cleanup Model
SADI Site-Specific Allowable Daily Intake
SAR Sodium-Adsorption Ratio (*also* Supplied Air Respirator)
SARA Superfund Amendments and Reauthorization Act
SASS Source Assessment Sampling System
SBR Sequencing Batch Reactor
SCBA Self-Contained Breathing Apparatus
scfm Standard Cubic Feet Per Minute
SCWO Supercritical Water Oxidation
SDWA Safe Drinking Water Act
SE Synchronous Excitation (Fluorescence Spectroscopy)
SEPC State Emergency Planning Committee
SERC State Emergency Response Commission
SFE Supercritical Fluid Extraction
SI Surface Impoundment
SIA Surface Impoundment Assessment
SIC Standard Industrial Code
SIM Selected Ion Monitoring
SITE Superfund Innovative Technology Evaluation
SLA Solvent-Laden Air

SLCRS Secondary Leachate Collection and Removal System

SMART Save Money and Reduce Toxics

SMCRA Surface Mining Control and Reclamation Act

SO_x Sulphur Oxides (SO_2, H_2S, etc.)

SOP Standard Operating Procedure

SPC Statistical Process Control

SPCC Spill Prevention, Control, and Countermeasures

SPDES State Pollutant Discharge Elimination System

SPL Spent Pot Lining

SQG Small Quantity Generator

S/S Solidification and Stabilization

SSHP Site Safety and Health Plan

STEL Short-Term Exposure Limit

STEV Short-Term Exposure Value

SVE Soil Vapor Extraction

SVPOHC Semivolatile Principal Organic Hazardous Constituents

SW Solid Wastes

SWDA Solid Waste Disposal Act

SWE Single-Wavelength Excitation (Fluorescence Spectroscopy)

SWICH Solid Waste Information Clearing House

SWMU Solid Waste Management Unit

t Metric tonne(s) [English ton is never abbreviated]

TAG Technical Assistance Grant

TAT Technical Assistance Team

TC Toxicity Characteristic (*also* Temperature Control)

TCCDs Tetrachlorodibenzodioxins

TCDFs Tetrachlorodibenzofurans

TCE Trichloroethylene

TCLP Toxicity Characteristic Leaching Procedure

TDI Toluene Diisocyanate

TDS Total Dissolved Solids

THC Total Hydrocarbons

THMs Trihelomethenes

TLC Total Lung Capacity

TLV Threshold Limit Value

TLV–C Threshold Limit Value–Ceiling

TLV–TWA Threshold Limit Value–Time Weighted Average

TLV–STEL Threshold Limit Value–Short-Term Exposure Limit

TNT Trinitrotoluene

TO Toxic Organic

TOC Total Organic Carbon

TOD Total Oxygen Demand

TOH Total Organic Halides

TPH Total Petroleum Hydrocarbons

TQM Total Quality Management

TRI Toxic Release Inventory

TRU Transuranic Waste (Radioactive)

TSCA Toxic Substances Control Act

TSD Treatment, Storage, and Disposal

TSDF Treatment, Storage, and Disposal Facilities

TSP Total Suspended Particles

TSS Total Suspended Solids

TUHC Total Unburned Hydrocarbon

TWA Time-Weighted Average

UAREP Universities Associated for Research and Education in Pathology, Inc.

UASB Upflow Anaerobic-Sludge Blanket

UEL Upper Explosive Limit

UF Ultrafiltration (*also* Urea-Formaldehyde Resin)

UFL Upper Flammable Limit

UIC Underground Injection Control

UL Underwriters Laboratories, Inc.

UMTRCA Uranium Mill Tailings Radiation and Control Act

UN/NA United Nations–North American

USC United States Code

USCG United States Coast Guard

USDA United States Department of Agriculture

USDOC United States Department of Commerce

USDOT United States Department of Transportation

USDWs Underground Sources of Drinking Water

USEC U.S. Ecology, Inc.

USEPA United States Environmental Protection Agency

USGS United States Geological Survey

USOWRT United States Office of Water Research Technology

USPCI U.S. Pollution Control, Inc.

USTs Underground Storage Tanks

UV Ultraviolet

VA Volatile Acid

VCM Vinyl Chloride Monomer

VOA Volatile Organic Analysis

VOC Volatile Organic Carbon

VOST Volatile-Organic Sampling Train

VPHOC Volatile Principal Organic Hazardous Constituents

VSS Volatile Suspended Solid

VTR Vertical-Tube Reactor

vv % Volume Percent

WAO Wet Air Oxidation

WC Water Column

WET Model Waste, Environment, and Technology Model

WMI Waste Management, Inc.

WPCF Water Pollution Control Federation

WQCB Water Quality Control Board

WRAP Waste Reduction Assessment Program (*also* Waste Reduction Always Pays)

WREAFS Waste Reduction Evaluations at Federal Sites

WRITE Waste Reduction Innovative Technology Evaluation

wt % Weight Percent

WWT Wastewater Treatment

WWTP Wastewater Treatment Process

6-9 DRE 99.9999 Percent Destruction and Removal Efficiency

3P Pollution Prevention Pays

µg/L microgram per liter (ppb; 10^{-9})

APPENDIX C

Environmental Agencies

FEDERAL GOVERNMENT

Environmental Protection Agency

Assistant Administrator for Solid Waste and Emergency Response
401 M Street S.W.
Washington, DC 20460
(202) 260-2090

Public Information (202) 829-3535
Library (202) 382-5921
General Counsel (202) 382-4134

Hazardous Waste Collection Database
EPA Headquarters Library
(202) 382-5922

EPA Alternative Treatment Technology Information Center (ATTIC)
Suite 210
4 Research Place
Rockville, MD 20850
System Operator (301) 816-9153
ATTIC Project Officer (202) 260-5747
ATTIC System Operator (301) 670-6294
Online Computer Access (301) 670-3808

National Response Team (NRT) for Oil and Hazardous Substances Incidents
Assistant Administrator for Solid Waste and Emergency Response
Environmental Protection Agency
Washington, DC 20460
(202) 475-8600
Fax: (202) 252-092

Hazardous Waste Superfund Research Committee
Office of Solid Waste Management and Emergency Response
Office of Research and Development, Environmental Protection Agency
401 M Street S.W., Room 308 N.E. Mall
Washington, DC 20460
(202) 382-7486

Regional Offices, EPA

U.S. EPA
Region 1
John F Kennedy Federal Building
Boston, MA 02203
(617) 565-3420
Fax: (617) 565-3346

U.S. EPA
Region 2
26 Federal Plaza
New York, NY 10278
(212) 264-2657
Fax: (212) 264-5433

U.S. EPA
Region 3
841 Chestnut Street
Philadelphia, PA 19107
(215) 597-9800
Fax: (215) 597-7906

U.S. EPA
Region 4
345 Court Land Street NE
Atlanta, GA 30365
(404) 347-4727
Fax: (404) 347-4486

U.S. EPA
Region 5
230 S. Dearborn Street
Chicago, IL 60604
(312) 353-2000
Fax: (312) 353-1155

U.S. EPA
Region 6
1445 Ross Avenue
Dallas, TX 75202
(214) 655-6444
Fax: (214) 655-2146

U.S. EPA
Region 7
726 Minnesota Avenue
Kansas City, KS 66101
(913) 551-7000
Fax: (913) 551-7467

U.S. EPA
Region 8
999, 18th Street
Denver, CO 80202
(303) 293-1603
Fax: (303) 294-1087

U.S. EPA
Region 9
215 Fremont Street
San Francisco, CA 94105
(415) 744-1305
Fax: (415) 744-1510

U.S. EPA
Region 10
1200 6th Avenue
Seattle, WA 98101
(206) 442-1200
Fax: (206) 553-8509

United States Nuclear Waste Technical Review Board
1100 Wilson Boulevard, Suite 910
Arlington, VA 22209
(703) 235-4473

National Technical Information Service (NTIS)
U.S. Department of Commerce
5285 Port Royal Road
Springfield, VA 22161
(703) 487-4600

Advisory Council in Hazardous Substances Research
and Training
Office of Program Planning and Evaluation
National Institute of Environmental Health Sciences
PO Box 12233
Research Triangle Park, NC 27709
(919) 541-3212

Department of Energy
Civilian Radioactive Waste Management
Washington, DC 20590
(202) 586-9116

Department of Transportation
Office of Hazardous Materials Safety
400 Seventh Street S.W.
Washington, DC 20590
(202) 366-0656

Defense Technical
Information Center
Attn: Online Support Office (DTIC-BLD)
Building 5, Cameron Station
Alexandria, VA 22304-6145
(703) 274-7709

Council on Environmental Quality
Associate Director
Pollution Control and Prevention
722 Jackson Place N.W.
Washington, DC 20503
(202) 395-5750

Nuclear Regulatory Commission
Director
1717 H Street N.W.
Washington, DC 20555
(301) 492-7000

Nuclear Regulatory Commission, Region 3
Regional Administrator
799 Roosevelt Road
Glen Ellyn, IL 60137
(312) 790-5500

Nuclear Regulatory Commission, Region 4
Regional Administrator
611 Ryan Plaza Drive, Suite 1000
Arlington, TX 76012
(817) 860-8100

United States Coast Guard
2100 Second Street S.W.
Washington, DC 20593-0001
(202) 267-0518

Chemical Transportation Advisory Committee
2100 Second Street, S.W.
Washington, D.C. 20593-0001
(202) 267-2967

United States Fire Administration
Firefighters' Safety Working Group
16825 S. Seton Avenue
Emmitsburg, MD 21727
(301) 447-1105

STATES

ALABAMA
Director
Department of Environmental Management
1751 Congressman Dickinson Drive
Montgomery, AL 36130
(205) 271-7761

ALASKA
Commissioner
Department of Environmental Conservation
PO Box O
Juneau, AK 99811
(907) 465-2600

ARIZONA
Commissioner
Department of Environmental Quality
2006 N. Central Avenue
Phoenix, AZ 85004
(602) 257-2300

ARKANSAS
Director
Pollution Control and Ecology
8001 National Drive
PO Box 9583
Little Rock, AR 72209
(501) 562-7400

CALIFORNIA
Secretary
Environmental Protection Agency
555 Capitol Mall
Sacramento, CA 95814
(916) 445-3846

COLORADO
Director
Health and Environmental Protection Office
Department of Health
4210 E. 11th Avenue
Denver, CO 80220
(303) 331-4510

CONNECTICUT
Commissioner
Div. of Environmental Quality
Dept. of Environmental Protection
165 Capitol Avenue Room 161
Hartford, CT 06106
(203) 566-2110

DELAWARE
Secretary
Natural Resources and Environment Department
89 Kings Highway
PO Box 1401
Dover, DE 19903
(302) 739-4403

FLORIDA
Secretary
Department of Environmental Regulation
Twin Towers
2600 Blairstone Road
Tallahassee, FL 32399
(904) 488-4805

GEORGIA
Director
Environmental Protection Division
Department of Natural Resources
205 Butler Street, S.W.
Atlanta, GA 30334
(404) 656-4713

HAWAII
Director
Environmental Health Administration
Department of Health
1250 Punchbowl Street
Honolulu, HI 96813
(808) 548-4139

IDAHO
Administrator
Division of Environment
Dept. of Health and Welfare
1410 N. Hilton
Boise, ID 83706
(208) 334-5840

ILLINOIS
Director
Environmental Protection Agency
2200 Churchill Road
Springfield, IL 62708
(217) 782-3397

INDIANA
Commissioner
Department of Environmental Management
105 S. Meridian Street
PO Box 6015
Indianapolis, IN 46225
(317) 232-8162

IOWA
Director
Environmental Protection Division
Department of Natural Resources
Wallace State Office Building
Des Moines, IA 50319
(515) 281-6284

KANSAS
Department of Health and Environment
Division of Environment
Forbes Field, Building 740
Topeka, KS 66620
(913) 296-1535

Health and Environment Department
Air Quality and Waste Management
Bureau
Landon State Office Building, Room 900
Topeka, KS 66612
(913) 296-1593

KENTUCKY
Commissioner
Department for Environmental Protection
Frankfort Office Park
18 Reilly Road
Frankfort, KY 40601
(502) 564-2150

LOUISIANA
Secretary
Department of Environmental Quality
PO Box 44066
Baton Rouge, LA 70804
(504) 342-1266

MAINE
Commissioner
Environmental Protection Department
State House Station #17
Augusta, ME 04333
(207) 289-2812

MARYLAND
Secretary
Department of the Environment
2500 Broening Highway
Baltimore, MD 21224
(301) 631-3084

MASSACHUSETTS
Commissioner
Department of Environmental Protection
1 Winter Street
Boston, MA 02108
(617) 292-5856

MICHIGAN
Director
Environmental Protection Bureau
Department of Natural Resources
PO Box 30028
Lansing, MI 48909
(517) 373-7917

MINNESOTA
Executive Director
Environmental Quality Board
State Planning Agency
658 Cedar Street, Suite 300
St. Paul, MN 55155
(612) 296-0212

MISSISSIPPI
Executive Director
Department of Environmental Quality
PO Box 20305
Jackson, MS 39289
(601) 961-5000

MISSOURI
Division of Environmental Quality
Department of Natural Resources
205 Jefferson
PO Box 176
Jefferson City, MO 65102
(314) 751-4810

Department of Conservation
Planning Section
Administrator
2901 W. Truman Boulevard, PO Box 180
Jefferson City, MO 65102
(314) 751-4115

Department of Public Safety
State Emergency Management Agency
Director
Adjutant General's Building
1717 Industrial Drive
PO Box 116
Jefferson City, MO 65102
(314) 751-9500

MONTANA
Director
Environmental Sciences Division
Cogswell Building
Helena, MT 59620
(406) 444-2544

NEBRASKA
Director
Department of Environmental Control
301 Centennial Mall S.
PO Box 94877
Lincoln, NE 68509
(402) 471-2186

NEVADA
Administrator
Division of Environmental Protection
123 W. Nye Lane
Carson City, NV 89710
(702) 687-4670

NEW HAMPSHIRE
Commissioner
Department of Environmental Services
6 Hazen Drive
Concord, NH 03301
(603) 271-3503

NEW JERSEY
Commissioner
Department of Environmental Protection
401 E. State Street , CN 402
Trenton, NJ 08625
(609) 292-2885

NEW MEXICO
Secretary
Department of Environment
PO Box 26110
Santa Fe, NM 87502
(505) 827-2850

NEW YORK
Commissioner
Department of Environmental Conservation
50 Wolf Road
Albany, NY 12233
(518) 457-3446

NORTH CAROLINA
Director
Division of Environmental Management
Environment, Health and Natural Resources
512 N. Salisbury Street
PO Box 27687
Raleigh, NC 27604
(919) 733-7015

NORTH DAKOTA
Chief, Environmental Health Section
Department of Health
1200 Missouri Avenue, PO Box 5520
Bismarck, ND 58502
(701) 224-2374

OHIO
Director
Environmental Protection Agency
1800 Watermark Drive, PO Box 1049
Columbus, OH 43266
(614) 644-2782

OKLAHOMA
Deputy Commissioner
Environmental Health Services
Department of Health
1000 NE 10th Street
Oklahoma City, OK 73117
(405) 271-8056

OREGON
Director
Department of Environmental Quality
811 SW Sixth Avenue
Portland, OR 97204
(503) 229-5300

PENNSYLVANIA
Department of Environmental Resources
Fulton Building, 9th Floor
PO Box 2063
Harrisburg, PA 17105
(717) 787-5028

RHODE ISLAND
Director
Department of Environmental Management
9 Hayes Street
Providence, RI 02908
(401) 277-2771

SOUTH CAROLINA
Environmental Quality Control Office
Department of Health and Environmental Control
2600 Bull Street
Columbia, SC 29201
(803) 734-5360

SOUTH DAKOTA
Secretary
Department of Water and Natural Resources
523 E. Capitol Avenue, Joe Foss Building
Pierre, SD 57501
(605) 773-3151

TENNESSEE
Bureau of Environment
Department of Environment and Conservation
150 Ninth Avenue, N., 1st Floor
Nashville, TN 37247
(615) 741-3657

TEXAS
Environment and Consumer Health Protection
Department of Health
1100 W. 49th Street
Austin, TX 78756
(512) 458-7541

UTAH
Director
Division of Environmental Health
Department of Health
288 N. 1460 W.
PO Box 16690
Salt Lake City, UT 84114
(801) 538-6121

VERMONT
Commissioner
Department of Environmental Conservation
Agency of Natural Resources
103 S. Main Street, 1 S. Building
Waterbury, VT 05676
(812) 244-8755

VIRGINIA
Administrator
Council on the Environment
903 Ninth Street Office Building
Richmond, VA 23219
(804) 786-4500

WASHINGTON
Director
Department of Ecology
M/S: PV-11
Olympia, WA 98503
(206) 459-6168

WEST VIRGINIA
Director
Division of Natural Resources
Commerce, Labor and Environmental
Resources Department
Capitol Complex, Building 3
Charleston, WV 25305
(304) 348-2754

WISCONSIN
Administrator
Environmental Quality Division
Department of Natural Resources
PO Box 7921
Madison, WI 53707
(608) 266-1099

WYOMING
Director
Department of Environmental Quality
Herschler Building
122 W. 25th Street
Cheyenne, WY 82002
(307) 777-7938

DISTRICT OF COLUMBIA
Administrator
Housing and Environmental Regulation
Administration
Consumer and Regulatory Affairs
614 H Street, N.W., Room 505
Washington, DC 20001
(202) 727-7395

AMERICAN SAMOA
Executive Secretary
Environmental Quality Commission
Office of the Governor
Pago Pago, AS 96799
(684) 633-2304

GUAM
Administrator
Environmental Protection Agency
130 Rojas Street
Harmon, GU 96911
(671) 646-8863

NORTHERN MARIANA ISLANDS
Chief
Environmental Quality Division
Public Health and Environmental Services
PO Box 409
Saipan, MP 96950
(670) 234-6114

PUERTO RICO
President
Environmental Quality Board
PO Box 11488
Santurce, PR 00910
(809) 767-8056

U.S. VIRGIN ISLANDS
Commissioner
Department of Planning and Natural Resources
Nisky Center, Suite 231
St. Thomas, VI 00802
(809) 774-3320

APPENDIX D

Environmental Journals, Magazines, and Databases

JOURNALS AND MAGAZINES

Alternatives
Alternatives, Inc.
c/o Faculty of Environmental Studies
University of Waterloo
Waterloo, ON N2L 3G1, Canada
(519) 885-1211
Fax: (519) 746-2031

Ambio
Royal Swedish Academy of Sciences/Pergamon Press, Inc.
Fairview Park
Elmsford, NY 10523

Atmospheric Environment
Pergamon Press, Inc.
Maxwell House
Fairview Park
Elmsford, NY 10523

BioCycle
JG Press, Inc.
419 State Avenue, 2nd Floor
Emmaus, PA 18049

Coastal Management
Taylor & Francis
1900 Frost Road, Suite 101
Bristol, PA 19007

Compost
Department of Environmental Protection
Division of Solid Waste Management
840 Bear Tavern Road
CN 414
Trenton, NJ 08625
(609) 530-8593

Critical Reviews in Environmental Control
CRC Press
2000 Corporate Boulevard N.W.
Boca Raton, FL 33431

Recycle
Environmental Action Coalition
625 Broadway
New York, NY 10012
(212) 677-1601

Defense Cleanup
Pasha Publications, Inc.
1401 Wilson Boulevard, Suite 900
Arlington, VA 22209-9970
(703) 528-1244
Fax: (703) 528-1253

Directory of Hazardous Waste Services
Southam Business Communications, Inc.
1450 Don Mills Road
Don Mills, ON M3B 2X7, Canada
(416) 445-6641
Fax: (416) 442-2200

E: the environmental magazine
Earth Action Network, Inc.
28 Knight Street
Norwalk, CT 06851

EI Digest
Environmental Information, Ltd.
4801 W. 81st Street, Suite 119
Minneapolis, MN 55437
(612) 831-6550
Fax: (612) 831-6550

Earth Island Journal
Earth Island Institute
300 Broadway, Suite 28
San Francisco, CA 94133

Earthwatch
Earthwatch Expeditions, Inc.
680 Mount Auburn Street
PO Box 403
Watertown, MA 02272

The Ecologist
Ecosystems, Ltd.
Corner House, Station Road
Sturminster Newton
Dorset DT10 1BB, UK

Environment
Heldref Publications
4000 Albemarle Street N.W.
Washington, DC 20016

Environment Advisor
J.J. Keller & Associates, Inc.
3003 Breezewood Lane
Box 368
Neenah, WI 54957-0368
(414) 722-2848
Fax: (414) 727-7516

Environmental Action
Environmental Action, Inc.
1525 New Hampshire Avenue N.W.
Washington, DC 20036

Environmental Toxicology & Chemistry
Elsevier Science, Ltd., Pergamon Press, Inc.
PO Box 800
Kidlington, Oxford
OX5 11DX, UK
44-865-843000 (UK)
Fax: 44-865-843010 (UK)
(914) 524-9200 (USA)
Fax: (914) 333-2444 (USA)

Environmental Conservation
Foundation for Environmental Conservation
Elsevier Sequoia SA
PO Box 564
1001 Lausanne 1
Switzerland

Environmental Ethics
Environmental Philosophy, Inc.
Chestnut Hall
University of North Texas
1926 Chestnut Street
Denton, TX 76203-3496

The Environmental Forum
Environmental Law Institute
1616 P Street N.W., Suite 200
Washington, DC 20036

Environmental Geochemistry and Health
Science and Technology Letters
PO Box 81
Northwood
Middlesex, HA6 3DN, UK

Environmental Geology and Water Sciences
Springer-Verlag New York, Inc.
175 5th Avenue
New York, NY 10010
(212) 460-1500
Fax: (212) 473-6272

Environmental History Review
American Society for Environmental History
Center for Technology Studies
New Jersey Institute of Technology
Newark, NJ 07102

Environmental Impact Assessment Review
Elsevier Science Publishing Co., Inc.
655 Avenue of the Americas
New York, NY 10010
(212) 989-5800

Environmental Management
Springer-Verlag New York, Inc.
175 5th Avenue
New York, NY 10010
(212) 460-1500
Fax: (212) 473-6272

Environmental Periodicals Bibliography
International Academy at Santa Barbara
800 Garden Street, Suite D
Santa Barbara, CA 93101-1552
(805) 965-5010
Fax: (805) 965-6071

Environmental Pollution
Elsevier Applied Science Publications, Ltd.
Crown House, Linton Rd.
Barking, Essex IG11 8JU, UK

Environmental Protection
Stevens Publishing Corporation
225 N. New Road
Waco, TX 76710
(817) 776-9000
Fax: (817) 776-9018

Environmental Research
Academic Press
One E. First Street
Duluth, MN 55802

Environmental Science & Technology
American Chemical Society
1155 16th Street N.W.
Washington, DC 20036

Environmental Waste Management
International Association of Environmental Managers
243 W. Main Street
Kutztown, PA 19530
(215) 683-5098
Fax: (215) 683-3171

The Environmentalist
Science and Technology Letters
PO Box 81
Northwood, Middlesex HA6 3DN, UK

Everyone's Backyard
Citizens Clearinghouse for Hazardous Waste, Inc.
Box 6806
Falls Church, VA 22040
(703) 237-2249

Garbage
Old House Journal Corp.
435 Ninth Street
Brooklyn, NY 11215

Greenpeace Magazine
Greenpeace USA
1436 U Street N.W.
Washington, DC 20009

HMCRI
Hazardous Materials Control Research Institute
7237 Hanover Parkway
Greenbelt, MD 20770-3602
(301) 982-9500
Fax: (301) 220-3870

HMCRI Focus
Hazardous Materials Control Research Institute
7237 Hanover Parkway
Greenbelt, MD 20770-3602

Hazardous Materials Management
CHMM Inc.
401 Richmond Street W., Suite 139
Toronto, ON M5V 1X3, Canada
(416) 348-9922
Fax: (416) 348-9744

Hazardous Materials Transportation
The Bureau of National Affairs, Inc.
1231 25th Street N.W.
Washington, DC 20037
(202) 452-4200
Fax: (202) 822-8092

Hazard Monthly
Research Alternative, Inc.
966 Hungerford Drive, Suite 1
Rockville, MD 20850
(301) 424-2803
Fax: (301)738-1026

Hazardous Substances & Public Health
U.S. Department of Health and Human Services
Agency for Toxic Substances and Disease Registry
1600 Clifton Road N.E.
Mailstop E-33
Atlanta, GA 30333
(404) 639-6206
Fax: (404) 639-6208

Hazardous Waste & Hazardous Materials
Hazardous Materials Control Research Institute
1651 Third Avenue
New York, NY 10128
(212) 289-2300
Fax: (212) 289-4697

Hazardous Waste Business
McGraw-Hill, Inc.
Energy & Business Newsletter
1221 Avenue of the Americas, 36th Floor
New York, NY 10020
(212) 521-6410

Hazardous Waste Consultant
McCoy and Associates, Inc.
13701 W. Jewell Avenue, Suite 202
Lakewood, CO 80228-4173
(303) 987-0333
Fax: (303) 989-7917

Hazardous Waste News
Business Publishers, Inc.
951 Pershing Drive
Silver Spring, MD 20910-4464
(301) 587-6300
Fax (301) 585-9075

Hazmat Transport
Business Publishers, Inc.
951 Pershing Drive
Silver Spring, MD 20910-4464
(301) 587-6300
Fax: (301) 585-9075

Hazmat World
Tower-Borner Publishing
800 Roosevelt Road, Building C, Suite 206
Glen Ellyn, IL 60137
(708) 858-1888
Fax: (708) 858-1957

Haztech News
Business Publishers, Inc.
951 Pershing Drive
Silver Spring, MD 20910-4464
(301) 587-6300
Fax: (301) 587-1081

Human Ecology
Plenum Press
233 Spring Street
New York, NY 10013

Industrial Waste Conference
Purdue University
School of Civil Engineering
W. Lafayette, IN 47907
(317) 494-2194
Fax: (317) 496-1107

Integrated Waste Management
McGraw-Hill, Inc.
Energy & Business Newsletters
1221 Avenue of the Americas, 36th Floor
New York, NY 10020
(212) 512-6410

International Environmental Affairs
University Press of New England
17 1/2 Lebanon Street
Hanover, NH 03755

Journal of Environmental Sciences
Institute of Environmental Sciences
940 E. Northwest Highway
Mt. Prospect, IL 60056

International Journal of Environmental Studies
Gordon and Breach Science Publishers
PO Box 786, Cooper Station
New York, NY 10276
(212) 206-8900
Fax: (212) 645-2459

Journal of Coastal Research
PO Box 368
Lawrence, KS 66044
Phone/Fax: (305) 565-1051

Journal of Environmental Education
Heldref Publications.
4000 Albemarle Street N.W.
Washington, DC 20016

Journal of Environmental Economics and Management
Academic Press, Inc.
1250 6th Avenue
San Diego, CA 92101
(619) 699-6825
Fax: (619) 699-6800

Journal of Environmental Engineering
American Society of Civil Engineers
United Engineering Center
345 E. 47th Street
New York, NY 10017-2398

Journal of Environmental Health
National Environmental Health Association
Delegation of the Commission of the
European Communities
2100 M Street N.W., Suite 700
Washington, DC 20037
(202) 862-9555

Journal of Environmental Management
Academic Press, Ltd.
Foots Cray
Sidcup, Kent DA14 5HP, U.K.

Journal of Environmental Quality
American Society of Agronomy
677 S. Segoe Road
Madison, WI 53711
(608) 273-8080
Fax: (608) 273-2021

Journal of Environmental Science and Health
Marcel Dekker Journals
PO Box 10018
Church Street Station
New York, NY 10249

Journal of Environmental Systems
Baywood Publishing Company, Inc.
26 Austin Avenue
Amityville, NY 11701

Journal of Soil Contamination
Lewis Publishers
CRC Press, Inc.
2000 Corporate Boulevard
Boca Raton, FL 33431
(800) 272-7737

Journal of Toxicology and Environmental Health
Hemisphere Publishing Corporation
1101 Vermont Avenue N.W., No. 200
Washington, DC 20005
(202) 289-2174
Fax: (202) 289-3665

Management of World Wastes
Communication Channels, Inc.
6255 Barfield Road
Atlanta, GA 30328-4369
(404) 256-9800
Fax: (404) 256-3116

Natural Resources Journal
University of New Mexico
School of Law
1117 Stanford N.E.
Albuquerque, NM 87131

Nature and Resources
The Parthenon Publishing Group
120 Mill Road
Park Ridge, NJ 07656

The New Crucible
De Young Press
RR 1, Box 76
Stark, KS 66775-9802
(316) 754-3203

The Northwest Environmental Journal
University of Washington
Institute for Environmental Studies
Engineering Annex, FM-12
Seattle, WA 98195

Pollution Engineering
Cahners Publishing Company
1350 E. Touhy Avenue
Des Plaines, IL 60018-3358

Resources, Conservation and Recycling
Elsevier Science Publishers
655 Avenue of the Americas
New York, NY 10010
(212) 633-3900

Water Research
Elsevier Science, Ltd., Pergamon Press, Inc.
PO Box 800
Kidlington, Oxford OX5 11DX, UK
44-865-843000 (UK)
Fax: 44-865-843010 (UK)
(914) 524-9200 (USA)
Fax: (914) 333-2444 (USA)

Water Science and Technology
Springer-Verlag, Inc.
175 Fifth Ave
New York, NY 10010

World Watch
Worldwatch Institute
1776 Massachusetts Avenue N.W.
Washington, DC 20036

Water, Air and Soil Pollution
Kluwer Academic Publishing Group
PO Box 358
Accord Station
Hingham, MA 02018-0358

Water Environment Research
Water Environment Federation
601 Wythe Street
Alexandria, VA 22314
(703) 684-2400
Fax: (703) 684-2492

DATABASES

CERCLIS
Bruce, Allen and Hamilton
Crystal City, VA
(800) 424-9346

Enviroline
Congressional Information Service
4520 East West Highway, Suite 800
Bethesda, MD 20814-3389
(800) 638-8380
Fax: (301) 654-4033
(coverage: 1971–present)

Environmental Bibliography
Environmental Studies Institute
800 Garden Street, Suite D
Santa Barbara, CA 93101-1552
(805) 965-5010
Fax: (805) 965-6071
(coverage: 1973–present)

GeoArchive
Geosystems
PO Box 40
Didcot, Oxon OX119BX, UK
0235 813913
(coverage: 1974–present)

GEOBASE
CD Publishing Co., Inc.
777-8 Avenue, SW
Calgary, AL T2P3R5, Canada
(403) 294-0080
Fax: (403) 294-0082
(coverage: 1980–present)

APPENDIX E

Universities and Colleges in the United States Offering Courses and Programs in Solid and Hazardous Waste Management

The majority of the colleges and universities in the United States offer courses in solid and/or hazardous waste management as part of their civil engineering and environmental engineering degree programs. The numbers of solid/hazardous waste management courses offered under the various degree programs are listed below.

- Environmental Engineering: 38
- Civil Engineering: 61
- Environmental Science: 9
- Environmental Health: 1
- Environmental Quality Engineering: 4
- Environmental Resources Engineering: 1
- Environmental Systems Engineering: 2
- Environmental Studies: 1
- Hazardous and Solid Materials Management: 3

Table E-1 is a list of colleges and universities that offer courses and/or degree programs in solid and/or hazardous waste management.

TABLE E-1 Colleges and Universities in the United States Offering Courses and/or Degree Programs in Solid/Hazardous Waste Management

COLLEGES AND UNIVERSITIES LISTED BY COURSES OFFERED

Course Offerings	Degree Program	College/University
Advanced Hazardous Waste Management	Environmental Quality Engineering	University of Alaska, Fairbanks
Air Pollution and Solid Waste Management	Civil Engineering	Montana State University
Environmental Restoration and Waste Management	Environmental Engineering	Florida Institute of Technology
Geology and Hazardous Waste Management	Urban Environmental Geology	University of Missouri–Kansas City
Geotechnical Aspects of Waste Disposal	Civil and Environmental Engineering	Marquette University
Geotechnical Aspects of Waste Management (undergraduate)	Civil Engineering	University of Missouri, Rolla
Hazardous Materials (undergraduate)	Civil Engineering	San Jose State University
Hazardous Waste Control	Environmental Engineering	University of Florida
Hazardous Waste Disposal and Solid Waste Management	Civil Engineering	University of Rhode Island

Course Offerings	Degree Program	College/University
Hazardous Waste Engineering	Civil Engineering	University of Washington
Hazardous Waste Engineering	Environmental Engineering	Illinois Institute of Technology
Hazardous Waste Management	Civil Engineering	Syracuse University
Hazardous Waste Management	Environmental Engineering	Iowa State University
Hazardous Waste Management	Environmental Science	Colorado School of Mines
Hazardous Waste Management	Civil Engineering	University of Texas, Austin
Hazardous Waste Management	Environmental Science	Washington State University
Hazardous Waste Management	Civil Engineering	University of Alabama
Hazardous Waste Management	Environmental Systems Engineering	Clemson University
Hazardous Waste Management	Civil and Environmental Engineering	University of Southern California
Hazardous Waste Management	Environmental Science	Rutgers, The State University of New Jersey
Hazardous Waste Management	Civil Engineering	Oklahoma State University
Hazardous Waste Management	Hazardous and Solid Material Management (AS)	North Dakota State College of Science
1. Solid and Hazardous Waste Management 2. Hazardous Waste Management	Civil and Environmental Engineering	Virginia Polytechnic Institute and State University
Hazardous Waste Management	Civil Engineering	University of Illinois, Champaign-Urbana
Hazardous Waste Management	Civil and Environmental Engineering	Georgia Institute of Technology
Hazardous Waste Management I and II	Civil and Environmental Engineering	Rensselaer Polytechnic Institute
Hazardous Waste Management I and II (undergraduate)	Civil Engineering	Bucknell University
Hazardous Waste Management and Control (advanced undergraduate and graduate)	Civil Engineering	Brigham Young University
Hazardous Waste Management and Toxicology	Civil Engineering and Environmental Science	University of Oklahoma
Hazardous Waste Management (graduate)	Environmental Science	McNeese State University
Hazardous Waste Management (undergraduate)	Civil Engineering	University of Missouri, Columbia
Hazardous Waste Management (undergraduate)	Civil Engineering	Case Western University
Hazardous Waste Management (undergraduate)	Environmental Engineering	University of Delaware
Hazardous Waste Management: RCRA and Superfund	Environmental Science	Duke University
Hazardous Waste Treatment and Disposal	Environmental Science and Engineering	Johns Hopkins University
Hazardous Waste: Regulation, Remediation, and Worker Protection	Civil and Environmental Engineering	University of Michigan

Course Offerings	Degree Program	College/University
Hazardous and Industrial Waste Management	Civil and Environmental Engineering	University of Colorado, Boulder
Hazardous and Toxic Waste Engineering	Environmental Engineering	Duke University
Hazardous and Toxic Waste Management	Environmental Quality Engineering	University of Alaska, Fairbanks
Hazardous and Toxic Waste Management	Civil and Environmental Engineering	Polytechnic University, Brooklyn Campus
Industrial Hazardous and Residual Waste Management	Civil Engineering	Pennsylvania State University
Industrial Waste Treatment and Disposal	Civil and Environmental Engineering	Georgia Institute of Technology
Industrial Waste Treatment and Disposal (graduate)	Civil Engineering	South Dakota School of Mines and Technology
Industrial and Hazardous Waste Management	Environmental Engineering	Michigan Technological University
Introduction to Solid/Hazardous Waste Management	Hazardous and Solid Material Management (AS)	North Dakota State College of Science
Liquid Waste Technology and Management (undergraduate)	Civil Engineering	California State University
Management of Hazardous Substances	Civil Engineering	Oregon State University
Management of Hazardous Waste	Environmental Science and Engineering	University of North Carolina, Chapel Hill
Management of Solid and Hazardous Wastes (undergraduate)	Civil Engineering	University of Illinois, Chicago
Municipal Solid Waste Management	Civil and Environmental Engineering	Marquette University
Municipal, Industrial, and Hazardous Waste Management	Civil Engineering	University of Massachusetts, Lowell
Municipal Solid Waste Management	Environmental Systems Engineering	Clemson University
Sludge Management and Disposal	Environmental Engineering	Duke University
Solid Waste and Residual, Management	Environmental Science	Antioch New England Graduate School
Solid Waste Collection and Disposal	Civil Engineering	Syracuse University
Solid Waste Disposal	Civil Engineering	University of Washington
Solid Waste Engineering	Environmental Engineering	Illinois Institute of Technology
Solid Waste Engineering (graduate)	Civil Engineering	California State University, Fresno
Solid Waste Management	Civil and Environmental Engineering	University of Southern California
Solid Waste Management	Environmental Engineering	Michigan Technological University
Solid Waste Management	Civil Engineering	University of Arkansas
Solid Waste Management	Environmental Quality Engineering	University of Alaska, Anchorage
Solid Waste Management	Civil Engineering	University of Illinois, Champaign-Urbana
Solid Waste Management	Environmental Quality Engineering	University of Alaska, Fairbanks
Solid Waste Management	Civil and Environmental Engineering	Polytechnic University, Brooklyn Campus

Course Offerings	Degree Program	College/University
Solid Waste Management	Environmental Resources Engineering	Humboldt State University
Solid Waste Management	Environmental Engineering	Iowa State University
Solid Waste Management	Environmental Science	Washington State University
Solid Waste Management	Environmental Engineering	University of Florida
Solid Waste Management	Environmental Engineering	Florida International University
Solid Waste Management	Civil Engineering	Oregon Institute of Technology
Solid Waste Management	Civil and Environmental Engineering	University of Michigan
Solid Waste Management	Civil Engineering	Vanderbilt University
Solid Waste Management I & II (undergraduate)	Civil Engineering	Bucknell University
Solid Waste Management and Control	Civil Engineering	University of Hawaii at Manoa
Solid Waste Management and Design	Civil Engineering	Washington State University
Solid Waste Management and Treatment	Environmental Science	Rutgers, The State University of New Jersey
Solid Waste Management (undergraduate and graduate)	Environmental Science	McNeese State University
Solid Waste Management (undergraduate)	Civil Engineering	University of Missouri, Columbia
Solid Waste Management (undergraduate)	Civil Engineering	Case Western University
Solid Waste Management (undergraduate)	Civil Engineering	University of Missouri, Rolla
Solid Waste Management	Civil Engineering	Pennsylvania State University
Solid Waste Management	Hazardous and Solid Materials Management (AS)	North Dakota State College of Science
Solid Waste Technology and Management (undergraduate)	Civil Engineering	California State University
Solid Waste Technology I and II	Civil and Environmental Engineering	Georgia Institute of Technology
Solid Waste and Resource Recovery Engineering	Environmental Engineering	Duke University
Solid and Hazardous Waste Disposal	Civil and Environmental Engineering	University of California, Berkeley
Solid and Hazardous Waste Engineering	Civil and Environmental Engineering	University of Wisconsin, Madison
Solid and Hazardous Waste Management	Civil Engineering	University of Massachusetts, Amherst
Solid and Hazardous Waste Management	Civil and Environmental Engineering	Utah State University
Solid and Hazardous Waste Management	Civil and Environmental Engineering	Rensselaer Polytechnic Institute
Solid and Hazardous Waste Management	Civil Engineering	Louisiana State University
Solid and Hazardous Waste Management	Environmental Engineering	Florida Institute of Technology
Solid and Hazardous Waste Management (undergraduate)	Environmental Studies	San Jose State University

Course Offerings	Degree Program	College/University
Solid and Hazardous Waste Management (undergraduate)	Environmental Health	Ohio University
Solid and Hazardous Waste Management (undergraduate)	Civil Engineering	Southern University A&M College
Treatment, Disposal, and Management of Hazardous Waste	Civil Engineering	South Dakota School of Mines and Technology
Treatment/Disposal of Residuals	Civil and Environmental Engineering	Georgia Institute of Technology
Waste Disposal Management	Civil Engineering	University of Notre Dame
Waste Management	Environmental Engineering	SUNY College of Environmental Science and Forestry
Waste and Hazardous Waste Management	Civil Engineering	University of California, Los Angeles

SOURCES

Career Guidance Foundation College Reference Guide, College Catalog Collection 1993–1994, vol. XX, issue 2, 1993.

Peterson's Guide to Graduate Programs in Biological and Agricultural Sciences 1991: Princeton, NJ, 2455 p.

Peterson's Guide to Graduate Programs in Engineering and Applied Sciences 1991: Princeton, NJ, 1356 p.

APPENDIX F

Fundamentals of Toxicology

INTRODUCTION

Toxicology is the science that deals with the adverse effects of poisons or toxins on living organisms. Exactly when and what kinds of adverse effects are produced is a function of

- The quantity or dose of poisonous substances
- The receptor, whether mice, rats, rabbits, dogs, or humans—their age, sex, and physical condition
- The exposure path
- Duration of exposure to the poison

All other factors remaining the same, dose is the most critical factor. For example, if one drinks a bottle of pure iodine, one may die, but applying a drop of diluted iodine on a cut produces healing effects. Caffeine in small doses is tolerable by most people, but large quantities produce harmful effects. The key in both examples is the quantity of the substance, or the *dose*. A substance may be safe at low levels of dose but may become poisonous at high levels. This fact was recognized by the German physician Theophrastus Bombastus von Hohenheim (1493–1541), otherwise known as Paracelsus. He stated "If you want to explain each poison correctly, what is there that is not poison—all things are poison and nothing without poison. Solely the dose determines that a thing is not poison."(Doull, Klaassen, and Amdur, 1980, p. 12)

In a strict sense, toxicology is not the science of identification of poisons, but the science that quantifies the levels at which substances become toxic.

A poison or toxicant is a substance that, *above* a certain level of exposure or dose, produces detri-

mental effects on tissues, organs, biological processes, or human health. Hazardous substances are of concern because of their effects on living organisms. The study of the adverse effects of such substances on life processes constitutes the basis of toxicology. Hazardous substances may occur in

- The work environment, where people can be exposed to toxic agents. The first production of a synthetic dye, aniline, took place in 1860. People working in dye-manufacturing factories developed cancer of the bladder. Chimney sweepers in the U.K. developed scrotal cancer, which is associated with coal tar and soot deposits in chimneys.

- The ambient environment itself—air, water, or soil—may have a high concentration of toxic substances, either naturally or from human activities. Examples of natural occurrence include elevated levels of radon (a carcinogen—causes lung cancer) in homes and high levels of UV radiations at high altitudes. Emissions of vapors and effluents containing toxic substances are examples of human-induced (anthropogenic) factors. Burning high-sulfur coal in power plants releases SO_x and NO_x, which cause formation of sulfuric and nitric acids, which are toxic substances.

- The food chain, where hazardous substances may take one of several paths and ultimately enter bodies of living forms. One of two things can happen: (a) the substance may be either detoxified by the body's immune system activities and excreted, for example, by the addition of oxygen to the toxicant by enzymes, which makes it water soluble and easy to excrete, or (b) it may accumulate and cause chronic effects.

TOXICITY

Toxicity is the capacity of a substance to produce adverse effects on living forms; it is not the effect of a poison or chemical. Toxicity is usually expressed in milligram/kg (mg/kg) of body weight. For air and water, mg/L (ppm) or µg/L (ppb) are the commonly used units.

Toxicity Rating

Chemicals have a wide range of doses at which they can produce detrimental effects, including death, in living forms. Some chemicals will cause death when administered even in minute quantities (micrograms; one millionth of a gram); such substances are extremely poisonous. Other chemicals can be tolerated in larger quantities (few grams); these are less toxic. Table F-1 shows the probable lethal doses of various chemicals on humans, and the relative toxicity class.

Types of Toxic Effects

Depending on the organism's response to toxic substances, one of the following four different kinds of toxic effects may be produced

1. *Reversible Toxicity.* Reversible toxicity occurs when a toxicant enters the body, but does not cause any permanent effects, either due to the organism's natural defense mechanisms or due to the administration of substances to counteract the action of the toxicants—antidotes. Reversible toxicity generally occurs in rapidly multiplying cells, such as those in the liver, bone marrow, and intestines. Damaged cells are quickly replaced by new cells. For example, even if 50 percent of the liver is lost, it can regenerate itself in about six months.

2. *Irreversible Toxicity.* Irreversible toxicity includes effects that last after the toxicant (causing the adverse effects) has been eliminated. This type of toxicity is common in cells that do not multiply rapidly. For example, when skin is burned with sulfuric acid, the immediate effect is pain and burn, which can be eliminated by using a neutralizing agent or washing with water. But there is still a permanent effect on the skin, which is the scar that the acid has produced. Carcinogenic effects of some chemicals are also examples of irreversible toxic effects.

3. *Acute Toxicity.* This type of toxicity results in responses that are observed soon after exposure to a toxic substance; acute toxicity involves short response time. Acute effects

TABLE F–1 Toxicity Rating Chart (modified from Doull, Klaassen, and Amdur, 1980)

Toxicity Rating/Class	Probable Oral Lethal Dose for Humans		
	Dose	For Average Adult	Example
Practically Nontoxic	>15 g/kg	More than one quart	Bis (2-ethylhexylphthalate
Slightly Toxic	5–15 g/kg	Between one pint and one quart	Ethanol
Moderately Toxic	0.5–5 g/kg	Between an ounce and a pint	Malathion (an insecticide)
Very Toxic	50–500 mg/kg	Between a teaspoonful and an ounce	Heptachlor (an insecticide)
Extremely Toxic	5–50 mg/kg	Between seven drops and a teaspoonful	Parathion (an insecticide)
Supertoxic	<5 mg/kg	A taste (less than seven drops)	Sarin (a military poison)

normally result from brief exposures to relatively high levels of a toxicant. Such effects are comparatively easy to observe, and can be attributed to a particular toxicant because of the pronounced symptoms displayed. For example, accidental release of methyl isocyanate at a chemical plant in India in December, 1984, caused blindness and eye injuries to 20,000 people who were exposed to the chemical for a brief interval of time.

4. *Chronic Toxicity.* This involves effects that take a long time to manifest. Chronic toxicity results from low exposures to a toxicant over long periods of time. Chronic responses to toxicants may have latency periods as long as several decades in humans. For example, persons working in nuclear weapons production plants during the 1940s and 1950s were exposed to nuclear radiation, but cancer was not detected until decades after the exposure.

Chronic effects are obscured by normal background symptoms and are much more difficult to study. Nonetheless, chronic effects are of great significance in understanding the long-term effects of toxic substances.

DOSE–RESPONSE RELATIONSHIP

Toxic effects of a substance are usually determined by conducting a series of laboratory experiments exposing selected organisms to varying doses of the toxicant and carefully monitoring its effect on their health. The experimental results are presented in the form of a graph called the *dose response curve.* Dose represents the degree of exposure of an organism to the substance; response is the observed effect of the substance on the organism.

Dose can be defined as the ratio of the mass of the toxicant to the body weight of the organism over the period of time for which it was administered. Dose is expressed in units of mass of toxicant/unit body weight; e.g., 0.5 g/kg or 10 mg/kg, etc.

The dose response curve is a plot of the percentage of organisms that exhibit a particular adverse response as a function of dose (high levels). The horizontal X axis is a log scale on which dose is plotted, and percent response is plotted on an arithmetic scale on the Y axis. Figure F-1 shows a dose response curve in which the response to a specific dose is the death of the organism.

Two distinctive features can be noted in the dose response curve.

- It is usually an S-shaped plot, which means that for incremental doses, the response does not increase in a linearly proportional manner
- It does not pass through the origin, which may be interpreted to mean that (a) the real response (effects) at low levels of dose are unknown; i.e. it has not been measured, or (b) the response is zero (no effect) for low doses.

The reasons for these ambiguities are related to the ways in which the dose response tests are carried out. Since tests involving an organism's response to varying doses of a toxicant are very expensive, as few tests as possible are conducted, and nearly all of them at higher dosage levels. This leads to extrapolation between the test data points. And, because tests are done at higher dosages, the curve is extrapolated down for lower dosage.

LD_{50} on the Y axis represents a statistical estimate of the dose that will cause death in 50 percent of the subjects; this is also called the *median lethal dose.* Obviously, depending upon the nature of the slope of the curve, dose corresponding to LD_{50} will be different for different organisms, and different toxicants will produce different effects on the same organism. Therefore, the nature of the toxicant and the organism on which the tests were conducted must be specified. Figure F-2 shows the dose response curve for two chemicals, A and B, on an organism. A is more toxic because a dose of 0.07 mg/kg will result in the death of 50 percent of the organisms. Chemical B, on the other hand, produces same mortality rate at a much higher dose of 2 mg/kg. Also, the slope of the curves is another indicator of the relative toxicity of the two chemicals. Chemical A's steeper slope means that a relatively small increase in the dose—from 0.07 to 0.093 mg/kg (33 percent)—will result in an 85 percent mortality rate. But for chemical B, the dose has to increase from 2 to 34 mg/kg, or 1700 percent, to cause the same mortality rate of 85 percent. Curve A is typical of highly toxic substances; B represents substances of low toxicity.

The dose–response relationships are based on laboratory experiments done on animals—typically mice or rats (cheap mammals with relatively short lifespans). The results are then extrapolated for higher life forms, including humans. This is done by using a factor of safety that ranges from 10^3 to 10^5 for humans. For example, if a 1000 ppm

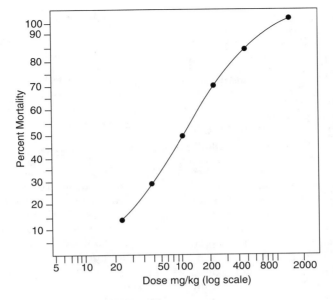

FIGURE F-1   Dose response curve

level of dose of a certain toxicant did not produce any adverse response in rats and is accepted as safe dose for rats, then the safe dose for humans will be

$$1000 \text{ ppm} \div 10^4 = 0.1 \text{ ppm}$$

This conservative value reflects weakness and uncertainty in dose response experimental results.

Other factors that control the nature of the dose response curve include

- Species habitat (for example, aquatic vs. terrestrial)
- Sex
- Genetic traits
- Nutrition level
- Age
- Exposure route
- Exposure duration

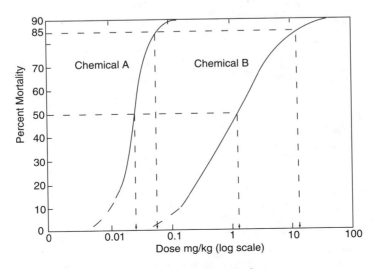

FIGURE F-2   Toxic rating chart

Even in closely related animals, such as mice and rats, one chemical may prove carcinogenic to, say, mice, but not to rats. Ames et al. (1987) tested 226 chemicals on both rats and mice and found that 96 were toxic to mice but not to rats.

Interspecies difference is by far the most serious problem in interpretation of test data. For example, aspirin is safe for people but is a teratogen for rabbits and causes birth defects in their offspring. Dioxin is 5000 times more toxic to hamsters than it is to guinea pigs. Effects on humans are not yet well understood; chloracne, particularly in children, is the only effect known so far.

Exposure path also plays a role; for example, nickel fumes are carcinogenic to humans if inhaled, but are non-carcinogenic if ingested. Another example is that very fine particles of asbestos are carcinogenic, because they can easily enter the lungs by inhalation, but larger particles may be ingested and excreted without causing any adverse effects.

The best way to determine human toxicity of a chemical is to subject human beings to the dose response test. But this is not allowed by law, and rightly so.

The next best option is to study the effects of chemicals on people working in chemical manufacturing plants. This too is limited, and not particularly effective because of the latency effect. These limitations have forced us to rely on laboratory data from tests conducted on animals.

Hypersensitivity and Hyposensitivity.

When an organism develops adverse effects at a very low dose level of a toxicant, it is called *hypersensitive*. When it takes an extremely high dose to produce a response, the organism is called *hyposensitive*. Most organisms, though, respond to toxicants at doses in the mid-range of the dose response curve; such subjects are called *normals*.

EXPOSURE PATHS OF TOXICANTS TO HUMANS

Toxic chemicals in hazardous wastes or in emissions from industrial plants can enter into the environment through several pathways. Figure F-3 shows the physical and biological pathways of transport of toxic substances and their potential for human exposure. Once the toxic substances are mobilized in the environment, they can enter the human body through the following pathways:

- Ingestion
- Inhalation
- Dermal contact
- Injection (rare)

In an industrial setting, inhalation has been found to be the primary route of exposure, followed by skin or eye contact, ingestion, and injection.

Ingestion

Toxic substances may be swallowed with food or drink, or alone. The substance then passes directly

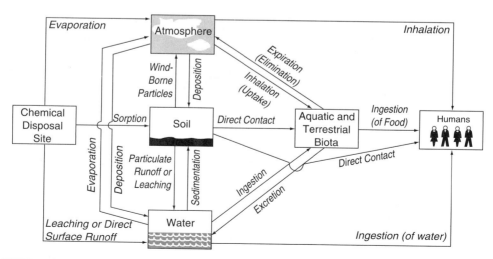

FIGURE F-3 Pathways of transport of harardous substances, their release from disposal sites, and potential for human exposure (adopted from National Research Council, 1991)

into the digestive tract, where it may eventually be absorbed into the blood. The most common ways through which we ingest toxicants are by eating or drinking in a contaminated area, or by consuming contaminated food or water.

Inhalation

Inhalation is absorbing toxic substances through breathing. Gases, vapors, fumes, or mist (all suspensions of fine solids/liquids/gas in one another), when inhaled, get into the respiratory system, where they are absorbed into the blood in the lung through alveoli (small air sac clusters).

Dermal Contact

Dermal contact involves direct contact of hazardous substances with the skin. Dermal exposure may affect only the contact point, or it may cause the toxicant to be absorbed into the skin and subsequently into the bloodstream.

Sometimes certain hazardous materials may come in direct contact with the eyes (ocular exposure), causing eye damage.

Injection

In injection, hazardous substances enter the bloodstream directly via breakage of skin or tissues—cuts and abrasions.

TERATOGENESIS, MUTAGENESIS, AND CARCINOGENESIS

Teratogenesis

Teratogenesis (birth defects) affects offspring while they are still in the fetal stage of development. Teratogenesis usually results from damage to embryonic or fetal cells, inhibiting their normal development. The first 8–12 weeks of pregnancy are the most critical in humans because most defects occur in this period. Chemicals that cause birth defects are called teratogens. Teratogenesis is caused by enzyme inhibition, deprivation of vitamins to the fetus, alteration of energy supply, or alteration of the permeability of the placental membrane.

Mutagenesis

Mutagenesis is the process of alteration of DNA, which causes genetic damage to reproductive cells. Birth defects are a common manifestation of muta-

gens. Such effects are usually felt in the next generation (e.g., mutagenesis due to radiation exposure).

Carcinogenesis

Carcinogenesis is the process of uncontrolled cell replication (cancer) due to the presence of a toxicant, called a carcinogen, in the body. Many hazardous substances are potentially carcinogenic. The problem is determination of the carcinogenic effects of a large number of chemicals. The disturbing fact is that in the EPA's list of 48,500 chemicals, only about one-fifth (<10,000) have been tested for acute effects, and fewer than one-tenth (4850) have been tested for chronic effects (for example, carcinogenesis, teratogenesis, or mutagenesis). This means that no information on the toxic effects of 38,300 chemicals (79 percent) is available (Postel, 1988).

It is believed that only 15 percent of cancer cases are related to genetic effects and 85 percent to environmental effects.

ASSESSMENT OF HEALTH RISKS OF HAZARDOUS WASTE

Because hazardous wastes contain all kinds of chemicals, they pose serious risk to all life forms, including human health. The first task in assessment of health risk is to find out what is present in the waste. If accurate records of what has been placed at the waste site are available, it is of some help. But complications arise because of the chemical and biological transformation of wastes, selective migration of waste, and interaction of various waste materials to generate other products. Toxicologists know that when two toxic substances combine, the toxicity of the product is not a simple function of the toxicity of the individual components. Terms like synergism, antagonism, and potentiation are used to describe the combined effects of two or more toxic substances.

Synergism relates to the situation in which the combined effect of two chemicals is much greater than the effect of each chemical alone; stated another way, 2 + 3 = 8. For example, both carbon tetrachloride and ethanol are hepatotoxic agents, but together they produce much greater damage to liver than the mathematical sum of their individual effects on the liver would suggest.

Antagonism can be described as the situation in which two chemicals, when given together, interfere with each other's actions, producing an effect that is less than the combined effect of the individual chemicals (for example, $4 + 3 = 5$ or $4 + 0 = 1$). Antagonism is of special interest in toxicology because it constitutes the basis of antidotes. The toxic effect of heavy metals, such as lead, mercury, and arsenic, can be reduced by using dimercaprol.

Potentiation is the situation where a substance does not produce a toxic effect on a certain organ or system, but, when added to another chemical, makes the latter more toxic (for example, $0 + 2 = 5$). Isopropanol, for instance, is not hepatotoxic by itself, but when it is added to carbon tetrachloride, the hepatotoxicity of carbon tetrachloride becomes much greater than when it is not given with isopropanol.

REFERENCES CITED

Ames, B.N., R. Magaw, and L.S. Gold, 1987, Ranking possible carcinogenic hazards: Science, v. 236, p. 271-280.

Doull, J., C.D. Klaassen, and M.O. Amdur, eds., 1980, Casarett and Doull's Toxicology (2nd ed.): New York, Macmillan Publishing Company, 778 p.

National Research Council, 1991, Environmental Epidemiology: Washington, DC, National Academy Press, p. 118.

Postel, S., 1988, Controlling toxic chemicals, *in* Brown, L.R., et al. eds., State of the World 1988: New York, W.W. Norton & Company, p. 120.

APPENDIX G

Hazardous Waste Exchanges in the United States, Canada, and Taiwan*

U.S.A.

Alkem
25 Glendale Rd.
Summit, NJ 07901
(201) 277-0060

California Waste Exchange
Department of Health Services
Toxic Substances Control Division
Alternative Technology Section
714/744 P Street
PO Box 942732
Sacramento, CA 94234-7320
(916) 324-1807

Enstar Corporation
PO Box 189
Latham, NY 12110
(518) 785-0470

Great Lakes Waste Exchange
400 Ann Street, N.W., Suite 201-A
Grand Rapids, MI 49504-2054
(616) 363-3262

ICM Chemical
20 Cordova Street, Suite 3
St. Augustine, FL 32084
(904) 824-7247

Idaho Waste Exchange
Idaho Department of Environmental Quality
Hazardous Materials Bureau
450 West State Street
Boise, ID 83720
(208) 334-5879

Indiana Waste Exchange
Purdue University
School of Civil Engineering
West Lafayette, IN 47907
(317) 494-5063

Industrial Materials Exchange (IMEX)
Seattle–King County Environmental Health
172 20th Avenue
Seattle, WA 98122
(206) 296-4633
Fax: (206) 296-0188

Industrial Material Exchange Service (IMES)
PO Box 19276
2200 Churchill Road, #24
Springfield, IL 62794-0276
(217) 782-0450
Fax: (217) 524-4193

Industrial Waste Information Exchange
New Jersey Chamber of Commerce
5 Commerce Street
Newark, NJ 07102
(201) 623-7070

Montana Industrial Waste Exchange
Montana Chamber of Commerce
PO Box 1730
Helena, MT 59624
(406) 442-2405

Northeast Industrial Waste Exchange (NIWE)
90 Presidential Plaza, Suite 122
Syracuse, NY 13210
(315) 422-6572
Fax: (315) 442-9051

Ore Corp.
2415 Woodmere Dr.
Cleveland, OH 44106
(216) 371-4869

Pacific Materials Exchange (PME)
S. 3707 Godfrey Blvd.
Spokane, WA 99204
(509) 623-4244

*Adapted from U.S. EPA, 1988, and Goldman et al., 1986. See "References Cited," Ch. 6.

Resource Exchange Network for
Eliminating Waste (RENEW)
Texas Water Commission
PO Box 13087
Austin, TX 78711-3087
(512) 463-7773
Fax: (512) 463-8317

San Francisco Waste Exchange
2524 Benvenue #435
Berkeley, CA 94704
(415) 548-6659

Southeast Waste Exchange (SEWE)
Urban Institute
Department of Civil Engineering
University of North Carolina
Charlotte, NC 28223
(704) 547-2307

Southern Waste Information Exchange (SWIX)
PO Box 960
Tallahassee, FL 32302
(800) 441-7949
(904) 644-5516
Fax: (904) 574-6704

Techrad Industrial Waste Exchange
4619 North Santa Fe
Oklahoma City, OK 73118
(405) 528-7016

Wastelink Division of TENCON, Inc.
140 Wooster Pike
Milford, OH 45150
(513) 248-0012
Fax: (513) 248-1094

Zero Waste Systems
2928 Poplar Street
Oakland, CA 94608
(415) 893-8257

CANADA

Alberta Waste Materials Exchange
Alberta Research Council

PO Box 8330, Postal Station F
Edmonton, AB T6H 5X2, Canada
(403) 450-5408

British Columbia Waste Exchange
2150 Maple Street
Vancouver, BC V6J 3T3, Canada
(604) 731-7222

Canadian Chemical Exchange
PO Box 1135
Ste-Adele, PQ J0R 1L0, Canada
(514) 229-6511

Canadian Waste Materials Exchange (CWME)
ORTECH International
Sheridan Park Research Community
2395 Speakman Drive
Mississauga, ON L5K 1B3, Canada
(416) 822-4111, ext. 265

Manitoba Waste Exchange
c/o Biomass Energy Institute, Inc.
1329 Niakwa Road
Winnipeg, MB R2J 3T4, Canada
(204) 257-3891

Ontario Waste Exchange
ORTECH International
Sheridan Park Research Community
2395 Speakman Drive
Mississauga, ON L5K 1B3, Canada
(416) 822-4111, ext. 512

Peel Regional Waste Exchange
Regional Municipality of Peel
10 Peel Center Drive
Brampton, ON L6T 4B9, Canada
(419) 791-9400

TAIWAN

Union Chemical Laboratories
Industrial Technology Research Institute
321, Kuang Fu Road, Sec. 2
Hsinchu, Taiwan (Republic of China) 30042

APPENDIX H

Hazardous Waste Information Exchanges in the United States*

California Waste Exchange
Department of Health Services
Toxic Substance Control Division
714/744 P Street
Sacramento, CA 95814
(916) 324-1818

Colorado Waste Exchange
Colorado Association of Commerce and
   Industry
1390 Logan Street
Denver, CO 80203
(303) 831-7411

World Association for Safe Transfer and Exchange
130 Freight Street
Waterbury, CT 06702
(203) 574-2463

Southern Waste Information Exchange
Box 6437
Tallahassee, FL 32313
(904) 644-5516

Georgia Waste Exchange
Business Council of Georgia
181 Washington Street SW
Atlanta, GA 30303
(404) 223-2264

Industrial Material Exchange Service
Illinois State Chamber of Commerce
IEPA-D LPC-24
2200 Churchill Rd.
Springfield, IL 62706
(217) 782-0450

Louisville Area Industrial Waste Exchange
Louisville Chamber of Commerce
1 Riverfront Plaza, 4th Floor
Louisville, KY 40202
(502) 566-5000

Great Lakes Regional Waste Exchange
Waste Systems Institute of Michigan, Inc.
3250 Townsend NE
Grand Rapids, MI 49505
(616) 363-7367

Midwest Industrial Waste Exchange
Regional Commerce and Growth Association
10 Broadway
St. Louis, MO 63102
(314) 231-5555

New England Materials Exchange
34 North Main Street
Farmington, NH 03835
(603) 755-4442
(603) 755-9962

New Jersey State Waste Exchange
New Jersey Chamber of Commerce
5 Commerce Street
Newark, NJ 07102
(201) 623-7070

Industrial Commodities
Bulletin, Enkarn Corp.
Box 590
Albany, NY 12210
(518) 436-9684

Northeast Industrial Waste Exchange
90 Presidential Plaza, Suite 22
Syracuse, NY 13202
(315) 422-6572
(315) 474-4201

Piedmont Waste Exchange
Urban Institute
University of North Carolina
Charlotte, NC 28223
(704) 597-2307

*Adapted from Goldman et al., 1986. See "References Cited," Ch. 6.

Tennessee Waste Exchange
Tennessee Manufacturers Association
501 Union Street, Suite 601
Nashville, TN 28223
(615) 256-5141

Chemical Recycle Information Program
Houston Chamber of Commerce
110 Milam Bldg., 25th Floor
Houston, TX 77002
(713) 658-2462
(713) 658-2459

Inter-Mountain Waste Exchange
W.S. Hatch Co.
643 South 800 West
Woods Cross, UT 84087
(801) 295-5511

APPENDIX I

Groundwater Monitoring Constituents
(Appendix IX, 40, CFR, 1990)

Common Name
Acenaphthene
Acenaphthylene
Acetone
Acetophenone
Acetonitrile (methyl cyanide)
2-Acetylaminofluorene (2-AAF)
Acrolein
Acrylonitrile
Aldrin
Allyl chloride
4-Aminobiphenyl
Aniline
Anthracene
Antimony
Aramite
Arsenic
Barium
Benzene
Benzo[a]anthracene (benzathracene)
Benzo[b]fluoranthene
Benzo[k]fluoranthene
Benzo[ghi]perylene
Benzo[a]pyrene
Benzyl alcohol
Beryllium
alpha-BHC
beta-BHC
delta-BHC
gamma-BHC; Lindane
Bis(2-chloroethoxy)methane
Bis(2-chloroethyl)ether
Bis(2-chloro-1-methylethyl) ether; 2,2'-Dichlorodiisopropyl ether
Bis(2-ethylhexyl)phthalate
Bromodichloromethane
Bromoform (tribromomethane)

Common Name
4-Bromophenyl phenyl ether
Butyl benzyl phthalate (benzyl butylphthalate)
Cadmium
Carbon disulfide
Carbon tetrachloride
Chlordane
p-Chloroaniline
Chlorobenzene
Chlorobenzilate
p-Chloro-m-cresol
Chloroethane (ethyl chloride)
Chloroform
2-Chloronaphthalene
2-Chlorophenol
4-Chlorophenyl phenyl ether
Chloroprene
Chromium
Chrysene
Cobalt
Copper
m-Cresol
o-Cresol
p-Cresol
Cyanide
2,4-D (2,4-dichlorophenoxyacetic acid)
4,4'-DDD
4,4'-DDE
4,4'-DDT
Diallate
Dibenz[a,h]anthracene
Dibenzofuran
Dibromochloromethane; chlorodibromomethane
1,2-Dibromo-3-chloropropane (DBCP)

Common Name	Common Name
1,2-Dibromethane (ethylene dibromide)	Hexachlorobutadiene
Di-*n*-butyl phthalate	Hexachlorocyclopentadiene
o-Dichlorobenzene	Hexachloroethane
m-Dichlorobenzene	Hexachlorophene
p-Dichlorobenzene	Hexachloropropene
3,3'-Dichlorobenzidine	2-Hexanone
trans-1,4-Dichloro-2-butene	
Dichlorodifluoromethane	Indeno(1,2,3-cd)pyrene
1,1-Dichloroethane	Isobutyl alcohol
1,2-Dichloroethane (ethylene dichloride)	Isodrin
1,1-Dichloroethylene (vinylidene chloride)	Isophorone
trans-1,2-Dichloroethylene	Isosafrole
2,4-Dichlorophenol	
2,6-Dichlorophenol	Kepone
1,2-Dichloropropane	
cis-1,3-Dichloropropane	Lead
trans-1,3-Dichloropropene	
Dieldrin	Mercury
Diethyl phthalate	Methacrylonitrile
O,O-Diethyl O-2 pyrazinyl	Methapyrilene
phosphorothioate (thionazin)	Methoxychlor
Dimethoate	Methyl bromide (bromomethane)
p-(Dimethylamino)azobenzene	Methyl chloride (chloromethane)
7,12-dimethylbenz[a]anthracene	3-Methylchlolanthrene
3,3'-Dimethylbenzidine	Methylene bromide (dibromomethane)
alpha, alpha-Dimethylphenethylamine	Methylene chloride (dichloromethane)
2,4-Dimethylphenol	Methyl ethyl ketone (MEK)
Dimethyl phthalate	Methyl iodide (iodomethane)
m-Dinitrobenzene	Methyl methacrylate
4,6-Dinitro-*o*-cresol	Methyl methanesulfonate
2,4-Dinitrophenol	2-Methylnaphthalene
2,4-Dinitrotoluene	Methyl parathion (parathion methyl)
2,6-Dinitrotoluene	4-Methyl-2-pentanone (methyl isobutyl ketone)
Dinoseb (DNBP or 2-sec-Butyl-4,6-dinitrophenol)	
Di-*n*-octyl phthalate	Naphthalen
1,4-Dioxane	1,4-Naphthoquinone
Diphenylamine	1-Naphthylamine
Disulfoton	2-Naphthylamine
	Nickel
Endosulfan I	*o*-Nitroaniline
Endosulfan II	*m*-Nitroaniline
Endosulfan sulfate	*p*-Nitroaniline
Endrin	Nitrobenzene
Endrin aldehyde	*o*-Nitrophenol
Ethylbenzene	*p*-Nitrophenol
Ethyl methacrylate	4-Nitroquinoline 1-oxide
Ethyl methanesulfonate	N-Nitrosodi-*n*-butylamine
	N-Nitrosodiethylamine
Famphur	N-Nitrosodimethylamine
Fluoranthene	N-Nitrosodiphenylamine
Fluorene	N-Nitrosodipropylamine (di-*n*-propylnitrosamine)
	N-Nitrosomethylethylamine
Heptachlor	N-Nitrosomorpholine
Heptachlor epoxide	
Hexachlorobenzene	

Common Name	**Common Name**
N-Nitrosopiperidine	1,2,4,5-Tetrachlorobenzene
N-Nitrosopyrrolidine	1,1,1,2-Tetrachloroethane
	1,1,2,2-Tetrachloroethane
5-Nitro-*o*-toluidine	Tetrachloroethylene (perchloroethylene
	or tetrachloroethene)
Parathion	2,3,4,6-Tetrachlorophenol
Polychlorinated biphenyls (PCBs)	Tetraethyl dithiopyrophosphate
Polychlorinated dibenzo-*p*-dioxins (PCDDs)	(Sulfotepp)
Polychlorinated dibenzofurans (PCDFs)	Thallium
Pentachlorobenzene	Tin
Pentachlorethane	Toluene
Pentachloronitrobenzene	*o*-Toluidine
Pentachlorophenol	Toxaphene
Phenacetin	1,2,4-Trichlorobenzene
Phenanthrene	1,1,1-Trichloroethane (methylchloroform)
Phenol	1,1,2-Trichloroethane
p-Phenylenediamine	Trichloroethylene (trichloroethene)
Phorate	Trichlorofluoromethane
2-Picoline	2,4,5-Trichlorophenol
Pronamide	2,4,6-Trichlorophenol
Propionitrile (ethyl cyanide)	1,2,3-Trichloropropane
Pyrene	O,O,O-Triethyl phosphorothioate
Pyridine	sym-Trinitrobenzene
Safrole	Vanadium
Selenium	Vinyl acetate
Silver	Vinyl chloride
Silvex (2,4,5-TP)	
Styrene	Xylene (total)
Sulfide	
	Zinc
2,4,5-T (2,4,5-Trichlorophenoxyacetic acid)	
2,3,7,8-TCDD (2,3,7,8-Tetrachlorodibenzo-*p*-dioxin)	

APPENDIX J

Sources of Geologic and Related Information

U.S. GEOLOGICAL SURVEY AND OTHERS

Public Inquiries Offices (PIOs)

ALASKA
PIO, USGS
108 Skyline Building
508 2nd Avenue
Anchorage, AL 99501
(907) 277-0577

CALIFORNIA
PIO, USGS
7638 Federal Building
300 North Los Angeles Street
Los Angeles, CA 90012
(213) 688-2850

PIO, USGS
504 Custom House
555 Battery Street
San Fransicso, CA 94111
(415) 556-5627

COLORADO
PIO, USGS
1012 Federal Building
1961 Stout Street
Denver, CO 80202
(303) 837-4160

TEXAS
PIO, USGS
Room 1 C45
1100 Commerce Street
Dallas, TX 75242
(214) 749-3230

UTAH
PIO, USGS
8102 Federal Office Building
125 South State Street
Salt Lake City, UT 84138
(801) 524-2524

WASHINGTON
PIO, USGS
678 U.S. Court House Building
West 920 Riverside Avenue
Spokane, WA 99201
(509) 456-2524

DISTRICT OF COLUMBIA
PIO, USGS
1028 GSA Building
19th and F Streets NW
Washington DC 20244
(202) 343-8073

Distribution Offices

U.S. Geological Survey
Branch of Distribution
Eastern Region
1200 South Eads Street
Arlington, VA 22202

U.S. Geological Survey
Branch of Distribution
Western Region
Building 41, Federal Center
Denver, CO 80225

U.S. Geological Survey

National Center
12201 Sunrise Valley Road
Reston, VA 22092
(703) 648-7411
Fax: (703) 648-4466

National Technical Information Service (NTIS)

U.S. Department of Commerce
Springfield, VA 22151

Superintendent of Documents

U.S. Government Printing Office
Washington, DC 20402

Regional Engineering Offices (USGS)

Atlantic Region Engineer
U.S. Geological Survey
1109 N. Highland Street
Arlington, VA 22210

Central Region Engineer
U.S. Geological Survey
Box 133
Rolla, MO 65401

Rocky Mountain Region Engineer
U.S. Geological Survey
Building 25, Federal Center
Denver, CO 80225

Pacific Region Engineer
U.S. Geological Survey
345 Middlefield Road
Menlo Park, CA 94025

Cartographic Division
Soil Conservation Service

Federal Center Building
Hyattsville, MD 20782

National Ocean Survey

Department of Commerce
Washington Science Center
Rockville, MD 20852

AERIAL PHOTOS AND REMOTE SENSING IMAGERY

ERTS-1, Landsat, Skylab, High-Altitude NASA Stereo Photos

EROS Data Center, USGS
Sioux Falls, SD 57198
(605) 594-6511

Landsat from Foreign Countries

Integrated Satellite Information
Box 1630
Prince Albert, Saskatchewan S6U 5T2
Canada

Instituto do Pesquisas Especiais (INPE)
Attn: Divisão de Banco de Dados
Av. dos Astronautas 1758
Caixa Postal 515
12.200 São Jose Dos Campos, SP
Brazil

Department of Forestry and Agriculture
Building 810
Pleasantville
St. Johns, Newfoundland A1A 1P9
Canada

Telespzaio—SES
Corso D'Italia
L-3 Roma
Italy

Aerial Photographs (Generally at 1:20,000) of the United States and Possessions

USGS: National Cartographic Information Center (NCIC) provides catalogs

NCIC-Reston (Headquarters)
USGS, 507 National Center
Reston, VA 22092
(703) 860-6045

NCIC-Mid-Continent
USGS, 1400 Independence Road
Rolla, MO 65401
(314) 364-3680, ext. 107

NCIC-Rocky Mountain
USGS Topographic Division
Stop 510, Box 25046
Denver Federal Center
Denver, CO 80225
(303) 232-2326

NCIC-Western
USGS, 345 Middlefield Road
Menlo Park, CA 94025
(415) 323-2427

SCS provides a catalog, "Status of Aerial Photograph"

U.S. Department of Agriculture
Soil Conservation Service
Cartographic Division
6505 Belcrest Road
Hyattsville, MD 20782
(301) 436-8756

Agricultural Stabilization Conservation Service (ASCS)
Aerial Photography Division
ASCS-USDA
2505 Parley's Way
Salt Lake City, UT 84109
(801) 524-5856

Aerial Photos with Some Color and Infrared Coverage in the United States

U.S. Department of Agriculture
Forest Service
PO Box 2417
Washington, DC 20013
(707) 235-8638

NOAA, National Ocean Survey
6001 Executive Boulevard
Rockville, MD 20852
Attn: Coastal Mapping Division
(301) 443-8601

Old Aerial Photos

National Archives (GSA)
Cartographic Records Division
Washington, DC 20408
(202) 962-0173

Outside the United States, photos may be available from the USAF Flights (country releases required from diplomatic representatives)

Photographic Records and Services Division
Aeronautical Chart and Information Center
2nd & Arsenal Street
St. Louis, MO 63118

Defense Mapping Agency
Building 56
U.S. Naval Observatory
Washington, DC 20305
(202) 254-4406

SLAR Imagery

Repository for Air Force radar imageries from certain programs:
Goodyear Aerospace Corporation
Litchfield Park, AZ 85340
(Attn: SLAR Imagery Depository)

National Cartographic Information Center
Reston, VA

Westinghouse Electric Corp.
Philadelphia, PA

Sanborn Maps and Information Service
629 5th Avenue
Pelham, N.Y. 10022

State Geological Surveys in the United States

ALABAMA
Alabama Geological Survey
PO Box O
Tuscaloosa, AL 35486-9780
(205) 349-2852
Fax: (205) 349-2861

ALASKA
Alaska Geological Survey
794 University Avenue
Suite 200
Fairbanks, AK 99709-3645
(907) 474-7147
Fax: (907) 479-4779

ARIZONA
Arizona Geological Survey
845 N. Park Avenue, #100
Tucson, AZ 85719-4816
(602) 882-4795
Fax: (602) 628-5106

ARKANSAS
Arkansas Geological Commission
3815 West Roosevelt Road
Little Rock, AR 72204
(501) 324-9165
Fax: (501) 663-7360

CALIFORNIA
Department of Conservation
Division of Mines and Geology
Division Headquarters
801 K Street, MS 12-30
Sacramento, CA 95814-3500
(916) 445-1923
Fax: (916) 445-5718

COLORADO
Colorado Geological Survey
1313 Sherman Street, Room 715
Denver, CO 80203
(303) 866-2611
Fax: (303) 866-2461

CONNECTICUT
Connecticut Geological and Natural History Survey
Department of Environmental Protection
Natural Resources Center
79 Elm Street
Store Level Floor
Hartford, CT 06106
(203) 566-3540
Fax: (203) 566-7292

DELAWARE
Delaware Geological Survey
University of Delaware
DGS Building
Newark, DE 19716-7501
(302) 831-2833
Fax: (302) 831-3579

FLORIDA
Florida Geological Survey
Gunter Building
903 W. Tennessee Street
Tallahassee, FL 32304-7700
(904) 488-4191
Fax: (904) 488-8086

GEORGIA
Georgia Geologic Survey
Environmental Protection Division
Room 400
19 Martin Luther King, Jr., Drive, S.W.
Atlanta, GA 30334
(404) 656-3214
Fax: (404) 651-9425

HAWAII
Department of Land and Natural Resources
Division of Land and Water Development
PO Box 373
Honolulu, HI 96809
(808) 587-0230
Fax: (808) 587-0283

IDAHO
Idaho Geological Survey
Room 332
Morrill Hall
University of Idaho
Moscow, ID 83843
(208) 885-7991
Fax: (208) 885-5826

ILLINOIS
Illinois Geological Survey
121 Natural Resources Building
615 East Peabody Drive
Champaign, IL 61820
(217) 333-4747
Fax: (217) 244-7004

INDIANA
Indiana Geological Survey
611 N. Walnut Grove
Bloomington, IN 47405
(812) 855-5067
Fax: (812) 855-2862

IOWA
Iowa Geological Survey Bureau
Department of Natural Resources
109 Trowbridge Hall
Iowa City, IA 52242-1319
(319) 335-1575
Fax: (319) 335-2754

KANSAS
Kansas Geological Survey
1930 Constant Avenue, West Campus
The University of Kansas
Lawrence, KS 66047
(913) 864-3965
Fax: (913) 864-5317

KENTUCKY
Kentucky Geological Survey
228 Mining and Mineral Resources Building
University of Kentucky
Lexington, KY 40506-0107
(606) 257-5500
Fax: (606) 257-1147

LOUISIANA
Louisiana Geological Survey
PO Box G
University Station
Baton Rouge, LA 70893
(504) 388-5320
Fax: (504) 388-5328

MAINE
Maine Geological Survey
Department of Conservation
State House Station #22
Augusta, ME 04333
(207) 289-2801
Fax: (207) 289-2353

MARYLAND
Maryland Geological Survey
2300 St. Paul Street
Baltimore, MD 21218-5210
(410) 554-5500
Fax: (410) 554-5502

MASSACHUSETTS
Massachusetts Geological Survey
Commonwealth of Massachusetts
Executive Environmental Affairs
100 Cambridge Street, 20th Floor
Boston, MA 02202
(617) 727-9800, ext. 213
Fax: (617) 727-2754

MICHIGAN
Michigan Geological Survey
Department of Natural Resources
Geological Survey Division
Box 30256
Lansing, MI 48917
    also
735 E. Hazel Street
Lansing, MI 48912
(517) 334-6907
Fax: (517) 334-6038

MINNESOTA
Minnesota Geological Survey
University of Minnesota
2642 University Avenue
St. Paul, MN 55114-1057
(612) 627-4780
Fax: (612) 627-4778

MISSISSIPPI
Mississippi Office of Geology
Department of Environmental Quality
PO Box 20307
Jackson, MS 39289-1307
(601) 961-5500
Fax: (601) 961-5521

MISSOURI
Missouri Geological Survey
Department of Natural Resources
Division of Geology and Land Survey
PO Box 250
Rolla, MO 65401
(314) 368-2100
Fax: (314) 368-2111

MONTANA
Montana Bureau of Mines and Geology
1300 West Park Street
Montana Tech
Main Hall
Butte, MT 59701-8997
(406) 496-4180
Fax: (406) 496-4451

NEBRASKA
Nebraska Geological Survey
Conservation and Survey Division
University of Nebraska
113 Nebraska Hall
901 N. 17th Street
Lincoln, NE 68588-0517
(402) 472-3471
Fax: (402) 472-2410

NEVADA
Nevada Geological Survey
Bureau of Mines and Geology
University of Nevada
Reno, NV 89557-0088
(702) 784-6691
Fax: (702) 784-1709

NEW HAMPSHIRE
New Hampshire Geological Survey
Department of Environmental Services
PO Box 2008
Concord, NH 03302-2008
(603) 271-3406
Fax: (603) 271-6588

NEW JERSEY
New Jersey Geological Survey
Division of Science and Research
Department of Environmental Protection and Energy
PO Box CN-427
Trenton, NJ 08625
(609) 292-1185
Fax: (609) 633-1004

NEW MEXICO
New Mexico Bureau of Mines/Mineral
Resources
Campus Station
Socorro, NM 87801
(505) 835-5420
Fax: (505) 835-6333

NEW YORK
New York Geological Survey
State Museum
Empire State Plaza
3136 Cultural Education Center
Albany, NY 12230
(518) 474-5816
Fax: (518) 473-8496

NORTH CAROLINA
North Carolina Geological Survey
Department of Environment Health
  and Natural Resources
Division of Land Resources
PO Box 27687
Raleigh, NC 27611-7687
(919) 733-3833
Fax: (919) 733-4407

NORTH DAKOTA
North Dakota Geological Survey
600 East Boulevard
Bismarck, ND 58505-0840
(701) 224-4109
Fax: (701) 224-3682

OHIO
Ohio Geological Survey
Department of Natural Resources
4383 Fountain Square Drive
Building B
Columbus, OH 43224
(614) 265-6988
Fax: (614) 447-1918

OKLAHOMA
Oklahoma Geological Survey
100 East Boyd
Room N-131
Norman, OK 73019-0628
(405) 325-3031
Fax: (405) 325-7069

OREGON
Oregon Department of Geology and Mineral Industries
Suite 965
800 N.E. Oregon Street #28
Portland, OR 97232
(503) 731-4100
Fax: (503) 731-4066

PENNSYLVANIA
Pennsylvania Geological Survey
Bureau of Topographic and Geologic Survey
Department of Environmental Resources
PO Box 8453
Harrisburg, PA 17105-8453
(717) 787-2169
Fax: (717) 783-7267

RHODE ISLAND
Office of the Rhode Island State Geologist
Department of Geology
University of Rhode Island
315 Green Hall
Kingston, RI 02881
(401) 792-2265
Fax: (401) 792-2190

SOUTH CAROLINA
South Carolina Geological Survey
5 Geology Road
Columbia, SC 29210-4089
(803) 896-7700
Fax: (803) 896-7695

SOUTH DAKOTA
South Dakota Geological Survey
Department of Environment and Natural Resources
University of South Dakota Science Center
414 Clark Street
Vermillion, SD 57069-2390
(605) 677-5227
Fax: (605) 677-5895

TENNESSEE
Tennessee Division of Geology
Department of Environment and Conservation
Division of Geology
401 Church Street
Life and Casualty Tower
Nashville, TN 37243-0445
(615) 532-1500
Fax: (615) 532-0231

TEXAS
Texas Bureau of Economic Geology
The University of Texas at Austin
Box X, University Station
Austin, TX 78713-7508
(512) 471-7721
(512) 471-1534
Fax: (512) 471-0140

UTAH
Utah Geological Survey
2363 South Foothill Drive
Salt Lake City, UT 84109-1491
(801) 467-7970
Fax: (801) 467-4070

VERMONT
Vermont Geological Survey
Agency of Natural Resources
103 South Main Street
Center Building
Waterbury, VT 05671-0301
(802) 241-3601
Fax: (802) 244-1102

VIRGINIA
Virginia Department of Mines, Minerals, and Energy
Division of Mineral Resources
Natural Resources Building
McCormick Road
PO Box 3667
Charlottesville, VA 22903
(804) 293-5121
Fax: (804) 293-2239

WASHINGTON
Washington Geological Survey
Department of Natural Resources
Geology and Earth Resources Division
1111 Washington Street, S.E.
PO Box 47007
Olympia, WA 98504-7007
(206) 902-1450
Fax: (206) 902-1785

WEST VIRGINIA
West Virginia Geological Survey
Mont Chateau Research Center
PO Box 879
Morgantown, WV 26507-0879
(304) 594-2331
Fax: (304) 594-2575

WISCONSIN
Wisconsin Geological Survey
3817 Mineral Point Road
Madison, WI 53705-5100
(608) 262-1705
Fax: (608) 262-8086

WYOMING
Geological Survey of Wyoming
Box 3008
University Station
Laramie, WY 82071-3008
(307) 766-2286
Fax: (307) 766-2605

GUAM
Department of Agriculture
PO Box 2950
Agana, GU 96910
(671) 734-3948

NORTHERN MARIANA ISLANDS
Natural Resources Department
Office of the Governor
Saipan, MP 96950
(670) 322-9830

PUERTO RICO
Department of Natural Resources
PO Box 5887
Puerta de Tierra Station
San Juan, PR 00906
(809) 722-2526
Fax: (809) 724-0365

U.S. VIRGIN ISLANDS
Department of Planning and Natural Resources
Nisky Center, Suite 231
St. Thomas, VI 00802
(809) 774-3320

CANADIAN GEOLOGICAL SURVEYS

National

Geological Survey of Canada
Room 215, 601 Booth Street
Ottawa, Ontario
K1A 0E8, Canada
(613) 992-5910

PROVINCIAL

ALBERTA
Geological Survey
Box 8330 Postal Station F
Edmonton, Alberta
T6H 5X2, Canada
(403) 438-7615
Fax: (403) 438-3364

BRITISH COLUMBIA
Geological Survey Branch
Fifth Floor, 1810 Blanshard Street
Victoria, British Columbia
V8V 1X4, Canada
(604) 952-0372
Fax: (604) 952-0371

MANITOBA
Geological Services Branch
535-330 Graham Avenue
Winnipeg, Manitoba
R3C 4E3, Canada
(204) 945-6559
Fax: (204) 945-1406

NEW BRUNSWICK
Geological Surveys Branch
Box 6000
Fredericton, New Brunswick
E3B 5H1, Canada
(506) 453-2206
Fax: (506) 453-3671

NEWFOUNDLAND AND LABRADOR
Geological Survey Branch
Box 8700
St. John's, Newfoundland
A1B 4J6, Canada
(709) 729-6487
Fax: (709) 729-3493

NORTHWEST TERRITORIES
Geology Division
Box 1500
Yellowknife, Northwest Territories
X1A 2RC, Canada
(403) 920-8210
Fax: (403) 873-5763

NOVA SCOTIA
Department of Mines and Energy
Box 689
Halifax, Nova Scotia
B3J 2T9, Canada
(902) 424-4161
Fax: (902) 424-7735

ONTARIO
Geological Survey
933 Ramsey Lake Road
Sudbury, Ontario
P3E 6B5, Canada
(705) 670-5866
Fax: (705) 670-5759

QUEBEC
l'Energie et des Resources
5700 4e Avenue Ouest
bur. A-211
Charlesbourg, Quebec
G1H 6R1, Canada
(418) 646-2707
Fax: (418) 643-2816

SASKATCHEWAN
Geology and Mines Division
1914 Hamilton Street
Regina, Saskatchewan
S4P 4V4, Canada
(306) 787-2476
Fax: (306) 787-7338

YUKON TERRITORY
Exploration and Geological Services
200 Range Road
Whitehorse, Yukon Territory
V1A 3V1, Canada
(403) 667-3201
Fax: (403) 668-2176

GLOSSARY

Abiotic   Without life; any system characterized by a lack of living organisms.

Absorption   Taking up, incorporation, or assimilation, as of liquids in solids or of gases in liquids.

Acclimate   Process requiring adjustment of microorganisms to the new environment after they are introduced at a remediation site.

Accumulator plants   Plants that accumulate high concentrations of certain metals in their systems.

Activated carbon   Finely powdered charcoal, having very large surface area, used for treatment of liquid waste.

Active exchange   An organization that arranges for the transfer of hazardous waste from a generator to a manufacturer to be used as raw material.

Active fault   A fault in which the latest movement(s) occurred anytime between the present and the past 10,000 years.

Acutely hazardous waste   Wastes listed in 40 CFR 261.31; are followed by the symbol H, including all P-code wastes, F022-F023 and F026-F028 wastes.

Adipose tissues   A type of connective tissue that stores fats (lipids); based on their gross appearance, there are two major types of adipose tissues: white and brown.

Adsorption   Adherence of ions or molecules in solutions to the surface of solids.

Advection   Process by which dissolved substances are transported along with the bulk of the moving fluid, such as groundwater.

Aerobic microorganisms   Microorganisms that grow in the presence of oxygen.

Agglomeration   Process that results in the collection of small solid particles to form a large mass.

Air sparging   A hazardous waste remediation technique in which air is introduced into the contaminated medium to remove hazardous constituents.

Air stripping   Use of air to transfer dissolved volatile chemicals from a liquid hazardous waste into a gaseous phase; a common method of treatment of groundwater contaminated with volatile organic compounds.

Aliphatic organic compounds   Hydrocarbon compounds having an open-chain structure, e.g., methane, ethane, and propane.

Aliquot   Definite proportion of a given quantity, e.g., a portion from a large sample.

Alkane   An organic compound composed of a straight chain of carbon atoms bound on all sides by hydrogen atoms and not containing any double bonds between carbon atoms, e.g., methane.

Alluvial   Sediment formed by water action.

Anaerobic bacteria   Microorganisms that grow in the absence of oxygen; some may even be killed by oxygen.

Analytes   Chemical elements and compounds of interest that should be determined in a chemical analysis procedure.

Andesite   Dark-colored, fine-grained extrusive rock, the extrusive equivalent of diorite.

Angle of internal friction   The arc tan value of the slope of the line on a graph showing the relationship between shear strength and the normal stress for a soil.

Anisotropism   Property of earth materials that results in different values for a parameter depending on the direction of measurement.

Anthropogenic   Anything related to human activity (as opposed to natural).

Anthropomorphic effects   Effects of human activities.

Aperture   Open space along a plane of discontinuity, such as a joint plane.

Apparent specific gravity   Ratio of the mass of oven-dried solid (in air) to the difference of mass of oven-dried solid in air and mass of saturated solid in water.

Aqueous phase   Of or pertaining to water.

Aquiclude   Earth material of low porosity and permeability, such that is has little or no capacity to hold or transmit groundwater.

Aquifer   Underground mass of saturated rock or soil material that is capable of yielding water supply.

Aquitard   Confining bed that retards but does not prevent the flow of water to or from an adjacent aquifer; a leaky confining bed. It does not readily yield water to wells or springs, but may serve as a storage unit for groundwater.

Aromatic compounds   Generally refers to aromatic hydrocarbons that have characteristic odor and benzene ring structure, e.g., benzene, nepthalene, toluene, etc.

Artesian well   A well in which the water rises above the top of the aquifer, whether or not it flows out at the land surface.

Atomize   Breaking of molecules of liquids into very fine particles.

Atterberg limits   The liquid limit, plastic limit, and shrinkage limit for soil. The water content at which the soil behavior changes from the plastic to the liquid state is the *liquid limit;* from the semisolid to the plastic state is the *plastic limit;* and from the semisolid to the solid state is the *shrinkage limit.*

Autotrophic   Microorganisms utilizing inorganic compounds as their source of nutrients.

Basalt   Dark-colored extrusive igneous rock, the fine-grained equivalent of gabbro.

Bedding planes   In sedimentary or stratified rocks, the division plane that separates each successive layer or bed from the one above or below. It commonly marks a visible change in lithology or color.

Best available technology   A general term for the best method currently available to treat/dispose of a hazardous waste.

Bioaccumulation   The retention and concentration of a substance by an organism.

Biochemical oxygen demand (BOD)   A measure of the amount of dissolved oxygen in a unit volume of water that is necessary for decomposition of organic materials. Water is considered polluted if the BOD drops to less than 5 milligrams per liter of water.

Biodegradation   Decomposition of a substance into more elementary compounds by the action of microorganisms such as bacteria.

Biogenic sedimentary rocks   An organic rock produced directly by the physiological activities of organisms, e.g., coral reefs, shelly limestone, pelagic ooze, or coal.

Biological air sparging   Introduction of air into contaminated groundwater to promote growth of microorganisms for eventual biodegradation of the hazardous constituents.

Bottomland   The part of the landscape that includes flood plains and the stream channel.

Bowen's Reaction Series   A schematic description of the order in which different minerals crystallize during the cooling and progressive crystallization of a magma.

Breakthrough curve   On a plot of solute concentration over time, it represents the time when a solute is first detected in the aqueous system.

Brittle solid   Solids that fail when the applied stress exceeds their elastic limit.

Brownian motion   Continuous irregular motion exhibited by particles suspended in a liquid or gaseous medium, generally in a colloidal suspension.

Buddy system   As part of the personnel health and safety plan at a hazardous waste site, each employee of a work group is to be observed by at least one other employee in the same work group to render assistance when needed or in case of an emergency.

Bulk specific gravity   Ratio of the mass of oven dried solid (in air) to the difference of the mass of the saturated solid (surface dried) in air and the mass of the saturated solid in water.

Bulk specific gravity, saturated surface dried   Ratio of the mass of the saturated solid (surface dried) in air to the difference of the mass of the saturated solid (surface dried) in air and the mass of the saturated solid in water.

Caliche   A solid, almost impervious layer of whitish calcium carbonate in a soil profile.

Capillarity   The action or condition by which a fluid, such as water, is drawn up in small interstices or tubes as a result of surface tension.

Capillary fringe   The lower subdivision of the zone of aeration, immediately above the water table, in which the interstices are filled with water under pressure less than that of the atmosphere, being continuous with the water below the water table but held above it by surface tension.

Carboniferous Period   The Mississippian and Pennsylvanian periods combined, ranging from about 345 to about 280 million years ago.

Carcinogen   An agent that has the potential to induce the abnormal, excessive, and uncoordinated proliferation of certain cell types, or the abnormal division of cells, i.e., a material that causes cancer cells to develop and proliferate.

Carcinogenicity   Cancer-causing potential of a substance.

Cation exchange   A reaction in which cations adsorbed on the surface of a solid, such as a clay mineral, are replaced by cations in the surrounding solution.

Cation exchange capacity   Process of replacement of a positively charged ion (cation) by another cation.

Centrifugation   A process that uses a centrifuge to separate hazardous constituents.

Chain-of-custody   A required document that contains information on the responsible parties that had the custody of analytical sample or samples during various stages of its sampling, transportation, analyses, and storage.

Channelled bedrock surface   Surface of rock that was scoured by water action to form channels.

Characteristic wastes   Wastes that fail any of the characteristics of hazardous waste that include ignitibility, corrosivity, reactivity, and/or toxicity.

Chemical weathering   The decomposition of rocks through chemical reactions such as hydration and oxidation.

Chlorinated hydrocarbons   Any organic compound that contains chlorine in its chemical structures besides carbon and hydrogen.

Chronic disease   Disease in which symptoms develop slowly over a long period of time.

Clastic sedimentary rocks   Rock composed principally of fragments derived from preexisting rocks or minerals.

Clay   Very small soil particles having a layered structure, formed as a result of chemical alteration of primary rock minerals. Clay particle dimensions are smaller than 0.002 mm (2 microns).

Clay marl   A sediment comprised of carbonate and clay minerals; also used for an impure limestone.

Claypan   A dense, heavy, relatively impervious sub-soil layer that owes its character to a high content of clay concentrated by downward-percolating waters. Cannot be excavated by using conventional earthmoving equipment and may need drilling and blasting.

Clean close   Removal or decontamination of all contaminated equipment, structures, liners, and soils, including records of sampling protocols, schedules, and the cleanup levels, to be used as standards for assessing whether removal or decontamination has been achieved.

Coagulation   A process using coagulant chemicals and mixing by which colloidal and suspended materials are destabilized and agglomerated into flocs.

Code of Federal Regulations (CFR)   A compilation of all final federal regulations in effect in the United States at the time of its publication. It contains the full text of all final regulations, excluding the preamble, promulgated by all federal government agencies. The CFR is updated each year in July. The U.S. EPA's regulations are included in Title 40 of the CFR, commonly referred to as 40 CFR.

Co-disposal (of wastes)   Household and industrial wastes disposed of in the same landfill.

Coefficient of compressibility   The potential of volume change (compression) of a soil when it is subjected to external loading.

Coefficient of cubical thermal expansion   Measure of expansion (or contraction) of a volume (three dimensions).

Coefficient of linear thermal expansion   Measure of expansion (or contraction) along the length of a material (one dimension).

Coefficient of surficial thermal expansion   Measure of expansion (or contraction) along an area (two dimensions).

Collision zone   A convergent plate margin where two plates collide.

Colloids   A particle-size range of less than 0.005 mm, i.e., smaller than clay size.

Compaction   The process of increasing the density or unit weight of a soil (frequently fill soil) by rolling, tamping, vibrating, or other mechanical means.

Composite liner   A barrier made of synthetic material, such as plastic, used to hold liquid.

Compressibility   The change, or tendency for change, that occurs in a soil mass when it is subjected to compressive loading.

Conductivity   The property of transmitting a fluid, heat, electricity, etc. through a material.

Cone of depression   The lowered surface of the groundwater table resulting from pumping of water; the maximum lowering or depression of the groundwater table occurs at the center of the pumping well and rises upward away from the center, forming a cone-like feature in three dimensions.

Consolidation   The process by which compression of a stressed clay soil occurs simultaneously with the expulsion of water present in its void spaces.

Containment   The process of confining a material at the place where it occurs or at a desired location to prevent environmental contamination.

Contaminant   Any unwanted physical, chemical, biological or radiological substance or matter in water, soil, rock, or air.

Contaminant plume   Mass of contaminated groundwater.

Contaminant reduction zone   The area at a hazardous waste site that has been set aside for the decontamination of equipment and personnel.

Controlled sample   A sample of known chemical composition that is introduced into a batch of samples sent for analysis to a laboratory to ensure quality control.

Controlled waste   Any waste that is capable of causing

harm to any living organism.

Convergent plate margins   The zone in which adjacent plates either collide with each other or one of the plates sinks beneath the other.

Corrosion potential   A measure of the ability of a liquid to corrode steel; such liquids have a pH of 2 or less or 12.5 or more.

Corrosivity   One of the criteria used to designate a waste as hazardous.

Covalent bond   A chemical bond in which electrons are shared equally between two atoms.

Cradle-to-cradle   A new concept in material conservation that aims at recycling of a manufactured product or its components after it reaches the end of its useful life into other products or components that can be used as input material in a process, thereby avoiding its disposal.

Cradle-to-grave   A phrase used for the provision of the Resource Conservation and Recovery Act that requires complete tracking of a hazardous waste from the time of its generation to its ultimate disposal.

Critical height   Maximum rise of a liquid in the pore spaces above the saturated zone due to capillary action.

Crust   The outermost and thinnest of the Earth's compositional layers, which consists of rocky matter that is less dense than the rocks of the mantle below.

Decontamination   Process that removes, destroys or renders innocuous hazardous constituents.

Deep well injection   Deep wells, drilled into bedrock, below aquifers, for disposal of hazardous liquid waste.

Degradation coefficient   A measure of the efficiency of an incinerator to destroy or remove principal organic hazardous constituent (POHC).

Denitrifying bacteria   Microorganisms that liberate elemental nitrogen from nitrogenous compounds, under anaerobic conditions.

Dense nonaqueous phase liquid (DNAPL)   An organic liquid, composed of one or more contaminants that is heavier than water and does not mix with water; e.g., chlorinated solvents.

Desiccation crack   A crack in sediment, produced by drying; especially a mud crack.

Destruction and removal efficiency (DRE)   The ratio of the difference in the mass of feed rate of POHC and mass emission rate of the same POHC to the mass feed rate of the POHC.

Devolatilization   Reverse of volatilization, i.e., conversion of vapors into liquids.

Diagenesis   All the changes undergone by a sediment after its initial deposition, exclusive of weathering and metamorphism.

Diffusion   The process that causes solutes to move from zones of higher concentration to zones of lower concentrations.

Diorite   Plutonic rocks intermediate in composition between acidic and basic; the approximate intrusive equivalent of andesite.

Discontinuities   An interruption in the continuity in the rock mass caused by the presence of planes of weakness.

Dispersion   Mixing of contaminated groundwater with uncontaminated groundwater as it moves through a porous medium.

Dispersivity   Measure of solute tendency for mixing.

Disposal waste pile   An accumulation of noncontainerized and nonflowing solid wastes.

Divergent plate margins   A fracture in the lithosphere where two plates move apart.

Domestic Substances List   A list of toxic substances manufactured or imported into Canada in a quantity of 100 kg or more per year used to determine if a toxic substance is new or already exists in the country.

Driving forces   In slope stability analysis, includes all forces that tend to cause the earth material to move downslope.

Due diligence   Any procedure or service used by a potential buyer of real estate to ensure that no hazardous waste is present at the property.

Dutch List   A list first developed in the Netherlands in which hazardous substances that include metals and organic and inorganic chemicals are grouped into seven categories.

Earth's plates   Rigid thin segments of the earth's lithosphere, which may be assumed to move horizontally and adjoin other plates along zones of seismic activity.

Economic geology   The study and analysis of geologic bodies and materials that can be utilized profitably by man, including fuels, metals, nonmetallic minerals, and water.

El Niño   Climactic changes attributed to redistribution of the circulation of warm water currents in the Pacific Ocean.

Endogenous phase   A phase of bacterial growth in which the microbes use the polysaccharide layer from their cell walls as nutrients.

Endothermic   A chemical reaction that is accompanied by absorption of heat (opposite of exothermic).

Engineering geology   The application of the geological sciences to engineering practice, to assure that the geologic factors affecting the location, design, and construction of engineering works are recognized and adequately provided for.

Environment   The physical and biological aspects of a specific area; includes all living things, soil, air, and water.

Environmental audit   Assessment of a real estate property or a site for signs of contamination.

Environmental geology   The application of geologic principles and knowledge to problems created by man's occupancy and exploitation of the physical environment.

Environmental impact statement   A document prepared by industry or a political entity on the probable environmental impact of its proposals before construction is allowed.

Epidemiologist   An expert who investigates the causes and control of epidemics.

Eukaryotic   Refers to higher life forms; eukaryotic cell has a true nucleus.

Evaporite deposits   Sediments which are deposited from aqueous solution as a result of extensive or total evaporation. Examples include anhydrite, rock salt, and various nitrates and borates.

Evapotranspiration   That portion of the precipitation returned to the air through evaporation and transpiration.

Exogenous   Not occurring in place; alien.

Exothermic   Chemical reaction that gives off heat.

Expansion cracking   Cracks in fine ground sediments produced as a result of drying of expansive clays.

Extrusive igneous (or volcanic) rocks   Result of cooling and crystallization of minerals from a molten material at or above the Earth's surface.

Fabric   The orientation in space of particles, crystals, or cement of which a rock is composed.

Facies change   Change in the characteristics of rocks in a lateral or horizontal direction, common in sedimentary rocks.

Facultative anaerobes   Microbes that can grow in the presence or absence of oxygen.

Fault   A fracture or fracture zone along which there has been displacement of the sides relative to one another.

Feedstock   Raw materials used in a manufacturing process.

Final cover   Layer of compacted silty clay used to close a landfill after it becomes full.

Fine grained soils   Soil containing particles smaller than 0.074 mm diameter.

Fission, nuclear   Splitting of an atomic nucleus with ejection of neutrons and heat.

Flash point   The lowest temperature at which a substance gives off enough combustible vapors to produce a momentary flash of fire when a small flame is passed near its surface.

Flexible membrane liner (FML)   A liner made of synthetic material used in landfills.

Flocculation   The process by which many minute suspended particles are held together in clot-like masses, or are loosely aggregated into small lumps or granules.

Fluff   An accumulation of nonflowing and nonmagnetic solid hazardous waste.

Folding   The process of bending of rocks or sediments.

Foliated metamorphic rocks   Metamorphic rocks that exhibit parallel arrangements of minerals giving the appearance of layering, e.g., schist, phyllite, and gneiss.

Foliation   The planar structure that results from flattening of the constituent grains of a metamorphic rock.

Foliation planes   Planar structures formed by parallel to subparallel alignment of mineral grains or structures in metamorphic rocks.

Fossil valley   An ancient valley in bedrock, filled with sediments.

Fractional distillation   Separation of liquids of different boiling points by distillation.

Fracture   A crack, joint, fault, or other break in rocks.

Frequency   The number of repeats.

Frost heave potential   The ability of a soil mass to expand (heave) upon freezing.

Fulvic acid   One of the several organic matters found in soils that are soluble in both acidic and basic solutions.

Gabbro   Dark colored basic intrusive igneous rock, the approximate intrusive equivalent of basalt.

Gas chromatography (GC)   A quantitative chemical analysis technique used to determine the nature and quantity of chemicals present in a sample.

Generator   A person whose act or process produces hazardous waste.

Geochemical cycle   Nature's way of causing concentration of chemical elements and compounds at one location or their removal at other locations by interaction of chemical elements through various paths.

Geohydrology   A term, often used interchangeably with hydrogeology, referring to the hydrologic or flow characteristics of subsurface waters.

Geologic structures   The attitude and relative positions of the rock masses of an area; the sum total of features resulting from such processes as faulting, folding, and igneous intrusion.

Geology   The study of the planet earth, the materials of which it is made, the processes that act on these materials, the products formed, and the history of the planet and its life forms since its origin.

Geomechanical   Pertaining to mechanical properties of rocks.

Geomorphology   The science that treats the general configuration of the earth's surface; specifically, the study of the classification, description, nature, origin, and development of landforms and their relationships to underlying structures, and the history of geologic changes as recorded by these surface features.

Geophysics   A branch of the Earth Sciences that deals with the study of the physics of the Earth; includes seismology, geomagnetism, volcanology, etc.

Geosynthetics   A general term for synthetic fabrics used as liners in landfills and other engineering applications.

Glaciated valley   A valley carved by glacial erosion, U-shaped in profile.

Gonzales Amendment   A 1986 amendment to the Safe Drinking Water Act, authorizing the U.S. EPA to designate aquifers that are especially valuable because they are the only source of drinking water in an area.

Gradient   A measure of the vertical drop over a given horizontal distance.

Grain size   Average diameter of sediments or soil particles.

Grain size distribution   Determines the range in size of mineral solids in a soil sample.

Granite   A coarse-grained igneous rock comprised primarily of feldspar and quartz.

Granular activated carbon   Finely powdered carbon (usually charcoal) used for treatment of hazardous waste.

Granular soil   Soil made up of grains of nearly the same size and in the range 2 to 10 mm.

Greenhouse effect   The heating of the earth's surface because outgoing long-wavelength terrestrial radiation is absorbed and re-emitted by carbon dioxide and water vapor in the lower atmosphere eventually returning to the earth.

Ground freezing   Temporarily freezing the water in loose soils to make them suitable for excavation and to stand steep to vertical cuts.

Ground Penetrating Radar (GPR)   A geophysical method that uses radar range electromagnetic radiation to delineate buried objects and bodies in the subsurface.

Groundwater   All the water contained in the spaces within bedrock and regolith.

Groundwater mound   Mass of groundwater that may occur above the general groundwater table and may influence overall groundwater flow pattern.

Groundwater table (GWT)   The uppermost surface of the zone of saturation.

Halogenated   Introduction of a halogen, usually chlorine or bromine, into a compound by substitution or addition.

Halophile   An organism requiring salt (NaCl) for growth.

Hazardous material   A substance or material capable of posing an unreasonable risk to health, safety, and property when commercially transported.

Hazardous substance   Any toxic material that is properly labelled and stored.

Hazardous waste   A waste material that has the potential to harm living forms and the environment.

Head   Pressure exerted by a column of water or by an elevated water source.

Heat capacity   Heat required to change the temperature of a substance by 1°C.

Heavy minerals   The accessory minerals of high specific gravity found in sedimentary rocks and sediments.

Heterotrophic   Microorganisms that use organic compounds as nutrients.

Historical geology   A branch of geology that deals with the evolution of the earth and its life forms from its origins to the present.

Holocene Period   The youngest period in earth's history, includes the time from the present to the past 10,000 years.

Homosphere   The domain of human activities and its impact on the environment.

Horst   Rock mass that occupies structurally high position due to faulting.

Horst-and-graben structure   Series of structurally high and low areas resulting from a number of faults.

Hot line   The boundary between the contaminant reduction zone and the support zone at a hazardous waste site.

Humic acid   Black acidic organic matter extracted from soils, low-rank coals, and other decayed plant substances. It is insoluble in acids but dissolves in basic solutions.

Humic substances   Solid products of microbial transformation of organic matter, generally dark in color and of complex chemical structure.

Humus   The organic matter of soil, so well decomposed that the original sources cannot be identified.

Hydraulic conductivity   Permeability coefficient.

Hydraulic dispersion   The process causing mixing of contaminated groundwater with uncontaminated groundwater when it moves through a porous material.

Hydraulic head   The height of the free surface of a body of water above a given reference point.

Hydrocarbon   Any organic compound (gas, liquid, or solid) consisting wholly of carbon and hydrogen.

Hydrodynamic dispersion   Process that causes a solute to spread by mixing with groundwater.

Hydrogen ion concentration (pH)   Potential for hydrogen; a measure of the acidity or alkalinity.

Hydrogeologist   An expert who has knowledge of the origin, movement, remediation, and exploration of groundwater.

Hydrogeology   The science that deals with subsurface waters and related geologic aspects of surface water.

Hydrometer method   A method used for determining the grain size of fine grained soils, smaller than 0.074 mm in diameter.

Hygroscopic force   Force that causes absorption of moisture from the air.

Hyperaccumulators   Plants capable of accumulating unusually high quantities of chemicals in their systems.

Hyperthermophiles   Microorganisms that can survive in high temperature environments, generally above $80°C$.

Igneous rocks   Rocks that have solidified from molten or partly molten material.

Ignitibility   The ability of a material to be set afire.

Immiscible liquid   A liquid that does not dissolve in another, e.g., gasoline and water.

Immobilization   Process that causes conversion of inorganic ions or compounds into organic form.

Incineration   A process that utilizes heat and oxygen to destroy organic constituents of a hazardous waste.

Index properties   Soil properties that indicate its type and condition, and provide an estimate of the engineering behavior of a soil.

Infilling material   Secondary materials found in open joints and other openings in rock mass.

Infiltration   The flow of a fluid into a solid substance through pores or small openings; specifically water into soil or porous rock.

Information exchanges   A matchmaking operation based on the idea that one manufacturer's waste may be another manufacturer's raw material.

Inner core   The innermost part of the earth that is a solid mass of materials rich in iron and nickel.

Input material   Raw materials used in a manufacturing process to produce the finished material.

Insular saturation   Relates to residual saturation of nonwetting fluids in the pore spaces, where they tend to occur as isolated blobs in the center of pores.

Intermittent stream   A stream that flows only at certain times of the year, as when it receives water from springs or from a surface source. A stream that does not flow continuously, as when water losses from evaporation or seepage exceed the available stream flow.

Intrusive igneous rocks   Any igneous rock formed by solidification of magma below the Earth's surface.

Ion exchange   The reversible exchange of ions of the same charge sign between a solution, usually aqueous, and an insoluble solid in contact with it.

Ionizing radiation   Radiation that either directly or indirectly displaces electrons from an atom, molecule, or ion.

Isotropism   The condition of having properties that are uniform in all directions.

Joint and several liability   A provision in the law which permits the U.S. EPA to recover cost of cleanup from parties responsible for the waste.

Joints   Fractures in rocks along which no observable movement has occurred.

Karst features   A type of topography that is formed over limestone, dolomite, or gypsum by dissolution, and is characterized by sinkholes, caves, and underground drainage.

Ketones   Any of a class of organic compounds containing a carbonyl group attached to two organic groups.

Labile   Rocks, minerals, or plant and animal products that are easily decomposed.

Land disposal restrictions   Bans land disposal of hazardous waste without pretreatment.

Landfarm   A general term for the practice of using the soil or the land for simultaneous treatment and disposal of hazardous waste.

Landfill   A common method of solid waste disposal where the waste is compacted to the smallest volume and then covered with suitable impermeable material.

Landfill gas (LFG)   Methane and related gases produced by decomposition of organic constituents in a landfill.

Land treatment   Use of soil or land to simultaneously treat and dispose of hazardous waste (synonyms: land-farming, land cultivation, land application).

Large quantity generator   Generators who produce more than 1000 kg of nonacutely hazardous waste or over 1 kg of acutely hazardous waste per month.

Latency effect   The time lag between exposure of an individual to a hazardous substance and the clinical manifestation of an adverse effect.

Lateral or transverse dispersion   Mixing of solute-solvent that occurs at right angles to the direction of flow of groundwater.

Leachants   A solution used to remove solids by dissolution.

Leachate   A highly mineralized liquid that comes out of landfills, rich in dissolved organic and inorganic materials, generally brownish red in color and a source of contamination.

Leachate management system   A provision in the design of a secured landfill to monitor, collect, and remove any leachate that may be produced over a period of time.

Light monaqueous phase liquid   Liquids that do not mix with water and are lighter than water, e.g., gasoline and fuel oil.

Limonite   A secondary material, formed by weathering (oxidation) of iron-bearing minerals. It occurs as coatings, earthy masses, and in a variety of other forms.

Lineation   A general term for any linear structures in a rock.

Liner   Clay or synthetic fabric used in waste containment structures.

Liquid limit   See *Atterberg limits.*

Listed wastes   A list of various hazardous wastes that are included in 40 CFR, part 261; these are listed in Appendix A.

Lithification   The process that converts sediment into a sedimentary rock.

Lithology   The systematic description of rocks in terms of mineral assemblage and texture.

Lithosphere   The outer 100 km of the solid Earth.

Loading rate   The rate at which contaminants are delivered into groundwater.

Longitudinal dispersion   Mixing along the streamline of fluid flow.

Low-level radioactive waste   Products of nuclear fission, characterized by low radiation level and short half life (up to a few decades).

Macronutrients   Chemical elements that are needed in large quantities by plants to maintain satisfactory growth.

Magma   Molten rock, together with any suspended mineral grains and dissolved gases, that forms when temperatures rise and melting occurs in the mantle or crust.

Manifest system   A system used to track hazardous waste.

Mantle   The zone of the earth below the crust and above the core.

Mass density   Ratio of the mass to volume of a material.

Mass transfer   Movement of material through chemical, biological, or nuclear reaction.

Mass transport   Movement of solute through porous medium along with the flow of groundwater.

Massive   Non-foliated rock.

Material exchanges   Organization that arranges physical exchange of hazardous waste from a generator to a manufacturer for use as feedstock; same as *active exchange.*

Maximum contaminant level (MCL)   Set by the U.S. EPA, to ensure water quality.

Medium-quantity generator   A generator who produces hazardous waste between 100 and 1000 kg per month.

Melt   Magma.

Metamorphic rocks   Rocks derived from preexisting rocks by mineralogical, chemical, and/or structural changes, essentially in the solid state, in response to marked changes in temperature, pressure, shearing stress, and chemical environment, generally at depth in the earth's crust.

Methanogenesis   Process of conversion of organic constituents in a hazardous waste into methane and carbon dioxide by anaerobic bacteria.

Methyl isocyanate   A chemical compound, CH_3NCO, having a boiling point of 39.1° C.

Microaerophilic   Microorganisms that require oxygen but at a level lower than atmospheric.

Micronutrients   Nutrients (chemical elements) needed by plants in very small quantities to maintain satisfactory growth.

Mineralization   Conversion of organic constituents in hazardous waste into CO_2 and H_2O by aerobic bacteria.

Mississippian   A period of the Paleozoic era (after the Devonian and before the Pennsylvanian) between 345 and 320 million years ago.

Mohr-Coulomb equation   Relates the shear strength of a solid to its cohesion, internal friction, and the normal stress acting on it.

Moisture content   Ratio of the mass (weight) of water to that of the solids present in the soil or rock.

Multimedia filter   A separation process utilizing layers of different materials of varying pore size and specific gravity.

Municipal solid waste   Common household and commercial solid waste.

Mutagenicity   The property of causing a permanent genetic change in a cell other than that which occurs during normal genetic recombination.

National Pollution Discharge Elimination System (NPDES)   A provision of the Clean Water Act that requires any party responsible for the discharge of a pollutant or pollutants into any waters of the United States from any point source, to apply for and obtain a permit.

National Priority List   A list of the hazardous waste sites in the United States targeted for cleanup under the Superfund.

Nonacute hazardous waste   Any waste that is not classified as acutely hazardous.

Nonaqueous phase liquid (NAPL)   Liquids that do not mix with water.

Non-foliated metamorphic rocks   (Massive) without banding, e.g., marble and quartzite.

Nuclear materials   Materials capable of giving off energy by fission or fusion of their atomic nuclei.

Nuclear waste   The spent fuel generated after fission of nuclear materials in reactors.

Obligate anaerobes   Microbes that grow only in the absence of molecular oxygen.

Occupational Safety and Health Administration (OSHA)   Created to enforce regulations of the Occupational Safety and Health Act of 1970 which deals with the protection of employees in the workplace.

Optimum moisture content   The moisture content at which a soil has the maximum dry density.

Organic substrate   Food (nutrient) source for microbes.

Organohalogen compound   An organic compound that contains one or more of the halogens (chlorine, bromine, fluorine, and iodine) in its structure, e.g., DDT, chlordane, and other pesticides.

Osmosis   The transport of solvent through a semipermeable membrane separating two solutions of different (solute) concentration. The solvent diffuses from the solution that is dilute to the solution that is concentrated.

Outer core   The outer or upper zone of the earth's core, extending from a depth of 2900 km to 5100 km. It is presumed to be liquid because it sharply reduces compressional wave velocities and does not transmit shear waves.

Ozone depletion   Reduction in the concentration of ozone in the stratosphere (upper atmosphere).

Ozone layer   Shields much of the earth from ultraviolet radiation.

Paleo-channels   Channels and depressions formed in rock formations during an earlier cycle of erosion; usually filled with sediments.

Parent material   Original rock or sediment from which a soil has developed.

Particulate matter (PM)   Substances that occur as very fine particles, generally become airborne, causing air pollution.

Passive exchange   Facilitates exchange of information on hazardous materials between the generator and potential user.

Pendular saturation   The fluid held in the narrowest part of the pore space by capillary force.

Pennsylvanian Period   A period of the Paleozoic era (after the Mississippian and before the Permian periods) between 320 and 280 million years ago.

Perched water table   The upper surface of a body of water perched atop an aquiclude that lies above the main water table.

Permeability   A measure of how easily an earth material allows a fluid to pass through it.

Permeation testing   Chemical analysis of protective clothing to ensure that the decontamination method has been effective and no contaminants have penetrated through the protective clothing.

Personnel protective equipment (PPE)   Clothing and other protective devices used by workers at a hazardous waste site.

Personnel safety and health plan (PSHP)   A written document describing the procedures to assure safety and health of workers at a hazardous waste site.

Petrology   A branch of geology dealing with the origin, occurrence, classification, structure, and history of rocks.

pH   The measure of acidity or alkalinity.

Phosgene   A deadly gas.

Phreatic zone   Zone of saturation in soil or rock.

Phytotoxicity   Study of the toxic effect of chemicals on plants.

Pinching and swelling   Thinning and thickening of geologic formations.

Pitting and trenching   Inexpensive method of obtaining information for geologic materials occurring in the subsurface.

Plasma -A highly ionized gas.

Plastic limit   See *Atterberg limits.*

Plasticity index   The numerical difference between the values of liquid limit and plastic limit.

Plate margin   Boundary of two adjacent earth plates.

Pleistocene Period   An epoch of the Quaternary Period, after the Pliocene or the Tertiary and before the Holocene period; it began 1.6 million years ago and lasted until the start of the Holocene some 10,000 years ago.

Pollutant   Anything that may result in the degradation of environmental quality. Includes dredged spoil, solid waste, incinerator residue, filter backwash, sewage, garbage, sewage sludge, munitions, chemical wastes, biological materials, radioactive materials, heat, wrecked or discarded equipment, rock, sand, dirt, and industrial, municipal, and agricultural waste discharge water.

Pollution   The process of contaminating air, water, or soil with materials that reduce the quality of the medium.

Pollution prevention   Reduction or elimination of hazardous waste in a manufacturing process.

Polycyclic aromatic hydrocarbon (PAH)   An organic compound comprising two or more benzene rings.

Porosity   Percentage of pore spaces in a unit volume of rock or soil.

Primary porosity   Percentage of the total volume of pore spaces that formed during rock formation.

Priority Substances List (PSL)   A list of substances that are toxic or are capable of becoming toxic.

Procaryotic   Simple organisms, e.g., bacteria.

Pump-and-treat remediation   A method of treatment where the contaminated groundwater is pumped to the surface and cleaned up.

Pyrolysis   A process of destruction of hazardous waste constituents, carried out at lower temperatures and in the absence of oxygen.

Quality assurance   Procedures used to ensure the quality of work, such as chemical analysis.

Quaternary Period   The second period of the Cenozoic era, following the Tertiary Period. It began 1.6 million years ago and extends to the present.

Radioactive element   An element capable of changing spontaneously into another element by the emission of charged particles from the nuclei of its atoms.

Radioactive material   Any material which spontaneously emits radiation.

Radionuclides   Isotopes.

Reactive contaminant   Undergoes chemical reaction with water.

Reactivity   One of the criteria used by the U.S. EPA to characterize a waste as hazardous.

Recalcitrant   Chemical with structure that is not analogous to any naturally occurring substance and will not biodegrade.

Reclamation   The recovery of valuable material from hazardous waste.

Redox reactions   A chemical reaction involving reduction and oxidation.

Relative permeability   The ratio of permeability of a fluid at a given saturation to its permeability at 100% saturation.

Remediation   Process of cleaning up contaminated water, soil, or rock.

Residual saturation   The volume of hydrocarbon trapped in the pores to the total volume of pores.

Resisting forces   Forces that offer resistance against sliding of an earth material and contribute to stability.

Resistivity   Inverse of (electrical) conductivity.

Retardation   Processes that delay the movement of solute in groundwater.

Reverse osmosis   Process that causes flow of solvent from a concentrated solution to a dilute solution.

Rhyolite   Extrusive igneous rock, typically porphyritic with phenocrysts; the extrusive equivalent of granite.

Rock bolting   A method of stabilizing rock slopes.

Rock cleavage   The tendency of a rock to break along preferred directions.

Rock cycle   The cyclic movement of rock material, in the course of which rock is created, destroyed, and altered through the operation of internal and external Earth processes.

Rock mass   Rock in place, includes rock outcrops and rock formations in the subsurface.

Rock quality designation (RQD)   A measure of the relative strength of rock mass.

Room-and-pillar method   A common method of underground mining in which void spaces, called *rooms,* created by extraction of geologic materials are supported by columns of the same geologic materials, called *pillars.*

Runoff   The fraction of precipitation, leachate, or other liquid that flows over the land surface or a hazardous waste facility.

Run-on   The portion of rainfall, leachate or other liquid that flows into a hazardous waste facility.

Salinity   The measure of saltiness.

San Andreas Fault   An example of the transform (fault) plate boundary.

Sand   A detrital particle smaller than a granule and larger than a silt grain, having a diameter in the range of 0.074 to 4.75 mm; a loose aggregate of such particles, most commonly of quartz.

Sanitary landfill   A method of solid waste disposal.

Saturated zone   A subsurface zone in which all the interstices are filled with water under pressure greater than that of the atmosphere. Although the zone may contain gas-filled interstices or interstices filled with fluids other than water, it is still considered saturated. This zone is separated from the overlying zone of aeration by the water table.

Schistosity   The foliation in schist or other coarse-grained, crystalline rock due to the parallel arrangement of mineral grains of the platy or prismatic types, usually mica.

Screening   A physical method of hazardous waste treatment to separate suspended solids from liquid waste.

Secondary minerals   Minerals formed later than the rock enclosing it, usually at the expense of an earlier-formed primary mineral, as a result of weathering, metamorphism, or solution action.

Secondary permeability   Ability of an earth material to allow fluid transport, resulting from secondary porosity.

Secondary porosity   Porosity developed in a rock after its deposition or emplacement, through such processes as solutioning or fracturing.

Secure landfill   A landfill with multiple barrier systems designed to contain the waste and with provision to monitor, collect, and remove leachate.

Sedimentary rocks   Layered rocks resulting from the consolidation of sediment.

Sedimentation   Separation of solids from the liquid by gravity.

Sediment   Solid material that has settled down from a state of suspension in a liquid.

Seismic waves   Earth waves generated by earthquakes.

Seventh Approximation   Soil classification for agricultural purposes based on physical and chemical properties.

Shear strength   The internal resistance of a body to shear stress, typically including a frictional part and a part independent of friction called *cohesion*.

Shear zone   A zone of rock that has been crushed and brecciated by many parallel fractures due to shear strain.

Sheeting   A type of jointing produced by pressure release.

Shock loading   Introduction of a large quantity or high concentration of contaminants that causes a marked slow down or termination of biodegradational activities at a remediation site.

Shotcrete   Pneumatically applied concrete, used for reinforcement.

Shrink-swell potential   Ability of an earth material to undergo volume change with changing water content.

Silicate minerals   Compounds whose crystal structure contains SiO_4 tetrahedra, either isolated or joined through one or more of the oxygen atoms to form groups, chains, sheets, or three-dimensional structures with metallic elements.

Silt   A detrital particle finer than fine sand and coarser than clay, commonly in the size range of 0.074 to 0.002 mm.

Sinkholes   Circular depressions in a karst area; its drainage is subterranean, size is measured in meters or tens of meters, and it is commonly funnel-shaped.

Slurry wall   A containment structure to prevent spreading of contaminated water or other liquid, constructed by injecting a mixture of clay and water into the rock or soil.

Small-quantity generator   Generating less than 100 kg of nonacutely hazardous waste or less than 1 kg of an acutely hazardous waste per month.

Soil horizon   A layer of a soil that is distinguishable from adjacent layers by characteristic physical properties such as structure, color, or texture, or by chemical composition, including content of organic matter or degree of acidity or alkalinity.

Soil profile   A vertical section through a soil mass that displays its component horizons.

Soil taxonomy   Soil classification.

Soil texture   Relates to the size of mineral grains making up the soil solids, e.g., sand, silt, and clay.

Soil washing   A method of remediation of contaminated soil.

Sole source aquifer   An aquifer designated as such under Section 1424(e) of the Safe Drinking Water Act.

Source reduction   Good operating practices, technology changes, materials or product changes to reduce amount of waste.

Specific gravity   The ratio of the weight of a given volume of a substance to the weight of an equal volume of water.

Specific heat   Ratio of the heat capacity of a substance to that of water.

Specific retention   The ratio of the volume of water that a given body of rock or soil will hold against the pull of gravity to the volume of the body itself.

Specific yield   The ratio of the volume of water that a given mass of saturated rock or soil will yield by gravity to the volume of that mass.

Standard Compaction Test   A standardized test commonly used to determine the maximum dry density of soil for construction purposes.

Storage coefficient   Volume of water that an aquifer releases or takes into storage per unit surface area of the aquifer per unit change in head.

Storativity   Storage coefficient.

Stratigraphy   (1) The science of rock strata. (2) The arrangement of strata especially at a geographic position and chronologic order of sequence.

Strength   The ability of a material to withstand differential stress, measured in units of stress.

Stress-relief joints   Joints or fractures that develop in rock mass when the confining stresses are removed.

Strict aerobes   Microorganisms that must have oxygen to survive.

Stripping   In hazardous waste treatment, it is the transfer of dissolved organic compounds present in a liquid into a vapor stream or gas flow.

Structural depression   Low spots caused by structural disturbance in an area, e.g., a horst or a basin.

Subduction zone   The linear zone along which a plate of lithosphere sinks down into the asthenosphere.

Subtitle C, D, and I   One of the major programs of the Resource Conservation and Recovery Act (RCRA). Subtitle C regulates hazardous waste; Subtitle D regulates (nonhazardous) solid waste; and Subtitle I regulates underground storage tanks that hold petroleum products and hazardous substances.

Suction head   Soil physicist's term for the capillary fringe.

Superfund   Common name for the Comprehensive Environmental Response Compensation and Liability Act (CERCLA).

Support zone   The area outside the contaminant reduction zone at a hazardous waste site; this area is free from contamination.

Surface impoundment   Any natural or man-made facility designed to hold an accumulation of liquid waste or waste containing free liquids.

Surface tension   The force that holds water in pores against the force of gravity.

TDS facility   Any facility that is involved in treatment, disposal, or storage of hazardous waste.

Tectonic cycle   A natural cycle responsible for the development of large-scale features of the earth, such as mountain ranges, oceans, and continents; also responsible for occurrence of earthquakes and volcanic activities.

Tension head   Pressure created due to capillarity in soil.

Tension-saturated (capillary) zone   The zone above the groundwater table from which water cannot flow out under the influence of gravity.

Texture   (1) The overall appearance that a rock has because of the size, shape, and arrangement of its constituent mineral grains. (2) The grain size of soil solids.

Thermal conductivity   The measure of the ability of a material to conduct heat.

Thermal diffusivity   Thermal conductivity of a substance divided by the product of its density and specific heat.

Thermal expansion   Relates to both expansion (upon heating) and contraction (upon cooling) of materials.

Thermoplastic materials   Materials that do not undergo any change in their inherent properties upon heating.

Throughput   Raw materials processed in a given time.

Total dissolved solids (TDS)   The total dissolved (filterable) solids as determined by use of the method specified in 40 CFR part 136.

Toxicity   The ability of a material to produce injury or disease on exposure, ingestion, inhalation, or assimilation by living organisms.

Transform fault plate margin   A fracture in the lithosphere along which two plates slide past each other, e.g., the San Andreas Fault.

Transmissivity   The rate at which water of the prevailing kinematic viscosity is transmitted through a unit width of an aquifer under a unit hydraulic gradient.

Transpiration   The process by which water absorbed by plants, usually through the roots, is evaporated into the atmosphere from the plant surface.

Unconfined compressive strength   A common measure of strength of geologic and other materials (synonyms: uniaxial compressive strength; strength).

Unconformity   A break or gap in the geologic record, such as an interruption in the normal sequence of deposition of sedimentary rocks, or a break between eroded metamorphic rocks and younger sedimentary strata.

Unconsolidated   Sediment that is loosely arranged or unstratified, or whose particles are not cemented together.

Unconsolidated aquifers   Uncemented and generally loose geologic materials, such as sand and gravel, that are capable of providing groundwater supply.

Uncontrolled site   Generally an abandoned hazardous waste site.

Underground source of drinking water (USDW)   Any aquifer or its portion that supplies or may supply water to any public water supply system, or for human consumption, or contains less than 10,000 mg/L of total dissolved solids.

Uniformly graded   In geotechnical engineering, the term refers to a soil that has a narrow range in the grain size of its solids.

Universal Hydrologic Equation   An empirical relationship between precipitation, runoff, infiltration, and evapotranspiration.

Unloading   Process of removal of pressure (stress) from the surface of a solid.

Upland   Part of the landscape that is above the bottomland.

Vapor pressure   The pressure exerted at any temperature by a vapor existing in equilibrium with its liquid or solid phases.

Viscosity   The property of a substance to offer internal resistance to flow; its internal friction.

Viscous fingering   Separation of the viscous part of the dense nonaqueous phase liquid into finger-like masses in the unsaturated zone.

Volatilization   Vaporization.

Waste exchange   Matchmaking operation based on the idea that one manufacturer's waste may be another manufacturer's feedstock (related terms: *active exchange; passive exchange*).

Waste pile   A noncontainerized accumulation of nonflowing solid hazardous waste.

Water cycle   Hydrologic cycle, the constant circulation of water from the sea, through the atmosphere, to the land, and its eventual return to the atmosphere by way of transpiration and evaporation from the land and evaporation from the sea.

Water potential   The potential energy of water in soil.

Weathering index   A number used to represent the degree of weathering of a rock.

Well graded   In geotechnics means a soil that shows a wide range in the grain size of its solids.

Wettability   Ability of a liquid to form a film on solids upon contact.

Whistleblower/bounty provision   The U.S. EPA's provision to pay a reward for information leading to criminal conviction.

Wipe sampling   A procedure that involves taking samples by wiping the contaminated material and chemically and/or physically examining the "wipe" for presence of hazardous substances.

Workability   Relative ease with which an earth material lends itself to handling during excavation and construc-

INDEX

CONVERSION TABLES

1. Fundamental Conversion Within the SI Units.

Multiplication Factor			Prefix	Symbol
1,000,000,000,000,000,000	=	10^{18}	exa-	E
1,000,000,000,000,000	=	10^{15}	peta-	P
1,000,000,000,000	=	10^{12}	tera-	T
1,000,000,000	=	10^{9}	giga-	G
1,000,000	=	10^{6}	mega-	M
1,000	=	10^{3}	kilo-	k
100	=	10^{2}	hecto-	h
10	=	10^{1}	deka-	da
0.1	=	10^{-1}	deci-	d
0.01	=	10^{-2}	centi-	c
0.001	=	10^{-3}	milli-	m
0.000 001	=	10^{-6}	micro-	μ
0.000 000 001	=	10^{-9}	nano-	n
0.000 000 000 001	=	10^{-12}	pico-	p
0.000 000 000 000 001	=	10^{-15}	femto-	f
0.000 000 000 000 000 001	=	10^{-18}	atto-	a

2. How Much Is Part per million/billion/trillion?

1 ppm = mg/kg, μg/g, ng/mg, pg/μg; mg/L, μg/mL, ng/μL; 0.00001%

1 ppb = μg/kg, ng/g, pg/mg; μg/L, ng/mL, pg/μL

1 ppt = ng/kg, pg/g, fg/mg; ng/L, pg/mL, fg/μL

Other ways to express trace concentration:

ppm	Equivalent to 1 *inch* in 15.8 *miles*, or 1 *minute* in 1.9 *years*, or 1 *cent* in $10,000
ppb	Equivalent to 1 *inch* in 15,782 *miles*, or 1 *second* in 31.7 *years*, or 1 *cent* in $10,000,000
ppt	Equivalent to 1 *inch* in 16,000,000 miles, 1 *second* in 320 *centuries*, or 1 *cent* in $10,000,000,000

3. Conversion Factors for Frequently Used Units.

To convert	To	Multiply by
acres	square feet	43,560
acre	hectare	0.405
acre-feet	gallons	325,900
acre-ft	cub m	1,233.5
barrels of oil	cub ft	5.61
barrels of oil	gallons	42
bars	lb/sq in.	14.504
BTU/hour	watts	0.2931
BTU/lb	cal/kg	0.556
cm	in.	0.3937
cm/s	ft/day	2,835
cm/s	gallons/day/sq ft	21,200
cm/s	m/day	864
cub cm	cub in.	0.06102
cub in.	cu cm	16.387
cub cm	pint (U.S.)	0.00211
cub m	acre-ft	0.000811
cub ft	cub m	0.028
cub m/s	cub ft/s	353.107
cub miles	cub km	4.167
cub km	cub miles	0.240
cub ft	barrels of oil	0.18
cub ft	cub cm	28,320